课题组成员

Research Grop

中方首席专家 Chinese Chief Expert

范必
Fan Bi

国务院研究室综合一司巡视员

DG, Department of Comprehensive Research, Research Office of the State Council, China

中国国际经济交流中心特邀研究员

Guest Researcher of China Center for International Economic Exchanges

外方首席专家 International Chief Expert

伊丽莎白・多德斯韦尔
Elizabeth Dowdeswell

加拿大学术委员会主席

President and CEO of Council of Canadian Academies

原联合国副秘书长和联合国环境署执行主任

Under-Secretary General of the United Nations，Executive Director of UNEP

课题组成员 Reserch Team Member

韩文科
Han Wenke

国家发展改革委员会能源研究所所长、研究员

Researcher, DG, Energy Research Institute of National Development and Reform Commission

刘应杰
Liu Yingjie

国务院研究室信息研究司司长

DG, Department of Information Research, Research Office of the State Council, China

夏光
Xia Guang

环保部环境与经济政策研究中心主任、研究员

Researcher, DG, Policy Research Center for Environment and Economy, Ministry of Environmental Protection

王敏
Wang Min

国务院研究室司长

DG, Department of Comprehensive Research, Research Office of the State Council, China

黄浩明
Huang Haoming

中国国际民间组织合作促进会副理事长兼秘书长、研究员

Researcher, Secretary General & Deputy Chairman,China Association for NGO Cooperation

莎拉·库克
Sarah Cook

联合国社会发展研究院院长

Director, United Nations Research Institute for Social Development (UNRISD)

杨・贝克斯
Jan Bakkes

荷兰环境评价局高级专家

Senior Export, PBL Netherlands Environmental Assessment Agency

安德莉亚・韦斯托尔
Andrea Westall

世界自然基金会同一地球项目负责人、高级研究员

Senior Researcher, Project Leader of the Sane Globe,Word Wildlife Fund

课题协调员 Coordinator

王飞
Wang Fei

国务院研究室综合研究司副巡视员

DDG, Department of Camprehensive Research, Research Office of the State Council, China

何秀珍
GørildHeggelund

联合国开发计划署高级顾问

Senior Climate Change Advisor, UNDP

王佩珅
Wang Peishen

前世界银行高级环境专家

Former Senior Environmental Specialist of the World Bank

中方支持专家 Chinese Support Expert

陈祖新
Chen Zuxin

国务院研究室综合研究司司长

DG, Department of Comprehensive Research, Research Office of the State Council, China

孙国君
Sun Guojun

国务院研究室宏观经济司巡视员

DG, Department of Macro Research, Research Office of the State Council, China

俞海
Yu Hai

环保部环境与经济政策研究中心战略部主任、研究员

Researcher Director, Policy Research Center for Environment and Economy, Ministry of Environment Protection

郭立仕
Guo Lishi

国务院研究室综合研究司处长

Director, Department of Comprehensive Research, Research Office of the State Council, China

王卓明
Wang Zhuoming

国务院研究室综合研究司处长

Director, Department of Comprehensive Research, Research Office of the State Council, China

张永亮
Zhang Yongliang

环保部环境与经济政策研究中心助理研究员

Research Assistant, Policy Research Centre for Environment and Economy, Ministry of Environmental Protection

芦丽莎
Lu Lisha

国家可再生能源中心助理研究员

Research Assistant，China National Renewable Energy Center

公正
Gong Zheng

国务院扶贫办外资项目管理中心社会发展专家

Social Development Specialist, Foreign Capital Project Management Center under State Council Leading Group Office of Poverty Alleviation and Development

CHINA'S ENVIRONMENTAL PROTECTION AND SOCIAL DEVELOPMENT RESEARCH

中国环境保护与社会发展
理论·改革·实践

《中国环境保护与社会发展》课题组

中国言实出版社

图书在版编目(CIP)数据

中国环境保护与社会发展：理论·改革·实践/中国环境保护与社会发展课题组著．—北京：中国言实出版社，2014.7

ISBN 978-7-5171-0691-3

Ⅰ.① 中… Ⅱ.①中… Ⅲ. ①环境保护－研究－中国 ②社会发展－研究－中国 Ⅳ. ①X-12②D668

中国版本图书馆 CIP 数据核字(2014)第 158001 号

责任编辑 郭江妮 曹庆臻

出版发行 中国言实出版社

地 址：北京市朝阳区北苑路 180 号加利大厦 5 号楼 105 室

邮 编：100101

编辑部：北京市西城区百万庄大街甲 16 号五层

邮 编：100037

电 话：64924853（总编室） 64924716（发行部）

网 址：www.zgyscbs.cn

E-mail: zgyscbs@263.net

经 销 新华书店

印 刷 北京温林源印刷有限公司

版 次 2014 年 10 月第 1 版 2015 年 12 月第 2 次印刷

规 格 787 毫米×1092 毫米 1/16 印张 21.5

字 数 409 千字

定 价 128.00 元 ISBN 978-7-5171-0691-3

本书编委会

目　录

下篇　分论

Contents

Pandect

Thesis

前 言

中国环境与发展国际合作委员会(简称“国合会”)是中国环境与发展领域的高级政策咨询机构，成立二十多年来很多研究成果和政策建议得到中国政府的重视和采纳。2012 年国合会决定，将“环境与社会发展研究”确定为 2012—2013 年的重点课题，这是中外学者首次合作，就这一主题进行框架性研究，对于完善中国的环境政策体系是一件很有意义的事。我和课题组的同事们，对于能够承担国合会这项研究任务，都感到十分荣幸。

在这里，首先介绍一下“环境与社会发展”课题的研究背景。人类进入工业化以来，虽然创造了超过以往所有时代的物质财富，但也面临人口膨胀、发展失衡、资源枯竭、环境恶化等重大挑战。40 年前，在斯德哥尔摩人类环境会议上通过了《人类环境宣言》；20 年前，在里约热内卢联合国环境与发展大会上通过了《21 世纪议程》。经过近半个世纪的研究和实践，国际社会普遍认识到，在处理人与自然的关系上，必须走可持续发展道路。可持续发展理论既强调保护环境，同时也强调环境、社会和经济三要素相互依存、相互制约，必须处理好彼此之间的关系。

中国对环境保护、环境与经济发展的关系研究较多，环境与社会发展内在关系的研究还没有得到足够的重视。中国的环境保护技术发展很快，政府每年环境保护投入越来越大，对污染环境的行为处罚也很严厉，企业的环境准入门槛越来越高。虽然这些举措取得了很大成效，但是环境恶化尚未在根本上遏制，公众对环境状况仍不满意，这其中的原因与社会发展政策的缺失不无关联。

课题组的研究中首先进行了问题识别。我们充分认识到，改革开放以来，我国一直将经济发展作为战略规划和决策的重点。在经济快速发展的同时，出现了环境污染引发群体性事件增多，环境恶化造成公共健康危害，环境破坏与贫困形成恶性循环，环境问题带来新的社会不公，城镇化快速发展使资源环境压力持续增加等突出问题。中国要实现生态文明的美好愿景，就必须要处理好环境保护与社会发展的关系。

课题组认真总结了国际国内对环境与社会关系的理论研究与实践成果。国际上对这一问题的研究开始于 20 世纪 70 年代。其时代背景也是由于当时环境污染引起的灾害事件频繁发生，对社会不同阶层的人群和社会结构造成冲击，甚至引发社会危机，各种环境运动在社会生活中不断涌现。在理论方面，主要是着眼于环境与社会的因果关系，环境问题与制度、社会结构的关联性进行研究，并形成了多个范式，如卡顿和邓拉普(Catton and Dunlap)的新生态范式，施耐伯格(Schnaiberg)的政治经济学范式、汉尼根(Jhon Hannigan)的建构主义范式等。在实践方面，主要关注环境与贫困，环境与人口，环境、移民和城市化，环境与健康，环境与就业，环境与社会公平，环境与可持续消费等问题。中国对于环境与社会发展问题的研究起步较晚，目前主要是引入和介绍国际上的前沿理论和成果。在国家战略和政策层面，如何通过环境社会政策，推动环境与社会发展和谐互动，改善环境可持续性，促进社会健康发展的实证性研究比较缺乏。

课题组综合国内外关于环境与社会关系研究的成果，提出了研究环境与社会关系“三个维度”的理论框架，它包括环境意识、环境行为、环境公共治理，这是本报告的一个创新点。我们主张，在价值观念上，推动全社会形成环境保护的主流价值观念；在环境行为上，倡导公众健康生活方式，落实企业的环境社会责任，促进和规范环保组织发展；在环境公共治理上，建立健全法规，完善环保社会风险的评估、沟通、化解、应急机制，提高环境基本公共服务水平。

基于三个维度的理论框架，我们提出了制定环境与社会政策的基本原则：一是多方参与原则。促进环境保护与社会发展涉及国家、企业、团体与个人的共同利益，所有这些社会主体不是旁观者或者批判者，而是要参与其中，发挥积极的影响。二是长期与短期目标相结合的原则。制定政策应当兼顾当前与长远。三是政策目标一致性原则。制定经济、社会和环境哪一方面的政策，都应综合考虑到其他两方面的政策，做到三种政策的相互衔接、相互配合。四是以法制为保障的原则。通过立法保障环境与社会相协调。五是公平正义原则。环境权利是公民的基本权利，良好的环境是一项公民的基本福利，而保护环境也是公民的基本义务。每个社会主体在享受环境权利的同时都应当履行保护环境的责任和义务。

研究报告提出的政策建议包括以下几个方面：

第一，报告提出了促进环境与社会相和谐的2050年愿景/2020年行动框架。在环境意识维度，建议提升和建立生态文明的主流价值；在环境行为维度，建议公众、企业、社会组织三个行为主体从自身特点出发，促进环境与社会相和谐；在环境公共治理维度，建议增强法律保障、建立独立的环境政策、提高社会风险控制和提高环境公共服务水平。

第二，形成生态文明的社会规范和价值观。课题组认为，应当使生态文明成为社会的主流价值和公序良俗。我们建议，一是制定教育和培训计划。将环境基础知识和可持续发展理论纳入学历教育、职业教育、继续教育、公务员培训，宣传倡导相关规范和行为。二是支持理论和政策研究。比如国合会在这方面可以进一步发挥作用，并不断扩大社会影响力。三是广泛传播生态文明价值观。运用新闻媒体、互联网等传播载体，开展形式多样的社会宣传活动。

第三，鼓励各类社会主体发挥作用。一是鼓励健康、可持续的生活方式。在这方面，社会组织、企业家、以及公众人物应当发挥示范带头作用。二是通过一定的制度安排，保持公众有条件参与决策过程，如保障公众的知情权，立法保证环境信息公开。三是促进企业履行环境社会责任。四是支持环保社会组织的进一步发展。当前中国有必要考虑改变社会组织注册相关政策，放松其开展环境、社会领域相关活动的限制。创造条件解决他们面临的注册难、经费难、社会参与难的问题。政府购买公共服务中，应当将环保组织纳入招标范围，弥补政府提供公共服务的不足。

第四，全面改进整个政府的公共治理。首先是规划名称和内容的调整。中国的发展计划，最早只有经济发展计划，从20世纪90年代开始，编制经济与社会发展计划，将社会发展政策与经济发展政策并列为重要的内容。我们希望从"十三五"规划开始再进一步，将每五年规划改为"国民经济、社会发展与环境保护规划"。在这一规划中，环境政策与经济和社会并列成为同等重要的内容。相应的，中国各级政府在每年"两会"上所提交的"国民经济与社会发展报告"也调整为"国民经济、社会发展与环境保护报告"。其次，建立重大政策的环境社会评估机制。完善政绩考核和政府绩效评价体系。改革政绩考核方法，逐年提高生态环境、社会发展等方面指标在评价体系中的权重，促进地方政府主动在生态环保上加大投入。

第五，建立环境保护社会风险评估、沟通和化解机制。我们建议，凡涉及公民环境权益的重大决策、重大政策、重大项目、重大改革，均纳入环境社会风险评估。政府应该建立一套全面的环境和社会风险评估方法。可以考虑采取的措施包括，一是对具有社会影响的重大项目、涉及公众环保权益的政策和改革进行"前置审批"，包括进行程序合法性评估、政策合理性评估、方案可行性评估、诉求可控性评估等。二是建立征求和吸纳民意的规范程序。包括座谈会、听证会、社会公示等多种形式，以获得民众的理解、信任和支持。三是建立环境社会影响问责制。对履行评估程序不严、不重视风险评估造成严重后果的决策者进行问责。四是构建突发环境事件的应急机制。五是提高环境信息的公开性和透明度。在应对环境事件过程中，发布及时、准确和实际的信息，以避免误导、失实报道、猜测和谣言。

第六，提高环境基本公共服务水平。一是合理确定环境基本公共服务的范围和标准如配备污水处理、垃圾处置等设施；保障公众清洁水权、清洁空气权及宁静权等。环境信息服务，如保障公众环境知情权和环境监督权。制定适当的协调机制，保证环境基本公共服务均等化。二是通过购买服务提高环境基本公共服务的水平。例如，调动社会组织开展环保监测、评估和提高环保意识的宣教活动。三是逐步提高环境基本公共服务在财政支出中所占比重。四是建立生态补偿机制。

当然，课题组开展的是初步的，框架性的研究。我们建议下一步重点开展以下三方面问题研究：一是如何改变公众的生活方式和行为方式；二是如何构建社会发展和环境保护的法律基础；三是如何解决环境保护和社会发展相协调所需的资金来源。

课题组的工作得到了国务院研究室宁吉喆主任、国合会李干杰秘书长的直接关心。课题组由中外双方的公共政策专家和环境专家组成，外方组长是原联合国副秘书长、联合国环境署执行主任，现加拿大学术委员会主席伊丽莎白·多德斯韦尔，中方组长由我担任。

中方课题组成员包括，国家发展改革委能源研究所所长韩文科、国务院研究室信息司司长刘应杰、环保部环境与经济政策研究中心主任夏光、国务院研究室王敏司长、中国国际民间组织合作促进会副理事长兼秘书长黄浩明。课题组外方成员有，联合国社会发展研究院院长莎拉·库克，荷兰环境评价局高级专家贝克斯，世界自然基金会同一地球项目负责人、高级研究员安德莉亚。国务院研究室综合研究一司副巡视员王飞、联合国开发计划署高级顾问何秀珍、前世界银行高级环境专家王佩珅担任课题联络员。

一大批支持专家参加了研究、讨论和撰稿，他们是国务院研究室的陈祖新司长、孙国君司长、郭立仕处长、王卓明处长，环保部俞海研究员、张永亮助理研究员，国家可再生能源中心芦丽莎助理研究员、国务院扶贫办外资项目管理中心社会发展专家公正女士。

在中方的研究团队中，有的长期从事环境政策研究，有的是从事宏观经济、社会政策的研究，大部分专家有政策制定的经验。我们充分认识到，环境与社会发展的研究，是在环境保护技术研究的基础上，借助社会学、经济学、公共政策等方面的研究力量，从多个视角提出如何把改善环境的诉求与政策制定的实践相结合。这一团队的知识结构有利于发挥多学科合作的优势。

在研究过程中，以沈国舫院士为首的中方专家组在我们研究过程中提供了宝贵意见。陕西西咸新区、亿利资源集团为课题组的调查研究、召开会议提供了大量支持和便利。国合会秘书处的同志们以极高的工作效率和敬业精神，为我们做了大量组织协调工作。在这里，谨向他们以及国合会捐助机构一并表示感谢。

中方首席专家　范必

2014 年 10 月

Preface

As a senior Chinese agency that provides advisory service on the policies of environment and development, China Council for International Cooperation on Environment and Development (CCICED) has worked out quite a few research findings and policy suggestions emphasized and adopted by the Chinese government since its establishment over two decades ago. In 2012, CCICED determined to take the "Studies on Environment and Social Development" as the key program for the period from 2012 to 2013. Then, the Chinese and foreign scholars started their first cooperation to carry out framework research on this program, which is of great significance in improving the environmental policies of China. My colleagues of the Research Group and I feel honored to take the research tasks.

At first, I'd like to present a brief introduction to the background of the program of "Environment and Social Development". Humans are faced with such major challenges as overpopulation, imbalanced development, resource exhaustion and environmental deterioration, though they have created unprecedented material wealth since the starting of industrialization. The *Declaration on the Human Environment* was approved on the Stockholm Conference on Human Environment 40 years ago; and *Agenda 21* was approved on the United Nations Conference on Environment and Development held in Rio De Janeiro 20 years ago. Having conducted studies and practices for half a century, the international community has realized generally that sustainable development is the only solution to the coexistence between human and nature. In the theories of sustainable development, equal importance is attached to the protection of environment and the interdependency and restriction among environment, society and economy.

More attention should be paid to the interrelations between environment and society, despite of the abundant studies on environmental protection and the relations between environment and economic development. With a rapid development of environmental protection technologies, the Chinese government has invested increasingly more in environmental protection every year, imposed strict punishments on environmental contamination and raised increasingly the environmental access for enterprises. Though these measures have shown a favorable effect, environmental deterioration has not been curbed radically and the general public are still unsatisfied with the environmental status. An important reason is the deficiency of social development.

Problem recognition is the first thing did by the Research Group. We have realized fully that China has been focusing on economic development in strategic planning and policy making since the reform and opening up. Along with the rapid development of economy, the country has encountered more mass disturbances incurred by environmental contamination, public health hazards due to environmental deterioration, the vicious circle of environmental destruction and poverty, new social injustice produced by environmental problems, increasing resource and environmental pressure stimulated by the rapid urbanization and other prudent problems. To realize the vision of ecological civilization, China must balance environmental protection and social development.

The Research Group has summarized the theoretical studies and practice results at home and abroad. International studies in this respect started from the 1970s, when environmental pollution caused frequent disaster events, impacted the populations of each social class and the entire social structure and even stimulated social crisis and thus environmental activities took place constantly. Theoretical studies were mainly aimed at the causality between environment and society and the relevance between environmental problems and system and social structure, for which several paradigms were formed, such as the New Ecological Paradigm of Catton and Dunlap, the Political Economy Paradigm of Schnaiberg and the Constructivism Paradigm of John Hannigan. On the other hand, practices focused mainly on the relevance between environment and poverty, population, migration and urbanization, health, employment, social equity, sustainable consumption and etc. Starting late on the research of environment and social development, China works presently to bring in and introduce internationally advanced theories and achievements. In relation to national strategies and policies, more material studies should be conducted about how to use environmental and social policies to promote the harmonious interaction between environment and social development, improve environmental sustainability and promote healthy social development.

An innovative point of the report is the "Three-Dimension" theoretical framework, composed of environmental awareness, environmental behavior and public management of environment, presented by the Research Group by summarizing both Chinese and foreign research findings. We suggest in relation to the view of value, the mainstream value of environmental protection be formed throughout the society; in relation to environmental behavior, public healthy lifestyles be advocated, environmental and social responsibilities of enterprises be performed, and the development of environmental organizations be promoted and standardized; in relation to public management of environment, sound laws and regulations be formulated, assessment, communication, solving and emergency mechanisms of environmental and social insurances be improved and fundamental public service of environment be advanced.

Based on the foregoing theoretical framework, we put forward the basic principles for the formulation of environmental and social policies. The first is multi-participation principle. Countries, enterprises, organizations and individuals should be participators, instead of bystanders or criticizers, and exert positive influence, for their common interest is involved in the promotion of environmental protection and social development. The second is the combination of long-term and short-term goals. Both short-term and long-term development should be taken into account in the formulation of policies. The third is compatibility of goals. In the formulation of economic policies, social policies or environmental policies, the other two shall be taken into consideration for the purpose of connecting and supporting one another. The fourth is security by law. Legislation should be adopted to secure the balance between environment and society. The fifth is the principle of fairness and justice. Environmental right is a basic civil right, favorable environment is basic welfare for civics and environmental protection is also a basic civil obligation. Every social subject shall perform the responsibility and obligation of protecting environment whilst exercising environmental rights.

Policy suggestions presented in the report:

Firstly, Vision 2050/ Action Framework 2020 is presented to propel the harmonious development of environment and society. In relation to environment awareness, it is suggested to improve and build the mainstream value of ecological civilization; In relation to environmental behavior, it is suggested the general public, enterprises and social organizations start from their characteristics to facilitate the harmonious development of environment and society; In relation to public management of environment, it is suggested to enhance legal security, establish independent environmental policies and improve the control of social risks and the public service regarding environment.

Secondly, social norm and value of ecological civilization should be formed. In the opinion of the Research Group, ecological civilization should be improved to serve as the mainstream value and public order and good custom of the society. At first, educational and training plans should be formulated. Basic knowledge of environment and theories of sustainable development should be covered by academic education, vocational education, continuing education and training of civil servants and applicable standards and behavior should be popularized and advocated. Furthermore, support should be given to theoretical and policy studies. For instance, CCICED may work better in this respect and raise its social influence constantly. Finally, the value of ecological civilization should be popularized widely in various social activities by virtue of news media, internet and other carriers of communication.

Thirdly, incentives should be given to social subjects of all kinds. At first, healthy and sustainable lifestyles should be encouraged, where social organizations, entrepreneurs and public figures should play the leading role. Then, regulations should be set forth so that the general public can take part in the policy-making process, such as securing their right to be informed and ensuring the opening of environmental information via legislations. Furthermore, enterprises should be urged to perform their environmental and social responsibilities. Finally, further development of social organizations shall be supported. It is necessary for China to consider changing the policies on the registration of social organizations and easing the restrictions on their operating environment and society-related activities to help them overcome the difficulties in registration, funds and social participation. Environmental organizations should be included in the bidders for government purchasing of public services to act as a supplementation.

Fourthly, the overall public management of governments should be improved. At first, the names and details of the plans should be adjusted. China's development planning covered economy merely at the very beginning and has formulated economic and social development plans since the 1990s, attaching equal importance to social development policies and economic development policies. We look forward for another progress in the Thirteenth Five-Year Plan, changing the five-year plan into "a plan of national economy, social development and environmental protection", where environmental policies will be regarded as important as social and economic policies. Accordingly, the "Report on National Economic and Social Development" submitted by governments at all levels at every NPC and CPPCC will be changed in the Report on National Economic and Social Development and Environmental Protection. Furthermore, an environmental and social assessment mechanism should be established for major policies. The performance evaluation and government evaluation systems should be improved. Methods of performance evaluation should be reformed and greater importance should be given to the indexes of ecological environment and social development year on year to promote local governments to spend more actively on ecological and environmental protection.

Fifthly, a mechanism should be built to assess, communicate and solve the social risks of environmental protection. We suggest the environmental and social risks of all major decisions, policies, projects and reforms that involve the environmental rights and interests of the civics should be evaluated. The government should establish a complete group of applicable measures, such as first, implementing pre-examination on major projects that impose influence on the society as well as the policies and reforms that involve the environmental rights and interests of the general public, including evaluating the legality of procedures, reasonability of policies, feasibility of plans and controllability of appeals; Second, establishing normative procedures for soliciting and absorbing public opinions through forums, hearings, social publicity and etc. to win the understanding, trust and support of the general public; Third, setting up an accountability system for environmental and social influence to investigate the duties of the decision makers who incur grave consequences by failing to perform the appraisal procedures strictly or taking no proper account of risks appraisal; Fourth, constructing a mechanism responding to environmental emergencies; Finally, making environmental information more open and transparent. In case of environmental incidents, accurate and real information should be announced timely to avoid misleading or false reporting, speculation or rumor.

Finally, basic public environmental service should be improved. First, the scope and standard of fundamental public environmental service should be identified, such as installing facilities for sewage treatment and rubbish disposal and securing the masses' right of access to clean water, clean air and tranquility. In relation to environmental information service, such as securing the masses' right to environmental information and environmental supervision, an appropriate coordination system should be established to secure the equalization of basic pubic environmental service; Second, improving the basic public environmental service through purchase, such as encouraging social organizations to monitor and assess and popularize and educate the awareness of environmental protection; Third, increasing gradually its proportion in the fiscal expenditure; Finally, establishing a mechanism for ecological compensation.

Of course, the studies of the Research Group are preliminary and to be enriched. We suggest the emphasis be paid to the research on how to change the lifestyle and behavioral pattern of the general public, how to construct the fundamentals of law for social development and environmental protection and how to collect funds for their balanced growth.

Ning Jizhe, Director General with the Research Office of the State Council, and Li Ganjie, Secretary General with CCICED, express immediate concern for the work of the Research Group, which consists of public policy and environmental experts from home and abroad. The lead of the foreign team is Ms. Elizabeth Dowdeswell, former Deputy Secretary General of the United Nations and Executive Director of the United Nations Environment Programme and present Chairman of Canadian Council of Academies and I am the lead of the Chinese team.

Members of the Chinese expert team are Han Wenke, General Director of Energy Research Institute of National Development and Reform Commission; Liu Yingjie, Director of Information Department of the Research Office of the State Council; Xia Guang, Director of Policy Research Center for Environment and Economy of the Ministry of Environmental Protection; Wang Min, Inspector of Comprehensive Research Department of the Research Office of the State Council; and Huang Haoming, Vice Chairman & Secretary General of China Association for NGO Cooperation. The foreign members include Sarah Cook, Director of UN Research Institute for Social Development; Becks, Senior Expert of Netherlands Environmental Assessment Agency (PBL); and Andrea, Responsible Person for the One Planet Program & Senior Researcher of World Wildlife Fund. Liaisons are Wang Fei, Deputy Inspector of Comprehensive Research Department I of Research Office of the State Council; He Xiuzhen, Senior Advisor of UN Development Programme; and Wang Peishen, former Senior Environmental Expert of the World Bank.

A number of experts have showed their support by participating in the research, discussions and writing, including Director Chen Zuxin, Deputy Director Sun Guojun, Division Head Guo Lishi, and Deputy Division Head Wang Zhuoming from the Research Office of the State Council, Researcher Yu Hai and Assistant Researcher Zhang Yongliang from the Ministry of Environmental Protection, Assistant Researcher Lu Lisha with China National Renewable Energy Center and Ms. Gong Zheng, a social development expert from the Foreign - Funded Programs Management Center of the State Council Leading Group Office of Poverty Alleviation and Development.

Among the Chinese members, some have been long engaged in the studies on environmental policies and the other work on the research of macroeconomic and social policies. Most of them are experienced in making policies. We have realized fully that the studies on environment and social development are aimed at finding out how to combining the appeals of improving environment and the making of policies from various perspectives on the basis of the research on environmental protection technologies and by virtue of the studies on sociology, economics and public policies.

I would like to deliver the earnest thanks to the expert team under the lead of Academician Shen Guofang for their valuable opinions, the Xixian New Area of Shanxi Province and ELION Group Limited for their tremendous support for the studies and meetings and the colleagues of CCICED Secretariat for their efficient and professional efforts in organization and coordination and the CCICED departments for their donations.

Fan Bi Chief Expert of Chinese Research Group for
Studies on Environmental Protection and Social Development of China
October, 2014

环境与社会关系研究的新探索

（序一）

国务院研究室主任　宁吉喆

在2013年国合会上，中国环境保护与社会发展研究课题组的报告得到了与会委员和专家们的高度肯定。肯特先生（国合会执行副主席、加拿大国会议员）代表国合会，将课题成果向中国国务院主要领导作了汇报。长期以来，国合会布置的研究课题主要是为制定中国环境政策提供参考。环境保护与社会发展研究报告基本达到了立题的初衷。

改革开放30多年来，中国经济发展日新月异，但环境制约发展、影响社会稳定的问题也日益凸显。随着雾霾遮日的天数增多，全社会对环境问题的重视达到了前所未有的高度。不惟雾霾一项，自来水水质不达标、土地重金属污染等与环境相关的问题往往会被置于聚光灯下、放大镜中，引发社会争议甚至群体事件。政府部门和研究机构越来越清楚地看到，必须把环境保护摆上更加重要的议事日程，深入认识环境与经济、社会的内在联系，着力建设人、自然、社会和谐相处的生态文明。

关于环境，人们首先想到的就是人类生存和发展的物质平台，如土地、空气、水等，这些都是环境的自然属性。从古代开始，中国的先哲就认识到，环境不是取用不尽的免费资源。《吕氏春秋》中说，“竭泽而渔，岂不得鱼，而明年无鱼；焚薮而田，岂不获得，而明年无兽。”甘地也曾经说过，“地球所提供的足以满足每个人的需要，但不足以填满每个人的欲壑”。

除此之外，环境还具有社会经济属性。20 世纪 60 年代，英国学者哈定在《科学》杂志上发表了《公地的悲剧》。他提出，“公地”作为一项公共资源，有许多拥有者。每一个拥有者都有使用权，但没有权利阻止其他人使用，从而造成资源过度使用和枯竭。这件事之所以叫悲剧，是因为每个人都知道资源将由于过度使用而枯竭，但每个人对阻止事态的继续恶化都无能为力。由于环境具有外部效应，企业和个体在这个问题上往往是短见的，只顾眼前效益，不管长远利益，甚至有人抱着“及时捞一把”的心态，从而使一些污染事件得以放大和加剧。一些文明因环境破坏、生态恶化而衰落的事例，在人类历史上发生过多次。

令人欣慰的是，目前国际上对于环境问题的研究，已经越来越多地跳出自然环境本身，而是从经济和社会的角度来认识。这已成为国际社会的共识，也是研究解决当代环境问题的重要方向。如今，环境已经成为中国经济社会发展的瓶颈制约，迫切需要站在社会经济的高度审视和剖析环境问题。国内学术界关于环境与经济关系研究已经有了大量成果，但对环境与社会关系仍缺乏研究。国合会将 2013 年的年度主课题确定为“中国环境保护与社会发展研究”，可以说恰逢其时，具有较强的现实针对性。

环境与社会课题组由中外两个专家团队构成。国务院研究室的几位同事，以及国家发改委宏观经济研究院、环保部环境与政策研究中心、中国民间组织合作促进会的几位专家组成了中方团队，由国务院研究室范必同志任组长。他们大都有二十多年政策研究经历，提出的不少政策建议已转化为政策实践。外方团队由前联合国副秘书长、现任加拿大学术委员会主席伊丽莎白·多德斯韦尔女士率领，联合国开发计划署、联合国社会发展研究院、世界自然基金会、世界银行、荷兰环境评价局的专家参加。他们对不少国家的环境政策提出过建议，拥有丰富的环境国际合作经验。应当说，这是一个国际化、高水准、充满朝气的研究阵容。在研究中，课题组第一次提出了分析中国环境保护与社会发展关系的理论框架，以及促进二者相协调的总体思路。报告提出了从现在起到 2050 年，经济、社会、环境相协调的四阶段愿景目标，建议在中长期发展规划中，将环境作为独立的政策体系与经济政策、社会政策并列。报告还提出了促进全社会形成生态文明的主流价值，制定有利于环境与社会相协调的公共政策体系等建议。这些研究对于未来制定相关政策，特别是制定“十三五”规划具有参考价值。

受时间、空间、财力的制约，住在世界各地的十几位课题组成员聚在一起不是一件容易的事。但在一年多的研究中，课题组成员除各自完成分担的任务外，他们克服种种困难，平均每两月就集中一次，既有会议，也有国内外调研，时间有的两三天，有的一两周。开始时，中外双方观点多有分歧，激烈的争论甚至令旁观者感到紧张。但是，经过相互切磋交流，最后达成了一致。可以说，正是不同意见的相互碰撞激起的火花，才使这份研究成果显得与众不同和更具创新性。这或许再次证明，政策研究需要中外互鉴的开放交流，需要有深度的理论探讨，需要从实际出发的务实精神，需要创新的勇气、智慧和勤奋。课题形成的具有探索性、创新性、突破性的研究成果，以及课题组成员科学严谨、务实合作的工作作风，都为今后进一步深入研究留下了宝贵财富。

2014 年 6 月

New Attempts on the Studies of Relations Between Environment and Society

The report given by China Environmental Protection and Social Development Research Group has been highly affirmed by the members and experts from China Council for International Cooperation on Environment and Development (CCICED) that attended the CCICED Roundtable Meeting 2013. On behalf of CCICED, Mr. Kent (Executive Vice Chairperson of CCICED & Member of Canada Parliament) gave a report on the achievements of the research to the major leadership of the State Council. The research programs arranged by CCICED have been long aimed at providing reference for China to formulate environmental policies. The research report on environmental protection and social development has adhered to the original intention basically.

Thanks to the reform and opening-up policy implemented for over thirty years, China has witnessed a rapid development in its economy and also been bothered by such increasingly prominent problems as environmental constraint and influence on social stability. Meanwhile, an unprecedentedly high attention has been paid to the environmental issues throughout the society. Besides haze, substandard quality of tap water, heavy metal pollution in land and other environment-related problems are often put in the spotlight or viewed through a "magnifying glass" and thus tend to trigger social disputes and even mass incidents. So, government organs and research institutes recognize clearly that greater importance must be attached to environmental protection, more knowledge should be obtained on the internal relations between environment and economy and society and more efforts should be made to construct the ecological civilization with a harmonious coexistence among human, nature and society.

When it comes to environment, we tend to think of the materials humans live and develop, such as land, air and water, all of which are natural attribute of environment. Chinese sages had realized the environment is not inexhaustible since the ancient times.

"How couldn't you catch fish if you pump the lake dry? But you won't get any fish in the following year. How couldn't you catch animals if you burn up the entire forest? But you won't get any animal in the next year", according to *Lu's Commentaries of History* (Lu Shi Chun Qiu, an ancient classic compiled in 239 B. C. by Lu Buwei). Gandhi also said, "There is enough on earth for everybody's need, but not for everyone's greed."

In addition, environment also has its social-economic attribute. Garret Hardin, a British scholar, published *The Tragedy of the Commons* on *Science*, remarking that as a common resource, the "common land" vests in many owners. Each owner is entitled to use but has no right to prevent others from using it, which thereby leads to overuse and exhausting. It is called a tragedy because everybody knows the resources will be exhausted by overuse, but nobody is capable of preventing the worsening. Due to the externality of environment, enterprises and individuals tend to be shortsighted in this respect, absorbed in the immediate benefits. Somebody even wants to reap some profits from it, intensifying some contamination accidents. There have been several civilizations declining for environmental destruction and ecological deterioration.

However, it is gratifying that more and more international studies on environment have been conducted from the perspectives of economy and society, other than the natural environment alone, which has been acknowledged by the international community and offers an important orientation for the research on environmental problems. Environmental problems should be surveyed and analyzed urgently from the social and economic perspectives, as they have been a bottleneck constraint on the economic and social development of China. The relationship between environment and society is to be studied further, though a large number of achievements have been obtained by the domestic academia. It is exactly appropriate and highly pertaining to the reality that CCICED has taken "Studies on the Environmental Protection and Social Development of China" as the main research of Year 2013.

The Environment & Society Research Group consists of the foreign expert team and the Chinese expert team. The latter is under the leadership of Fan Bi, Director of the Research Office of the State Council and composed of the colleagues from the Research Office and the experts from the Academy of Macroeconomic Research of National Development and Reform Commission, Policy Research Center for Environment and Economy of Ministry of Environmental Protection and China Association for NGO Cooperation. Most of them have been engaged in policy research for over two decades and many of their suggestions have been converted into policies. On the other hand, led by Ms. Elizabeth Dowdeswell, Former Deputy Secretary General of the United Nations and present Chairman of Canadian Council of Academies, the foreign expert team is composed of the experts from the United Nations Development Programme, United Nations Research Institute for Social Development, World Wildlife Fund, World Bank and Netherlands Environmental Assessment Agency (PBL). It is a world-oriented, high-standard and vigorous research team, thanks to its members who have put forward suggestions for the environmental policies of many countries and are quite experienced in environment-related international cooperation.

During the studies, the Research Group figured out the theoretical framework for analyzing the relations between environmental protection and social development of China as well as the overall thinking for promoting the coordination between them for the first time. In the report, the Research Group sets forth the visions and goals of four stages for balancing economy, society and environment from now to 2050. It is suggested environment be classified as an independent policy system in parallel with economic policies and social policies in the medium and long-term planning. Meanwhile, it is also suggested the mainstream value of ecological civilization be popularized throughout the society and public policy systems be formulated in favor of the coordination between environment and society. These studies may serve as reference for the formulation of applicable policies, especially designing the Thirteenth Five-Year Plan.

It is hard for the members of the Research Team to get together from around the world, due to the time, spatial and financial restrictions. However, they gathered for two or three days or one or two weeks for meetings or investigations at home and abroad every two months averagely during the studies for over one year, besides completing their tasks assigned. In the beginning, the violent debates between Chinese and foreign members on their discrepancies, which occurred frequently, even made the bystanders nervous. But a consensus was always reached at last through communications and exchanges. So to speak, it is the sparks ignited by the controversies that make the research findings unique and innovative. Again, evidence is shown that policy research should be based on the open exchanges between China and foreign countries, profound discussions on theories, pragmatic spirit, courage and intelligence and diligence of innovation. Valuable experience is left for the further studies, thanks to the exploratory, innovative and ground-breaking research findings as well as the members' scientific, rigorous, pragmatic and cooperative attitude of work.

Ning Jizhe, Director of the Research Office of the State Council
June, 2014

中国环境保护与社会发展的理论、改革与实践

（序二）

中国环境与发展国际合作委员会秘书长
中国环境保护部副部长　　李干杰

世界环境与发展委员会 1987 年在《我们共同的未来》中正式提出“可持续发展”的概念，并成为全球长期发展的指针。作为人类探索科学发展的理念，“可持续发展”内涵不断得到丰富和发展，尤其是明确了经济发展、社会进步和环境保护作为“可持续发展”的三大支柱。在 1992 年首次召开的联合国可持续发展大会上，各国领导人讨论通过了旨在指导实施促进三大支柱平衡发展的《21 世纪议程》。时隔十年后的 2002 年，各国领导人聚首南非并通过了约翰内斯堡执行计划，开始了探索提供促进三大支柱协调发展的更具针对性的办法和具体步骤，以及可量化的、有时限的指标和目标的新征程。然而，新十年经济全球化浪潮并未朝着人们预想和设计的轨道上发展，反而加剧了经济发展不平衡，南北差距迅速扩大；相当部分国家和地区环境恶化、生态破坏、气候变化趋势明显并侵蚀着经济发展取得的成果；社会问题非但没有减少，反而衍生出新的极端变化，特别是

恐怖主义、民族主义和极端主义泛滥。在此背景下，2012 年联合国召开了“里约+20”全球峰会，旨在谋求为所有人创造更加清洁、更加公平和更加繁荣的世界提供一个历史性机会。峰会提出基于可持续发展“三大支柱”框架下的两大主题：一是在可持续发展和消除贫困的背景下发展绿色经济；二是建立可持续发展的体制框架和行动措施框架。两大主题的根本目标是促进可持续发展“三大支柱”的融合。

如今，世界已迈入后“里约+20”时代，全球金融危机依旧发生着深刻影响，世界经济在缓慢中企稳向好但不确定性风险仍在加大。全球经济治理体系、环境治理体系和社会治理体系远未形成，世界仍然处于难以调和的状态。与此同时，中国全面进入了改革与发展的深水区，传统的人口、资源与环境等红利和优势逐渐消失，而长期累积和潜在的问题则凸显为影响经济社会进步的障碍和瓶颈，特别是近年频频发生的群体性事件，诸如四川什邡钼铜项目、广东江门核电项目等均与环境有关。环境问题已经成为影响中国社会发展和政治稳定的问题。在这一新的发展形势下，国合会于 2012 年启动并开展了关于环境保护与社会发展的课题研究，集中了大批中外高级专家，通过历史性梳理和分析国际社会特别是西方发达国家处理环境与社会关系问题成功经验基础上，从环境保护与社会发展关系理论分析入手，构建了推进环境保护与社会协调发展的框架，勾画了未来 2020 年和 2050 年中国环境保护与社会发展的路径和宏伟蓝图。

本书不仅开创了关于环境保护与社会发展系统性研究的先河，而且在分析方法和理论上具有重要的创新。我相信，这本集中外专家智慧的报告的编译出版，不仅对中国政府决策者们处理当前复杂的环境与社会关系具有重要的借鉴和参考意义，同时也为世界理解和支持中国为推进环境保护与社会发展付出的巨大努力提供了很好的素材。作为世界上拥有十几亿人口的负责任大国，中国在环境保护与社会发展实践取得的成绩、积累的经验可为其他发展中国家提供有价值的参考。尽管本书为我国深入探索和构建适应我国国情的环境保护与社会发展新型关系提供了广阔的思路和研究框架，但关于环境保护与社会发展研究还远远不够，我们开展的这项工作仅仅是一个开始，希望有更多的中外专家加入到推进环境保护与社会发展的研究中来，将全球的可持续发展事业推向更高的发展水平。

感谢中外专家为此作出的辛勤努力。

2014 年 1 月

Preface to Environmental Protection and Social Development of China Theories, Reforms and Practices

The concept of "sustainable development" has served as the guidance for the long-term development of the world since it was put forward in *Our Common Future* in 1987 by World Commission on Environment and Development (WCED). As a human philosophy on exploration of scientific development, "sustainable development" has been enriched and expanded constantly. Particularly, economic development, social progress and environmental protection have been identified as the major pillars that secure "sustainable development". On the first UN Conference on Sustainable Development held in 1992, state leaders approved *Agenda* 21, with the aim of guiding the balanced growth of economic development, social progress and environmental protection. Ten years later, the state leaders gathered again in South Africa and approved the Johannesburg Plan of Implementation (JPOI), starting the attempt of providing more targeted measures and specific measures as well as measurable and time-defined indexes and objectives for propelling the balanced growth of economic growth, social progress and environmental protection. However, the wave of economic globalization failed to run along the expected and designed route. Instead, it has exacerbated the unbalanced economic development by expanding rapidly the gap between the south and the north. A certain number of countries and regions have witnessed environmental deterioration, ecological damage and obvious trend of climate change that erodes the achievements of economic growth. Social problems have produced new extreme changes, especially the inundated terrorism, nationalism and extremism. RIO+20, the global summit, was held by the United Nations in 2012 in this context, for the purpose of providing a historic opportunity for creating a cleaner, fairer and more prosperous world for all. There, two principal themes under the "Three Pillars" of sustainable development were put forward—developing green economy under sustainable development and poverty elimination, and systematic framework and framework of action policies for sustainable development—with the fundamental aim of facilitating the integration of economic development, social progress and environmental protection.

Nowadays, the post-RIO+20era has started, the global financial crisis has been producing profound influence and world economy is tending to recover steadily, though it is still faced with increasing uncertain risks. The global governance systems of economy, environment and society are far more from taking shape and the world stays still unbalanced. Meanwhile, China has started comprehensively the deep reform and development and is losing gradually its conventional dividends and advantages with regard to population, resources and environment. Besides, the accumulative and latent problems have grown prominently into a barrier and bottleneck that hinders the progress of economy and society—all of the mass disturbances that took place in recent years, such as those relating to Sichuan Shifang Molybdenum-Copper Project and Guangdong Jiangmen Nuclear Power Project, have something to do with environment, particularly. So to speak, environmental issues have stretched their impact on the social development and political stability of China. In this new context, China Council for International Cooperation on Environment and Development (CCICED) launched and carried out the research on environmental protection and social development in 2012, when a large number of Chinese and foreign senior experts were gathered to construct a framework that balances environmental protection and social development on the basis of the theories on their relations and figure out the route and plan for the country by 2020 and 2050.

Besides initiating the systematic studies on environmental protection and social development, the book achieves significant innovations with regard to analytical methods and theories. I believe the publication of the compiled and translated report, which absorbs the intelligence of both Chinese and foreign experts, will be of great referential significance for the policy makers of the Chinese government to deal with the complicated relations between environment and society and also will serve as an excellent material for the world to understand and support China's tremendous efforts for promoting environmental protection and social development. As a responsible great power with a billion-plus population, China can provide valuable references for other developing countries by virtue of its practical achievements and accumulative experience in this respect. However, more studies are needed for environmental protection and social development, in spite of the extensive ideas and research frameworks for China to deepen and construct new environmental protection and social development patterns that meet its national conditions. As this research is just a beginning, we hope more Chinese and foreign experts will take part in the studies on how to propel environmental protection and social development and improve the global undertaking of sustainable development.

Thanks to the experts at home and abroad for their hard work.

Li Ganjie, Secretary General of China Council for International Cooperation on Environment and Development Deputy Minister of the Ministry of Environmental Protection, the People's Republic of China

January, 2014

致力于促进可持续发展的新成果

（序三）

国合会副主席、联合国副秘书长、联合国环境规划署执行主任
施泰纳(Achim Steiner)

当今中国频繁受到国际社会的瞩目：中国的经济正处于高速发展阶段；城市蓬勃发展、日新月异；人民渴望踏上新的征程。与其他处于工业化进程的国家和新兴经济体一样，“环境”主题——联合国环境规划署(UNEP)工作的核心和灵魂——在中国这个充满活力、快速发展的国家业已成为一大焦点问题。

“中国环境保护与社会发展研究”由中外专家联合开展，其研究成果受到了国合会2013年年会的广泛认可。我本人十分荣幸自2012年以来担任国合会副主席一职。

UNEP致力于促进可持续发展，尤其是反对错误的环境可持续性—经济增长二元模型的各项工作。令人欣慰的是，在当今的全球化信息时代，越来越多的中国和国际学者及相关决策者开始从更微观和社会化的视角来更好地了解环境。越来越多的人开始意识到与环境退化有关的高昂社会成本。尽管可以对环境退化进行测量，但由此造成的社会成本却难以量化，且有可能是毁灭性的。

“中国环境保护与社会发展研究”包括围绕环境与社会，尤其是环境与贫困的关系、人口、健康、就业、社会公平和生活方式等方面的前沿讨论，强调了绿色经济是环境与全球经济的未来发展方向。事实上，我们不能将环境问题与经济发展孤立起来；资源和生态系统的有限性决定了一个国家经济发展的潜力和规模。UNEP 的研究表明，与传统措施相比，1 美元的绿色能源和清洁技术投资将会增加 3—4 倍的就业机会。

中国等快速城镇化的国家将面临独特的挑战，中国有一百多个城市的人口规模超过百万。然而，就我个人来看，令人欣慰的是中国的特大城市正在重新考虑可持续发展的路径。全国各地涌现出更多的绿色出行方式和节能建筑，城市居民也正在积极的践行可持续生活方式，所有这些挑战确实令人感到备受鼓舞。

私营部门应该且必须在实现中国包容性的绿色经济中发挥更重要的作用。其中的一个最佳实践范例就是多家中国企业联合发起的“零公里行动”项目。该项目旨在清除世界最高峰——珠穆朗玛峰的垃圾，受到了广泛的关注，同时激励了数百万中国市民参与垃圾分类和回收等活动。

中国的其他最佳实践范例还包括 UNEP 参与的 2008 年北京奥运会和 2010 年上海世博会前期和期间开展的环境评估活动，其评估结果是令人满意的。秉承着包容、可持续发展的原则，北京和上海两个城市严格遵守环保法规、强调环保责任，为世界各地树立了新的标杆。

“中国环境保护与社会发展研究”详细的介绍了这些成果，并且承认中国在走上绿色、可持续发展的道路上仍然任重而道远。例如，北京和上海仍然遭遇着雾霾和烟雾的污染，这一现象已经蔓延至沿海及其它周边地区，约占整个中国国家领土的六分之一。该研究报告还提出了中国自然资源引起的社会冲突的增加及非政府组织和个人在实现包容性绿色经济中的作用等关键问题。

“中国环境保护与社会发展研究”在此背景下无疑显得十分及时和迫切。很高兴许多中国高级别专家和国际专家参与此项研究，相信相关研究成果不仅将纳入中国的国家政策，也将转化为切实的行动。

2014 年 4 月

Preface of *China's Environmental Protection and Social Development Researth*

Today, China is frequently at the centre of the world's attention. Its economy has been developing at a high speed, its cities flourish wish each passing day, and its people are eager to set out on new journeys. In this dynamic and rapidly evolving country, the topic of environment- the heart and soul of UNEP's work- has become a major issue, just as it has been for other industrializing countries and emerging economies.

China's Environmental Protection and Social Development Research, which was jointly researched and written by Chinese and international experts, was widely recognized by the 2013 Annual General Meeting of the China Council for International Cooperation on Environment and Development (CCICED), a body for which I have had the honour of serving as Vice Chairperson since 2012.

UNEP is committed to all efforts that promote sustainable development, in particular those that combat the false binary model of environmental sustainability versus economic growth. Fortunately, in today's globalized information age, more scholars and policymakers in China and beyond are beginning to better understand the environment from a more nuanced, socially-minded perspective. More and more people are becoming aware of the high social costs associated with environmental degradation. While the latter may be measurable, the social costs can be both difficult to quantify and devastating.

China's Environmental Protection and Social Development Research includes cutting-edge discussions on environment and society- in particular, the relationships between environment and poverty, population, health, employment, social equity and lifestyle. It underscores the fact that the Green Economy is the way forward for both the environment and the global economy. In fact, we cannot isolate environmental problems from economic development; the limits of resources and ecosystems define the potential and scale of a country's economic development. UNEP studies indicate that every dollar invested in green energy and clean technology could generate 3-4 times more job opportunities, compared with traditional measures.

Unique challenges arise in a rapidly urbanizing country such as China, which has more than one hundred cities with a population of over one million people. However, from my personal experience, I am pleased to see that China's mega-cities are rethinking the pathway to sustainable development. There are more green travel alternatives and energy-saving buildings across the country, and urban residents are exercisingincreasingly sustainable lifestyles-all of which challenges are truly inspiring.

The private sector should- and must- play a larger role in bringing about an inclusive Green Economy in China. One example of this best practice was the "Zero-Kilometer Action" project, which was initiated jointly by several Chinese enterprises. The project aimed to remove litter from Mount Everest, the Earth's highest peak. Having received wide attention, the project then inspired millions of Chinese individuals to take action such as waste sorting and recycling.

Other best practices emerging from China include environmental assessments undertaken before and during the 2008 Beijing Olympics and the 2010 Shanghai World Expo-in which UNEP was involved-and which found satisfactory results. Upholding the principle of inclusive, sustainable development, the two cities complied with environmental protection regulations and emphasized environmental responsibility, setting a new benchmark for the world.

China's Environmental Protection and Social Development Research both details these achievements and acknowledges that more remains to be done to move China on a green, sustainable path. For example, both Beijing and Shanghai still suffer from haze and smog, which have spread to coastal areas and beyond, covering an estimated one-sixth of the country. The report also raises critical questions about the rise of social conflicts in China that can be linked to natural resources, as well as about the roles of NGOs and private individuals in bringing about an inclusive Green Economy.

In this context, *China's Environmental Protection and Social Development Research* is both timely and urgently needed. I am delighted that many high-level Chinese and international experts are involved in the study, and I firmly believe that its implications will be translated not only into Chinese national policy, but into true action on the ground.

Achim Steiner
Vice Chairperson of CCICED
Under-Secretary-General, United Nations
Executive Director, United Nations Environment Programme
April, 2014

上篇
总论

第一章 引 言

国际社会普遍认为，经济、社会和环境是可持续发展三个不可或缺的组成部分。要实现可持续发展，需要系统地协调和平衡三者之间的复杂关系。然而，当今世界上还没有国家能够完全做到这一点。

改革开放以来，可持续发展的经济、环境和社会三维支柱中，经济一直是中国最为关注的中心维度，被置于发展战略、规划和政策制定的优先位置。然而，长期以来粗放式的经济发展方式也带来了一系列资源环境压力，大气、水和土壤等环境污染、生物多样性破坏、部分资源濒临枯竭、环境事故频发等问题不仅成为制约经济社会进一步发展的瓶颈，也对公众正常的生产、生活和身体健康造成了严重影响，这些影响比气候变化①带来的影响更为直接和紧迫。

生态破坏和生物多样性破坏很难直观地感受到，但是无形价值的损失，比如自然景观的消失、标志性物种濒临灭绝（如熊猫和江豚）对中国都有很大的影响。它使我们警醒：只有保护环境和生物物种，才能为建设"美丽中国"带来希望与灵感。

经济迅速发展在带来环境问题的同时，也带来了很多社会问题。严峻的资源环境形势不仅引发了公众对环境健康的严重担忧，还加剧了资源分布不均和贫富差距。农村贫困人口生活的地区大多生态富足但脆弱，依赖环境生存的人们为了获得生产生活资料而不得不破坏环境。贫困人口几乎没有选择在哪生活的权利，且大多面临着空气和水污染。在城市中，较贫穷和新兴中等收入阶层的生活条件尚未达到富裕水平，而那些高学历和富裕的中等收入群体已经开始对环境状况提出了更高的期望与要求。

以上状况加剧了与环境相关的社会矛盾。土地用途变更、工业项目选址以及突发性环境事件（如化学品污染）或慢性污染问题（如空气质量问题，这些问题对贫困人口造成过多影响，并有可能加剧已经存在的不平等或贫困现象），都有可能引起社会问题。环境问题往往是公众抵制大型项目的核心原因，尤其是该大型

① IPCC 工作小组评估报告，气候变化 2013：物理科学基础认定人类对气候系统的影响很明显，这在全球大多数地区都很明显。

项目对社会和环境产生的影响尚未确定之时。通过目前的信息公开制度，人们很难准确、自由地获取环境信息，这为谣言或不准确信息的散播提供了平台，而这会导致矛盾加剧。

改革开放以来，中国逐步意识到环境保护的重要性，环境与经济的关系日益受到重视。1983 年召开的第二次全国环境保护会议将环境保护明确为基本国策，强调经济建设和环境保护必须同步发展。1992 年，中国政府发布的《环境与发展十大对策》，首次提出要将环境保护与经济发展统筹考虑，可持续发展战略开始成为国家战略。进入 21 世纪，中国对环境与经济关系的认识进一步深化，2003 年中国政府提出了“科学发展观”，2005 年发布的《关于落实科学发展观加强环境保护的决定》明确提出，要加快从重经济增长轻环境保护转变为保护环境与经济增长并重，从环境保护滞后于经济发展转变为环境保护和经济发展同步推进，从主要用行政办法保护环境转变为综合运用法律、经济、技术和必要的行政办法解决环境问题。尤其是“十一五”以来，中国充分认识到了粗放式的经济发展带来的环境污染问题以及未来经济发展将面临的资源环境约束，通过出台和实施一系列的法律法规、规划、政策加快调整产业结构、推动节能减排，协调环境保护与经济发展的关系，取得了积极成效。然而，相比较对环境与经济关系的认识，中国对于环境保护与社会发展关系的认识和理解仍然较弱，主要存在以下问题：

- 环境与社会的关系：主要包括环境与人口、贫穷、公平、健康、社会福利、消费、防灾减灾、基本公共服务等的关系。
- 环境与经济的关系：包括环境与经济增长，可持续农业与农村发展，工业发展与环境污染防治，可持续交通，可持续能源生产与消费等。
- 资源利用与生态系统关系：包括自然资源保护与可持续利用，空气、水土生态系统保护，生物多样性保护，海洋资源可持续利用，荒漠化控制等。

自可持续发展的理念提出以来，人们对可持续发展中经济、社会和环境三者相互关系的认识和理解不断深化，如图 1-1 所示。事实上，在经济、社会和环境的政策行动中，保持彼此的协调和平衡是十分困难的。2012 年召开的里约可持续发展会议提出，需要制定一套全球目标来推动和实现可持续发展①。

中国于 2007 年首次提出了生态文明的理念，强调建设和谐社会，共享发展

① http：//sustainabledevelopment. un. org/index. php？ menu＝1300.

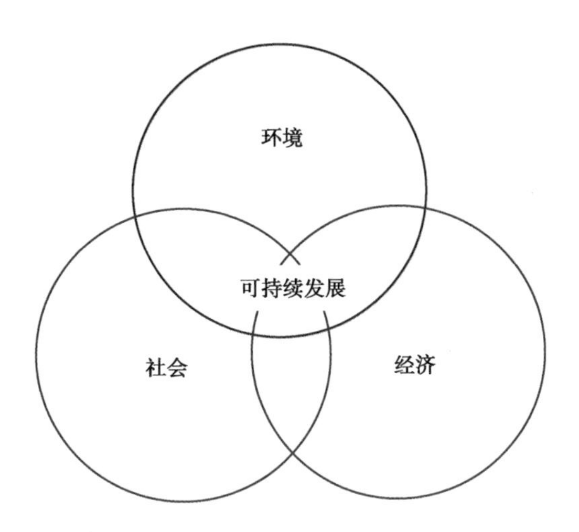

图 1-1 可持续发展环境、经济与社会三个维度的关系示意图

成果，维护社会公平和正义[①]。2012 年 11 月召开的中国共产党第十八次全国代表大会，将生态文明建设写入决议，提出将生态文明纳入经济、政治、社会和文化建设"五位一体"总体布局[②]。实现这些目标必须在原有基础上，进一步处理好环境保护与社会发展的关系。

本摘要报告由中外专家组共同撰写。报告借鉴了国内外相关经验，提出了促进环境保护与社会协调发展的总体思路、指导原则、阶段目标、行动框架、重点领域和政策建议。需要指出的是，这一研究开辟了新的领域，但相关结论与建议是初步的，需要在下一步的研究中进一步深化和拓展。

课题组工作重心围绕三个问题展开(专栏 1-1)。摘要报告以课题组的中国成员在总报告[③]中的研究成果，以及中外团队 2012 年 8 月到 2013 年 9 月进行的一系列会议和实地考察得出的结论和经验为基础。课题组尤其关注城镇化问题，因为城镇化问题能够说明经济、社会和环境的可持续发展关系，并且能够强化政策与实践的相互作用。本摘要报告中引用的主要观点和结论参见课题总报告。

---

① 见国合会 2008 年政策报告：和谐社会的环境与发展，第 26 页。

② http://china.org.cn/china/18th_cpc_congress/2012-11/15/content_27118842.htm.

③ 参见国合会课题组报告：中国环境保护与社会发展研究报告。

专栏 1-1　本报告研究的问题

- 怎样认识当今和未来一段时期中国环境保护与社会发展之间的关系？
- 制定兼顾环境保护与社会发展的政策时，要考虑哪些最为重要的机遇和挑战？
- 当今世界正处于大变革、大调整之中，中国将如何统筹长期愿景、中期目标和短期行动，实现社会公平与可持续发展？

报告其他章节内容如下：第二章，重点阐述中国环境与社会发展领域取得的成就和面临的挑战，识别当前环境与社会发展领域存在的主要问题，分析当前问题产生的深层次原因。第三章，梳理发达国家环境与社会发展的历史经验，总结国内外有关环境与社会发展问题的理论研究成果，构建基于环境价值观念、环境行为和环境公共治理的环境与社会发展理论框架；第四章，提出协调中国环境与社会发展的总体思路，指导原则和 2015 年、2020 年、2030 年、2050 年 4 个阶段的行动目标。第五章，提出政策建议。本报告中使用的术语见专栏 1-2。

专栏 1-2　关键术语

- 社会发展指实现一个社会期望目标的发展过程，也指这些目标最终可实现、可衡量的成果。社会发展可以从三个维度去理解，一是指社会发展，包括教育、科技、文化、卫生等领域；二是指社会组织，包括政府、企业、社会团体、社区、家庭、媒体、个人等社会主体；三是指社会行为，即公民和社会组织的行为方式。社会发展由社会制度及其行为主体共同推动。
- 环境保护指以保护和谨慎使用人类社会赖以生存和发展的环境资源的活动、策略和政策工具。生态或环境破坏（如经济活动导致的污染或气候变化）可影响现在和将来一段时期内人类生活、健康和幸福。法规、个人和集体道德规范以及教育等因素都能影响环保行为。
- 公共服务指由政府直接提供，或通过政府融资、社会资本或组织提供的服务，是整个社会不同群体平等获得体面生活所必需的基本服务。不同国家的公共服务内容虽不相同，但大多包括教育、社会保障、国防、环境保护、公共卫生以及其他社会服务。
- 公共产品指一个人使用或消费后不影响其他人再次使用（非竞争性）、但没有人能够独自占有（非排他性）的产品，它们通常都由公共部门提供和维护，污水处理、公园等是较为典型的公共产品。

- 环境公正强调环境权益和责任的公平分配，以及获得决定和认可社区生活方式和当地知识和文化及能力差异的平等权利。因此，环境公正重在平等。一般认为，大范围的政治或经济不平等可能导致环境不公正，权力更大、更富有的人群往往能够通过使环境恶化的经济活动获得利益，而穷人则需为这种经济活动承受过于沉重的代价。
- 环境权利指确保人类能够获得生存保障所必须的自然资源，包括土地、住所、食物、水和空气等的权利，以及个人的生存环境不受破坏的权利。环境权利成为人类的基本权利，因为人类生活资料、健康甚至存在都依赖于周围环境的质量以及是否有权使用周围环境。环境权利还包括获得获取环境信息、参与环境管理，获得环境安全和赔偿的权利。

第二章　中国环境保护与社会发展现状

自上世纪70年代以来，中国的环境保护工作取得了积极进展。尤其是“十一五”以来，环境保护在国家宏观决策和公共治理中的地位不断提升，环境保护公共基础设施的投入力度不断加大，相关环境法律法规、标准和政策体系不断完善和丰富，中国在协调环境与社会发展领域也取得了一定的成效，但同时也面临着诸多问题。

上世纪80年代开始，中国逐渐转向资源密集型经济。然而，面对随之而来的环境恶化和污染问题，中国社会并没有做好充足的应对准备。在农村，环境污染对健康和生计的直接影响迅速成为农村人口的主要担忧。经济和工业的大发展，以及铁路、公路和管道等基础设施的修建对环境和社会结构造成了严重影响。当前，随着城市富有中等收入阶层的出现，越来越多的市民开始意识到环境问题的重要性，并积极参与到环保行动当中。预计未来十年，将有上亿人涌向城市，这将大大增加能源和自然资源的使用压力和社会服务需求①。

改革开放早期，中国把经济增长放在首位，环境保护与社会发展方面的努力和支出存在不足。“十一五”和“十二五”规划推出后，情况有所改善。这两个五年规划更清晰地将环境与社会目标结合在一起，同时考虑了社会和经济发展对环境的影响以及环境保护对可持续均衡发展的贡献。十八大将生态文明建设纳入“五位一体”总体布局，为强化环境和社会发展之间的共生关系带来了新的可能性。

一、环境保护与社会发展领域取得的成就

上世纪70年代开始的环境保护工作，开启了中国探索可持续发展的先声。1973年召开的第一次全国环境保护工作会议，将环境保护工作列入国家重要的议事日程，标志着中国政府环保意识的觉醒。随着推行清洁生产、环境标志认

① 2013中国人类发展报告，《可持续与宜居城市——迈向生态文明》，北京：中国对外翻译出版有限公司。

证、企业环境信息公开等政策，中国企业的环保意识有所提升。一些企业开始发布企业社会责任报告（CSR），一些银行和金融机构也将环保标准纳入贷款评估体系。

此外，公众、媒体和社会组织对环境保护的关注程度和影响力也日益提高。媒体近年关于雾霾天气的报道，对于提高公众意识和参与度，推动《空气质量标准》的修订产生重大和深远影响①。公众开始越来越多的参与到环境保护法规、政策和标准的制定，参与到环境保护的社会监督和管理中，参与到环境信息的披露和重大项目的环境影响评价过程中。

随着环境恶化带来的直接影响不断增强，公众的环境意识水平不断提高，他们对环境议题的认识、关注度和直接参与度也愈来愈高②。

在1998年的一份热点问题调查中，环境议题成为第五大公众最关心的问题，位于社会安全、教育、人口和就业之后。在2008年的一份民意调查中，环境污染上升为第三大公众关注热点③。近期，媒体对空气污染的报道也为提升公众意识和鼓励公众支持《环境空气质量标准》的修订发挥了积极的作用。

越来越多的非政府组织也参与到环境保护工作当中。1994年3月，中国第一个民间环保组织“自然之友”成立。此后，一大批民间环保组织陆续成立。截至2012年底，全国在民政部登记的生态环境类民间组织数量达到7928个（图2-1）④。

（一）各类社会主体环境行为不断改善

当前，中国公众和社会组织以及一些大型企业正在积极参与环境保护工作。环境法规、政策、标准、社会监督管理、环境信息披露以及重大项目环境影响评估的发展为这些行为主体的参与提供了便利。公众参与力度也不断加大，2013年关于北京雾霾事件以及昆明的PX项目的社会讨论，充分反映了当前中国公众对自身环境利益的诉求、对环境质量改善的迫切要求，以及对参与国家和地方相关环境决策的强烈要求。

（二）企业社会责任逐步增强

随着媒体、公众和政府对环境的关注度不断提高，中国企业更加积极地履行

---

① 闫国东，康建成等，中国公众环境意识的变化趋势，中国人口·资源与环境，2010，(10).

② 参见《企业社会责任：认知的力量》，2011年国际商业报告第24页 http://www.internationalbusinessreport.com/files/IBR_2011_CSR_Report_v2.pdf

③ 同②。

④ 同②。

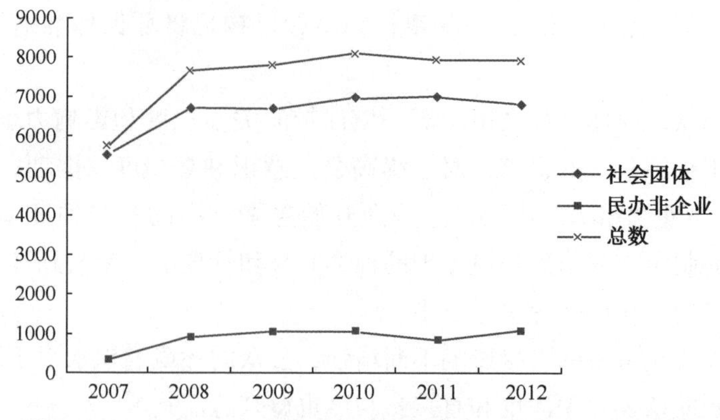

图 2-1　2007-2012 年在民政部登记的生态环境类民间组织数量

各自的企业社会责任。“十二五”规划将改善民生和节能减排设为工作重点，在其影响下，一些企业积极制定了环境管理计划，将战略重点转向可持续发展。根据《2011 年均富国际商业报告》①，在舆论、税收和监管政策等外部因素的驱动下，中国大陆企业的社会责任感正在不断增强。对中国规模以上企业的电话调查显示，84%的企业积极关注职工健康和福利；75%的企业为减少对环境和社会造成的负面影响，对其产品或服务作了改善；69%和 63%的企业分别在节能和减排方面付出了努力；39%的企业开始计算碳足迹。

本课题组考察了亿利资源集团有限公司。作为一个知名企业集团，亿利资源在中国第七大沙漠区的恢复和可持续发展工作中综合考虑了社会和环境目标，是沙漠绿化企业的典范。该企业在履行社会责任，处理保护生态环境与当地民生的关系方面进行了积极探索，并取得了显著成绩。

（三）环境保护法规框架初步建立

长期以来，尤其是“十一五”以来，环境保护在国家宏观决策体系中的地位不断增强。2008 年，国家环境保护总局升格为环境保护部，成为国务院的组成部门之一，环境保护工作的综合管理、污染物防控、监察执法体系逐步完善。截至目前，中国全国人大常委会已经制定了环境保护法等法律 10 部、资源保护法律 20 部。刑法规定了“破坏环境资源保护罪”专章，侵权责任法规定了“环境污

① 参见《企业社会责任：认知的力量》，2011 年国际商业报告第 24 页 http://www.internationalbusinessreport.com/files/IBR_2011_CSR_Report_v2.pdf.

染责任”专章。地方人大和政府制定了地方性环保法规和规章700余件，国务院有关部门制定的环保规章数百件，其中环保部的部门规章69件。2006年和2007年，原国家环境保护总局相继发布了《环境影响评价公众参与暂行办法》和《环境信息公开办法(试行)》。近年来，中国还开展了环境保护法的修订工作，对环境与社会相关条款进行了修订。2012年2月新《环境空气质量标准》和2013年9月《大气污染防治行动计划》的发布标志着中国将环保工作重点转向环境健康危害。

上述举措为中国公众表达环境诉求提供了更多的保障。此外，全国许多地区还在其他领域实施了多种新举措，如绿色信贷、环境污染责任保险、脱硫电价、燃煤发电技术改进、流域和矿产开发生态补偿和阶梯电价等，还有一些地区正在试行当中。这些举措可能将产生巨大的社会效益，然而，只有在进行了影响评估，并且有效地实施了相应的法律法规后才能看到这些举措的真正效益。

二、环境保护与社会发展存在的问题

虽然中国在环境与社会发展领域取得了一定的成就，但依然面临诸多重大挑战。环境问题已经成为影响社会发展与稳定(例如，健康、民生和公平等)的重要因素，还可能对未来经济社会的发展造成不利影响。

(一)环境污染引发群体性事件增多

自1996年来以来，环境信访和投诉数量保持年均29%的增速(图2-2)，主要反映的是食品与饮用水安全、持续性有机污染、危险化学品、危险废物等问题。1995—2010年间中国共发生突发性水污染、大气污染、固体废物污染、噪声污染及振动危害以及其它类型突发性环境事件(环境事故)21985起，自2005年以来，环保部直接接报处置的环境事件共927起。

2004年的云南怒江水电建设规划环境影响评价争议案、2007年的厦门PX项目事件[②]和北京六里屯垃圾焚烧厂事件 、2009年的湖南浏阳镉污染事件、2012年的启东王子造纸大型污水排海工程事件和四川什邡钼铜项目事件[③]等因环境问题导致的争议或群体事件便是其中的典型案例。

---

① 数据来源：1995—2011年《全国环境统计公报》。

② Wanxin Li, Jieyan Liu, Duoduo Li. Getting their voices heard: three cases of public participation in environmental protection in china . Journal of Environmental Management, 2012, 98: 65-72.

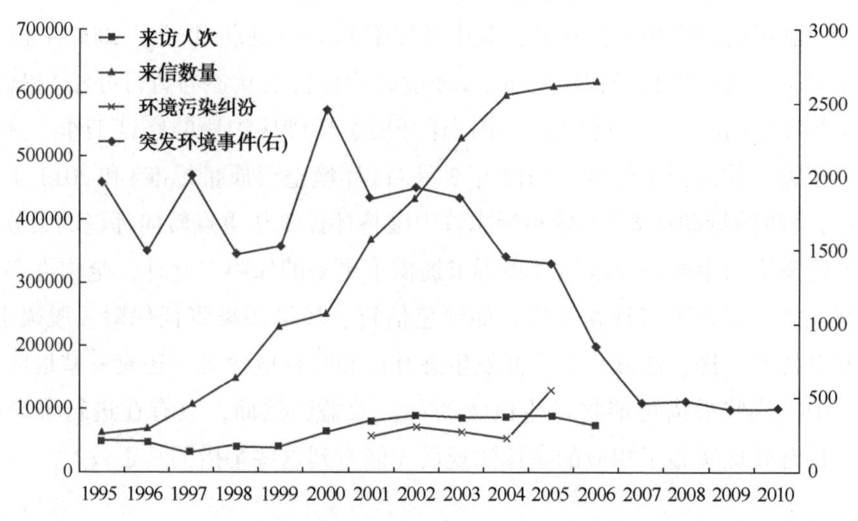

图 2-2 1995—2010 年中国环境信访人数、环境污染纠纷和突发环境事件①

上述事件发生后，有的建设项目因民众激烈反对而停建或迁址，一些垃圾焚烧厂之类的政府主导项目因为群众对于污染的强烈担忧而无法按原计划进行，更有一些是由于居民的身体健康受到伤害，继而引发群体性抗议，出现了围堵政府、司法机关、相关公司及打砸行为。这些事件可能构成社会动乱的诱发因素。另一方面，这些事件也说明了加强对环境问题社会影响的关注和监测、针对有可能引发争议的项目建设建立良好的群众意见反馈机制的必要性。

（二）环境恶化造成公共健康危害

环境污染对人类健康的影响是环境与社会发展中不可忽视的重要问题。与发达国家不同，由于中国的工业布局多集中人口密集的城市与地区，地表水的污染遍及主要经济发达与人口密集区域，以及采矿业和有色金属冶炼导致的农田污染，导致中国的人群直接暴露于环境污染物，并且这些人群长时间以来对污染物的摄入量和暴露水平要远高于发达国家。环境污染浓度高、暴露人口众多、暴露时间长、暴露途经复杂多样、历史累积污染问题的健康影响难以短期消除、城乡差异显著等特点，导致了传统型的环境健康问题尚未得到解决，各种新型环境污染物的健康危害和风险日趋严重。同时，由于复合性污染重，污染区内的主要污

① Glibert, N. Green protests on the rise in China. Nature, 2012, 488(7411): 261-262.

染物、污染源及其健康危害很难确定，加大了人群健康损害效应的调查和干预的难度。

世界卫生组织公布的《2010年全球疾病负担评估》数据显示，中国由于$PM_{2.5}$污染导致的中风、心脏病死亡率有所上升，1990—2010年，由室外空气污染导致的疾病负担增长了33%，2010年，中国20%的肺癌由$PM_{2.5}$引起①。相关研究还显示，$PM_{2.5}$浓度每升高100μg/m^3，总死亡率、呼吸系统疾病、心血管疾病、冠心病、中风、慢性阻塞性肺病（COPD）的死亡率将分别增加4.08%、8.32%、6.18%、8.32%、5.13%、7.25%②。此外，环境内分泌干扰物（EEDs）、持久性有机污染物（POPs）以及新材料和新化学污染物等新型污染物导致的健康损害使问题更加复杂化。

（三）环境破坏与贫困形成恶性循环

环境质量的好坏对人们的健康水平、生产力或收入能力、安全、能量供给及生活条件等方面都起到决定性的作用。特别是在农村地区，贫困人口对自然资源依赖性强，而环境退化又使得这种依赖性表现得更加脆弱。在一些地区，环境问题往往与贫困问题并存，环境恶化往往会进一步加剧地区贫困。反之，农村贫困问题可能导致有限资源的过度利用，加速生态恶化，并最终形成恶性循环。但中国的发展政策还没有充分意识到这些影响之间的相互联系。

（四）环境问题带来新的社会不公

中国不同地区、城乡、性别和民族之间存在着很多不平等现象，环境与社会发展的不公正问题日益突出。这些不平等体现为收入不平等、获得环境资源和服务的权利不平等、所受环境伤害以及承担的环境压力不平等，以及健康和社会保障不平等。在资源环境领域，主要体现在不同区域、群体间占有和支配环境资源的显著差异，以及不同区域、群体间享有环境公共服务的显著差异等。在已有主要强调教育与医疗等公共服务均等化的议题中，将推进环境基本公共服务发展纳入政府的社会目标，是发展绿色经济、实现经济结构转型以及向“服务型政府”转变的重要手段。在中国“十二五”规划纲要中，环境保护已经成为基本公共服务均等化的重要领域。固体废物处理及城乡饮用水安全被纳入环境保护工作重

---

① 世界卫生组织．2010年全球疾病负担评估.

② 赵可，曹军骥，文湘闽．西安市大气$PM_{2.5}$污染与城区居民死亡率的关系．预防医学情报杂志，2011，27(4)：257-261.

点，不同地区、不同群体的公众能否享有公平的环境公共服务，同样是衡量环境公正的重要指标。

然而，相关研究显示，中国的环境基本公共服务绩效存在显著的省际差异（图 2-3）。东部地区大部分省份的公共环境服务水平显著高于中部省份和西部省份，经济发达地区与经济落后地区、城市地区与农村地区的公众所享受的环境基本公共服务明显不均等。很显然，如果在加快落后区域经济发展的过程中，忽视了这些地区环境公共服务水平的提升，将会进一步导致区域、群体间的环境不公正，以及环境与社会关系的进一步恶化。

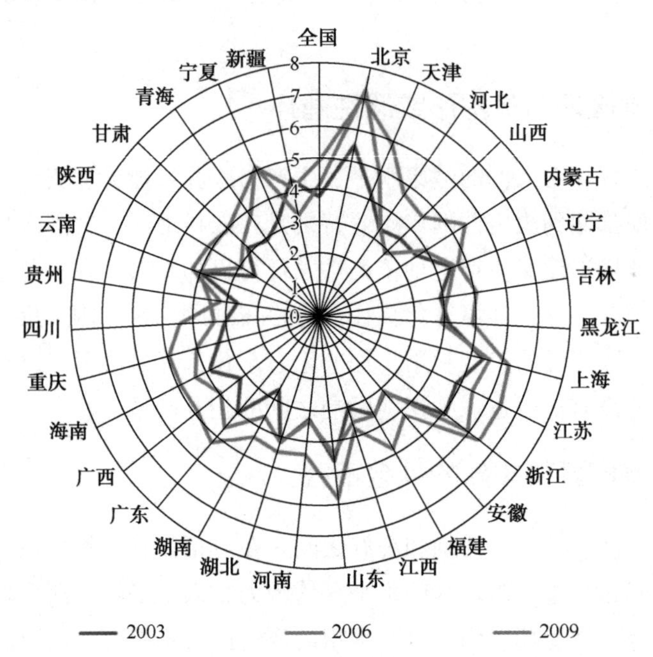

图 2-3　2003—2009 年中国大陆各省份环境基本公共服务评估①

（五）城镇化快速发展使资源环境压力持续增加

全国人口普查数据显示，中国城市人口从 1949 年的 11%②升至 2000 年的

① 卢洪友，袁光平，陈思霞等．中国环境基本公共服务绩效的数量测度．中国人口·资源与环境，2012，22(10)：48-54.

② 1982 年第三次全国人口普查，2000 年第五次全国人口普查数据。

30%，现在已超过 50%①。中国的城镇化率不断加速，每年有超过 1000 万人口涌向城市，为地方政府带来了巨大的挑战。根据预计，截至 2030 年，城镇化率将达到 70%，届时 3 亿左右人口将从农村移居到城市(图 2-4)。

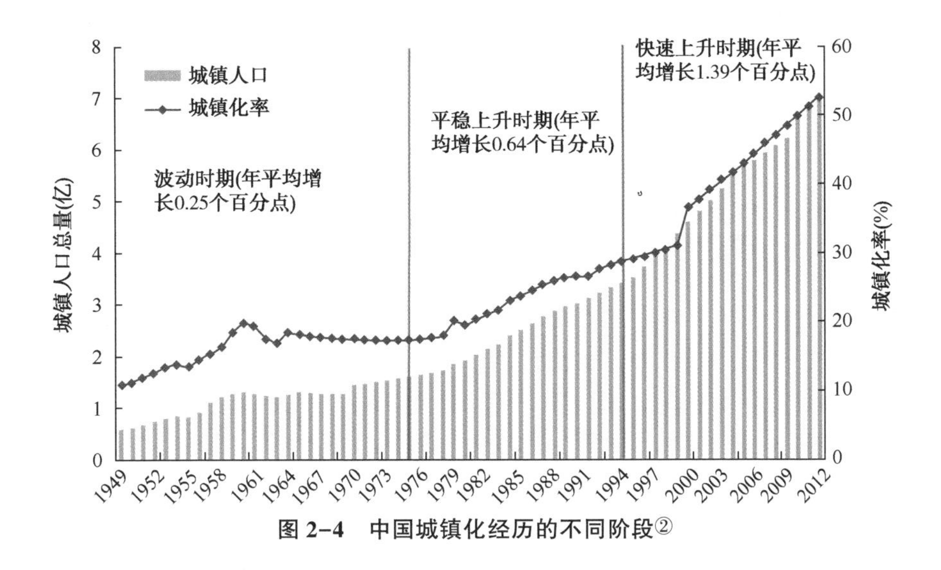

图 2-4　中国城镇化经历的不同阶段②

中国城镇化面临的主要挑战有：协调资源密集使用和环境承载能力；将城市中心和周边地区的发展与生态保护相结合；协调城乡发展，保证各地区公平享有公共服务等。

2007 年和 2008 年，中国国务院分别批准成立了武汉城市圈和长株潭城市群为全国资源节约型和环境友好型社会建设综合配套改革试验区。不仅这些地区，许多地方政府都在探索新的城镇化模式。

然而，中国城镇化用地使用效率不高，城镇空间分布与资源环境承载能力不匹配。2000—2010 年，城市建设用地年平均增长 6.04%，远高于城镇人口 3.85%的年平均增长速度，城镇空间和建设用地快速扩张导致耕地数量不断减少。650 多座城市中有 400 多个缺水，其中约 200 个城市严重缺水。此外，城市群布局不够合理，城市群内部分工协作、集群效率不高，部分特大城市人口压力偏大，特别是城镇建设过程中忽视了对原有自然生态系统的保护，大部分城镇中原生自然生态系统加速萎缩。城镇环境基础设施成本巨大、环保工作薄弱，导致

① 国家统计局，《中国统计年鉴 2012》。

② 中国人类发展报告 2013. 北京：中国对外翻译出版有限公司，2013：14.

区域性城市环境问题不断增加①。

目前的城镇化模式面临用水保障、城市建设用地保障、能源保障(图 2-5)、生态环境质量等瓶颈。城镇化带来的环境问题日益与社会问题交织，给中国的可持续发展带来巨大挑战。相关测算表明，未来城镇化进程对能源的需求将净增加 1. 89 倍，对水的需求将净增加 0. 88 倍，对建设用地的需求将净增加 2. 45 倍，对生态环境超载的压力将净增加 1. 42 倍②③。

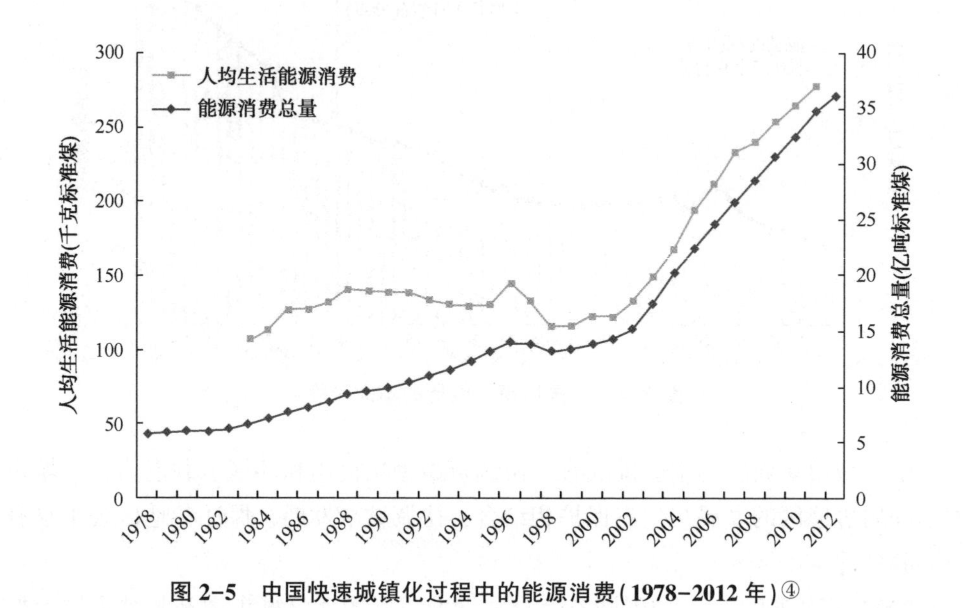

图 2-5　中国快速城镇化过程中的能源消费(1978-2012 年)④

三、环境保护与社会发展面临的障碍

要解决上述环境与社会发展中面临的问题，需要关注一系列主要障碍。我们认为，当前环境与社会协调发展主要面临三大瓶颈，即对环境与社会发展问题认识不足；行动主体的角色不明确，行动责任履行不到位；治理体系不完善。

① 参见国合会课题组报告：中国环境保护与社会发展研究报告，218-243.

② 方创琳，方嘉雯．解析城镇化进程中的资源环境瓶颈．中国国情国力，2013，(4) 33-34.

③ 方创琳，王德利．中国城市化发展质量的综合测度与提升路径．地理研究，2011，30(11)：1931-1946.

④ 中国统计年鉴 2012.

（一）环境意识与环境信息缺乏

当前，中国对环境与社会之间内在联系的认识还远远不够。因此，环境保护与社会发展尚未成为中国政府关注的政策重点。各级政府仍把经济增长放在首位，要设计一条能同时满足社会与环境目标的发展之路依然十分困难。

很多环境问题已经严重影响中国公众的日常生活，再加上他们掌握的信息或赔偿渠道不足，导致抗议活动增加。如果政府继续忽视这些问题，在组织经济活动和进行城市规划时，就会加剧与公众的矛盾，甚至引发群体性事件。

除此之外，公众对政府和厂商的信任缺失也对环境与社会相协调提出了严峻挑战。当前中国的环境群体性事件多发，即反映出社会出现了“共识断裂”，大多数人对重大问题的观点不一致，甚至尖锐对立，严重影响了社会稳定。因此，要在环境保护和社会发展的政策措施上达成一致意见，或让公众接受环境保护和社会发展的政策将变得愈加困难，尤其是当这个过程需要进行妥协或谈判时。

（二）社会主体的行动责任履行不到位

可持续发展要求所有行为主体履行其各自的责任和义务。当前，企业、公众和社会组织都还没有充分履行环境与社会协调发展的责任。提高个人和组织践行绿色生活方式的决心，让所有社会主体共同支持可持续发展是一大难题。

中国政府已经意识到，要推动生态文明的发展，就要让群众积极参与进来，共同制定和落实可持续发展政策。然而，实际运作机制还很缺乏。目前，对社会与环境相互影响及变化还缺少系统的调查研究，相关民间团体难以发挥应有的作用。同样，企业履行社会责任在中国还处于起步阶段。

（三）公共治理体系存在不足

中国现行的财税体制、法律法规、政策制定不利于实现环境保护与社会发展相协调的目标。例如，虽然近年来环境保护的投入有所增加，但由于中央与地方财权与事权不一致，一些地方政府的资金依然不足，政府提供的环境和社会公共服务在各地很不均衡。

在环境法规领域，环境与社会的相关法治建设仍不完善，实施力度不够，覆盖面不足。群众环境维权意识增强、跨界损害事件增多与环境民事赔偿法律制度不健全、环境民事赔偿和调处能力滞后的矛盾日益凸现。环境违法现象普遍、环境纠纷群体性事件增多与环保法规操作性不强、执法不力的矛盾日益加剧。群众环境信访案件走向复议、复议案件走向诉讼的快速变化趋势与环保部门行政复议工作重视不够、国家环境司法诉讼渠道不畅通的矛盾日益明显。

在政策领域，当前社会发展与环境保护的政策和行动，以及相关的经济政策，往往是分开制定、独立实施。这种分离弱化了这些政策在社会发展与环境保护之间建立关系的能力，降低了通过实施这些政策避免社会冲突演变为重大问题或避免冲突的可能性。在政策设计、制定、实施及评估中，缺乏公众以及其他利益相关方的参与。

最后，中国不同的区域之间、城乡之间、不同性别之间以及不同民族之间存在着不平等现象，主要表现在收入、环境资源和服务分配、环境危害程度、健康威胁和社会保障等多个方面。要想消除不平等现象，造福当代和后代人，就要在充分认识环境与社会之间关系的基础上落实相关政策和行动。

四、国际上对中国环境保护与社会发展的研究

经合组织、世界银行、联合国开发计划署等三个大型国际多边机构对中国环境、经济与社会发展三者之间的关系及政策协调的必要性进行了研究。专栏 2-1 对这些研究报告作了简要介绍。值得关注的是，经济合作与发展组织在两个规划周期以前提出的建议在今天依然适用，而且这些建议的实施变得更加紧迫。

专栏 2-1　国际上关于中国环境与社会发展问题的研究

1. 经合组织的《中国环境绩效评估研究报告》(2006 年)指出，为改善环境与社会发展的协调性，中国需要从六个方面进行改进：一是提升获得良好环境服务的人口比例(包括安全饮用水、基本的卫生和电力设施等)；二是加强对环境健康和风险信息的搜集；三是提高环境健康风险信息披露的质量、频率和范围；四是健全公众获取一般环境信息的渠道；五是进一步加强环境教育和宣传；六是政府需要加强与公众、非政府组织的合作和伙伴关系，推动企业履行社会责任。
2. 世界银行的《中国 2030 年发展研究报告》[①]指出，到 2030 年，中国有潜力成为现代、和谐、有创造力的高收入社会，实现该目标需要实施新的发展战略。主要包括，推进结构性改革以增强市场经济的基础，加快创新步伐和高

① 世界银行. 2030 年的中国：建设现代、和谐、有创造力的高收入社会 . 2012.

效创新型体系建设，通过市场激励、监管、公共投资、产业政策和制度建设等措施，抓住“绿色”机遇，在就业、融资、高质量社会服务和可转移的社会保障等领域提供相同的可及性，从而使所有人享有均等机会和社会保障，确保地方政府有充足财力履行责任等。

3. 联合国开发计划署的《2013 中国人类发展报告》[①]从人类发展和长期可持续发展的角度对中国的城镇化进程进行了深入研究。该研究认为，在城镇化过程中，要把人的发展作为落脚点和出发点，加强社会领域的治理将是未来成功的关键，如果没有强有力的治理结构和机制，中国将难以面对未来城镇化的巨大挑战。

① 联合国开发计划署．人类发展报告，可持续与宜居城市——迈向生态文明．2013.

第三章 国际实践与理论视角

在本章，我们将从三个方面阐述环境与社会问题的国际实践和理论视角。第一，工业革命以来环境与社会的关联和互动发展情况。第二，国际上关于环境与社会关系的理论研究，包括一些替代性的研究方法和观点。第三，用可持续发展的理念构建环境与社会的一体化政策框架。

一、工业革命以来环境与社会的互动发展

纵观人类发展的历史，现代的生产和生活方式是在工业革命之后逐步形成的。与之相伴的是人类与自然环境之间的关系发生了深刻变化。人类从完全依赖水、土地和生物多样性的农业社会过渡到城市和后工业社会。在这个进程中，人们有时会忘记人类的生产和健康仍然必须完全依赖于自然环境，甚至认为人类可以控制和取代来自于自然环境中的物品和服务。但是，自然环境正在给我们敲响警钟，甚至爆发了诸多“自然灾害”，而这些也正是人类行为带来的后果①。

（一）历史回顾

专栏3-1，用简要的办法，回顾和阐述了工业革命之后环境、社会和经济发展过程中的重要历史事件。其中最突出的变化是人口由农村大量流向城市，大规模地改变了人类的生产和消费方式；同时，工业化的发展增强了人类利用和改造自然环境的能力，各种工业活动大规模地改变了环境，导致人类所依赖的生态环境开始恶化。人们在尽情享受工业化生产和城镇化生活带来的发展成果时，各种环境问题也从天而降，给人们的日常生活甚至生存带来了巨大威胁，引发了各种

① 一些科学家认为人类对环境的影响太大，我们已经进入“人类世”的时期。http：//www. anthropocene. info/en/home.

社会问题。环境与社会问题的关系越来越密切，已成为现代社会的重要特征①②。

专栏 3-1　影响环境与社会的主要事件和行动时间表

时间	主要事件和行动
18 世纪 60 年代	·第一次科技革命，蒸汽机作为动力机并被广泛使用，社会财富随着生产工具的机器化暴增，社会发展进入大工业经济时代。 ·人口大量流向城市和工业城镇，导致生活环境变差并造成局部污染。 ·自然资源，特别是发展中国家的自然资源，被大规模地开发和利用。
19 世纪 20 年代	·全球人口、人类活动和环境压力急剧增长。 ·欧洲人口快速增长，大量人口移民到美洲、大洋洲和非洲，全球人口分布发生变化③。
1870—1900	·第二次科技革命，以电力的广泛应用、内燃机和新交通工具的使用、新通讯手段的发明和化学工业的建立为标志，世界进入“电气时代”，石油成为主要能源。 ·西方国家采取了大量措施，颁布了一系列污染防治法规，如英国《碱业法》和《河流防污法》；日本大阪府《工厂管理条例》。美国、法国等国也陆续颁布了防治污染的法规。建立国家公园等对生态环境进行保护的措施越来越多。 ·城市规划进入了新时期，重点关注水供应和公共卫生。
1900—1920	·地方性的社会组织积极参与自然和景观保护，努力获得自然和文化遗产的长期所有权。 ·在城市规划中，通过改善室内外生活环境，将公益住房的推广与环境、社会发展目标联系起来。
1920—1950	·工业化国家煤炭、冶金、化学等重工业以及战时工业大规模地建立和发展，由于缺少针对性的污染防治措施以及对生态效益的忽视，大气、土壤和水域遭受到持续不断的污染；战后城市化也带动了郊区得到发展。 ·人类历史上第一次环境问题爆发高潮，包括日本足尾铜矿区废水污染农田事件；比利时马斯河谷工业区大气污染事件；美国洛杉矶光化学烟雾事件和多诺拉烟雾事件等。

① 20 世纪 60 年代早期到 2012 年间更详细的关键环境事件和行动可向可持续发展国际研究所(IISD)咨询 http：//www. iisd. org/pdf/2012/sd_ timeline_ 2012. pdf.

② 世界范围内地区差异评估(考虑环境和社会问题之间的时间轴和关系)见 www. unep. org/GEO <http：//www. unep. org/GEO> 和 Kok，M，等人：发展环境-从全球环境评估中的政策教训。荷兰环境评估局。荷兰：Bilthoven；2009 - www. pbl. nl <http：//www. pbl. nl>.

③ Vries，Bert JM de：可持续性科学。剑桥大学出版社，2013 年。编入全球环境历史数据库中的数据 http：//themasites. pbl. nl/tridion/en/themasites/hyde/.

续表

时间	主要事件和行动
1950—1970	·第三次科技革命爆发，原子能、电子计算机、宇航工程、生物工程等领域出现了重大发明和突破。 ·二战后西方大国竞相发展，工业化、城市化进程加快。工业活动和私人汽车快速增长，健康问题随之急剧增加。 ·各种大气、土壤和食品污染公害事件加剧，包括1953—1965年的日本水俣病事件；1955—1972年日本富山县的骨痛病事件以及1968年在日本九州爱知县等23个县府发生的米糠油事件等。 ·1969年美国制定国家环境政策法令之后，西方国家相继成立环境保护专门机构，并颁布和制定了一些环境保护的法规和标准，以加强法治。 ·国际层面对环境变化(跨境空气污染和水污染，区域和全球问题)的意识得到提升，比如加拿大和美国共同建立的国际联合委员会。
1970—1990	·非法采伐和开荒、过度捕捞、平流层臭氧耗竭、化学污染、气候变化等一系列问题逐渐被公众所了解并要求采取相应的行动。 ·工业化国家重要流域逐渐得到修复，城市空气污染也得到控制。 ·环保部门以及非政府部门开始用专业化、集成化和系统化的手段解决公共卫生、自然景观和资源等领域的问题。 ·重大工业事故频发，导致了法律法规的强化以及社会自发志愿行动的增加，比如针对化工企业的企业责任等。1970年，上千个组织参与了第一个地球环境日活动。 ·1972年斯德哥尔摩召开联合国人类环境会议。全球环境变化的人文因素计划得以建立。世界自然保护联盟提出“可持续发展”，1987年联合国环境与发展世界委员会对此再次进行重申，可持续发展逐渐被世界各国人民所接受。 ·联合国环境规划署建立，通过联合国及其机构发挥协调和促进的作用。 ·20世纪70年代早期，人们发现并开始关注全球性的环境变化问题。国际非政府组织(如罗马俱乐部)联合国际的、国家的和当地的公民社会组织参与环境行动。 ·形成了一些区域性多边环境保护协议，如欧洲《远距离越境空气污染公约》，以及一些针对荒漠化和化学物质的协议。

续表

时间	主要事件和行动
1990—2010	·人类进入全球化时代，计算机网络技术、信息技术、生物技术、基因工程技术、微电子集成技术等学科高度融合并产业化。 ·1991 年，环境与发展部长级会议在中国召开，通过并发表了《北京宣言》。次年，中国环境与发展国际合作委员会在北京成立。 ·1992 年在巴西里约热内卢召开联合国环境与发展大会。会议通过了《里约热内卢宣言》和《21 世纪议程》两个纲领性文件，标志可持续发展被全球持不同发展理念的国家普遍认同。 ·2002 年，联合国可持续发展世界首脑会议在约翰内斯堡举行，将注意力转到消除贫穷和环境可持续发展上，形成千年发展目标。 ·联合国开发计划署、联合国环境规划署、世界卫生组织和经济合作与发展组织等对国家、地区、全球生态进行系统、综合的评估和展望。 ·通过互联网等各类社会媒体，环境和社会相关问题的传播范围和关注度明显扩大。 ·形成一些国际环保公约，如《维也纳公约》、《蒙特利尔议定书》，达成了一些环境全球治理的共识和原则，如“共同但有区别的责任”等。 ·2009 年在哥本哈根召开世界气候大会，商讨至 2020 年的全球减排协议。
2010 年至今	·2012 年联合国可持续发展大会——里约峰会（Rio+20），集中讨论两个主题：①绿色经济在可持续发展和消除贫困方面作用；②可持续发展体制框架（包括其确立可持续发展目标）。

从以上历史回顾看出，工业社会以来环境与社会问题的关系变得越来越复杂，两者相互依赖相互作用。环境政策及措施的制定越来越多地受到社会问题的影响和制约。下面列举了若干体现两者复杂关系的国际实例。

关于经济和社会的发展关系。在 2002 年约翰内斯堡可持续发展声明中提出，可持续发展的首要目标和基本要求是消除贫穷、改变消费和生产模式、保护管理经济社会发展的自然资源基础，这已经达成共识。①”

关于绿色就业。在里约 2012 期简报中提出，为实现一系列可持续发展目标，全球共同行动，投资大约 1.8 万亿美元，到 2050 年，每年将产生 1300 万个新绿色职位。因为能源供给方面成本较高，通过降低消费，可替代其他一些工作，所

① 2002 年约翰内斯堡可持续发展声明。http：//www.un-documents.net/jburgdec.htm.

以，全球提供的就业机会有所减少，到2050年，每年将产生6300万个体面的新工作，绿色职位将成为解决全球就业问题的方法之一①。

关于可持续与宜居城市。2013年中国国家人类发展报告提出，在短期和长期能够满足人们的基本需求和城市的舒适性要求的城市地区。指标包括：完善的城市规划和设计，城市形态、公共场所维护良好、服务充分，覆盖面积广、文化和传统保存良好、鼓励文化服务、基础设施及文化产业、天空晴朗、水干净以及自然资源的有效利用②。

（二）经验和启示

以上，我们对工业革命以来环境与社会的主要事件进行了回顾(专栏3-1)，这些事件说明，经济快速发展对环境保护和社会发展的影响有利有弊，并加深了环境与社会之间复杂性和相互依赖性。与此同时，科技创新和革命性技术的产生，也带来了两面性的影响，一方面对旧的问题提供了新的解决办法，另一方面也带来了新的问题。通过专栏3-1，我们不难发现，一是城市在现代发展中发挥了重要作用，二是统筹经济、社会和环境协调发展的重要性和迫切性。

通过对这些全球重要事件的总体回顾，我们可以得出一些对中国制定未来可持续发展方案及决策有重要参考的结论：

• 以煤炭为主的资源依赖型工业加速了工业化和城市化进程，但同时也刺激了不可持续的生活方式和消费模式上升，导致了严重的环境问题和社会问题。

• 对处于工业化过程中的国家而言，早在19世纪末期和20世纪30年代之前，一些政府便开始采取措施解决环境问题和社会问题。促使这些行动的主要原因包括：工业化对人类健康产生的直接影响，保护生态系统的需要，以及公众和社会卫生运动使得人们对贫穷、疾病和环境之间的关系有了更加深入的认识。美国和日本采取了很多相应措施，如清理公共水道、制定工厂管理法规、实施公共卫生控制措施和社会政策项目、增加对基本公共服务的投资等。然而，由于缺乏解决新出现问题的有效政策手段，这些行动受到了很大限制。

• 随着科技水平的提高，公众越来越认识到环境污染与人类健康之间的紧

---

① 里约2012期简报：绿色职业和社会包容。http：//www.uncsd2012.org/content/documents/224Rio2012%20Issues%20Brief%207%20Green%20jobs%20and%20social%20inclusion.pdf.

② UNDP中国办事处。2013年中国国家人类发展报告。可持续的和宜居城市：建设生态文明。http：//www.undp.org/content/china/en/home/library/human_ development/china-human-development-report-2013/.

密联系，促使人们对环保问题更加关注。20世纪下半叶出现了更为复杂的公共应对政策，以及各类污染治理和环境保护法规。从20世纪50年代开始，已实现工业化的国家大力开展受污染水道清理和烟雾减排工作，它们制定了新的环境法律并建立起许多环境保护机构，政府用于环境保护的开支也不断增加，如美国、日本环保支出占国民生产总值的1%—2%。20世纪60年代末期出现了环保运动，环境问题逐渐成为全球关注的焦点，1972年，联合国在瑞典首都斯德哥尔摩召开了人类环境会议。

• 许多传统的区域性、地方性环境问题演变成为全球性问题，如非法采伐、空气污染、气候变化、过度消费等，开展环境讨论的国际背景发生了巨大的变化。值得注意的是，发达国家的科技进步，以及在环境保护方面上所形成制度框架、治理机制和社会运动并没有减轻对发展中国家的影响程度，而对于经济发展、人类福祉与环境权利的讨论在发展中国家变得越来越激烈。

• 人们越来越广泛地认识到长期的人类活动所产生的问题，如污染物排放、自然资源过度使用，以及生物多样性和栖息地的消失等，都是不可逆的。科学研究表明，全球人类活动对环境产生的影响已经超出了生态系统所能承受的极限。因此，我们要转变风险分析方法，将更多的注意力放在预防措施上。

• 保护环境的行动与其他各种政治、经济和社会事件相互交织。例如，20世纪20年代欧洲的工人解放运动；20世纪六七十年代越战时期西方国家的反战运动；20世纪80年代苏联、东欧等中央计划经济国家的动荡；20世纪90年代以来，土耳其大型水源和采矿项目中的库尔德民族问题。

• 过去四十年中，社会组织和非政府组织在环境保护政策的制定和行动上，以及其他社会问题上发挥了重要作用。目前，发达国家、发展中国家和国际社会都出现大量的公众运动。跨国组织在制定策略行动方面发挥了重要作用，并常常在社会和环境问题研究上位于前沿。

• 在一些发展中国家，如中国，受过良好教育、有钱、有知识的中等收入阶层具有更高的环保意识，因此对政府提出了全新的要求。这些要求超出了简单的健康、生计和从眼前利益出发的环境问题的范畴，还提出了一些更高的诉求，包括参与决策、将治理过程透明化、拥有信息知情权以及要求政府重视环保问题等。因此，经济富足和期望值的不断增加也可能使环境问题更加复杂，因为这些因素导致人们对污染产品和不可持续消费提出了新的要求。同时，在信息传播速度越来越快和越来越透明的情况下，不同社会利益群体（社会组织、企业和政府）之间的紧张关系更容易引发公众群体性事件。

● 推动绿色经济转型、建立国际合作关系成为国际潮流。短期看，在促进经济增长和就业方面，绿色经济与传统经济平分秋色，但从长期看，重新分配投资比例可以加强社会和环保效益。根据联合国环境规划署发布的《绿色经济报告》①，从现在起至2050年，每年将全球国内生产总值的2%投资于10个主要经济部门，便可推动全球向低碳绿色经济转型。根据目前的数据，2%的全球国内生产总值约为1.3万亿美元。在各国国内和国际上采取的绿色经济政策的引导下，这笔资金将投资于农业、建筑、能源、渔业、林业、制造业、旅游业、交通等10个经济部门。这样，制定的相应的"绿色发展"战略不仅能促进经济增长，催生大量就业机会，还可减轻对水资源及其他关键资源的压力，而且有助于消除极端贫困和调节气候变化。

● 当前的全球生态危机和环境保护危机使人们重新思考如何协调环境、社会和经济政策，以便在21世纪探索出一条不同的发展道路。世界各国普遍认识到，全球生态环境是关系到所有人的共同问题。人口增长和人均消费的增加使人们开始关注气候变化和地球承载能力等②。根据过去二十年举行的联合国地球高峰会议和对可持续发展目标的探讨，探寻可持续发展道路成为国际社会的首要任务。然而，由于以经济增长为中心的总体格局和对责任分工的争执，很多政策措施事实上无法实施，也使得可持续发展面临严峻挑战。

二、环境保护与社会发展关系的理论及其实践

从目前国内外的研究看，还没有形成用某一个独立的理论或模型能够说明或解释环境与社会问题的复杂和多元关系，以及产生环境和社会问题矛盾的机理，因此很难制定一个能够平衡各不同因素的"综合性"框架目标。本课题组试图制定出一个相对完整的综合性框架，在平衡环境保护和社会发展需求时，可以帮助确定需要优先考虑的重要因素，最终为决策者提供参考和辅助。框架见下文。

（一）重点政策问题和研究领域

我们认为，任何团体或组织在其转型的过程中，都无法回避一些基本的矛盾和问题，如经济、环境与社会目标的设定、资源的再分配或使用、既得利益格局

---

① http：//www.unep.org/greeneconomy/greeneconomyreport/tabid/29846/default.aspx.

② 政府间气候变化专门委员会（IPPC）、斯德哥尔摩应变中心（Stockholm Resilience Centre）和世界野生动物基金（WWF）已经制定了这些方面的近期报告。

的打破等。这使得各种社会主体和团体之间的矛盾会不断上升，加之一些长期存在的不公平和不确定因素，有时甚至会使这些矛盾愈演愈烈。所以，在制定公共政策和管理机制以及机构设置中，必须考虑如何去解决这些矛盾和冲突。

以下是研究环境和社会关系需要重点关注的领域：

- 环境与贫困问题。环境恶化、水短缺和气候变化的压力通过一系列机制传导至社会弱势群体，并加重社会分化。农村贫困人口通常被视为他们赖以生存的自然资源的管理者，但因为缺乏替代性生活资料，不得不破坏环境，反而成为环境的破坏者。城市贫困人口则更可能受到来自其生活和工作环境的危害。与此同时，贫困人口可能因为居住位置或应对能力所限，更容易面临自然灾害的威胁。

- 环境与人口问题。由于技术进步，世界许多地区在减少污染、降低环境恶化和提高资源利用效率方面取得了重要进展。人们对全球人口急剧增长的早期担忧由于人口增长率的下降而逐渐减弱。然而，仅仅通过技术进步，是否能在有限环境资源范围内满足日益增长的人口需求，或是否需要更深入地改变消费和生活方式方面仍然存在争议。许多国家，包括中国正在经历人口的老龄化。这增加了政府预算和社会政策压力，护理成本高昂并不断上升。老年人更易受环境问题（如城市空气污染）的影响，环境导致的死亡率和发病率（及相关费用）随着人口年龄的增加而上涨。大城市人口高度聚集、人口老龄化已经引起世界卫生组织（WHO）、世界气象组织（WMO）和政府间气候变化专门委员会（IPCC）的注意。

- 环境、移民和城市化问题。人口流向城市和城市近郊地区时，一般都会引发很多问题。未来的20到30年里，移民对整个世界都非常重要，尤其是亚洲和非洲①。但是，中国还缺少与城镇化规模相配套的措施。因此，城镇化已成为中国领导人需要密切关注的问题②。

- 环境与健康问题。环境与健康的关系推动了主要公共卫生和环境治理活动，一方面，环境条件的恶化逐渐破坏了人们的健康和处理灾难或冲击的能力，反过来，疾病也使人们更易受到环境危害的影响。与环境有关的公共基础设施和服务，如水和公共卫生、固体废弃物管理等是提高公共卫生的关键因素，一般由国家通过公共卫生程序提供。

---

① 联合国人居署估计世界城市人口比例到2050年有可能从50%增加到70%。http://www.unhabitat.org/documents/GRHS09/FS1.pdf.

② 李克强总理主张新型城镇化：即“人的城镇化”，应以人为中心，确保民族繁荣和促进中国经济增长。http://www.chinadaily.com.cn/china/2013npc/2013-03/18/content_16314958.htm.

• 环境与就业问题。提高就业率是解决生计和贫困问题的重要途径，但是为此也需要提高工作场所的安全性，并改善工作环境。因此，当前普遍要求创造“绿色”就业，为能够推动可持续经济的部门提供技能培训，这将有益于获得社会和环境收益。同时，生态效益增加和越来越多的创新能帮助实现工作场所的环保目标，并且提高企业的竞争力和盈利能力。

• 环境与社会公平问题。在收入较低地区，环境和社会的紧张关系主要指为实现基本生计和福祉，在利用资源(矿产、土地、水)、森林、草地及其他生态系统时引发的冲突。贫困人口的健康更易受到危害，且更容易受到工业及工作场所产生污染的影响。然而，对富裕人口和生活在较发达地区的人们而言，环境问题可能更多地与生活质量或消费与行为模式、对生活方式的期待、对信息的需求等方面有关。

• 环境和可持续消费问题。自20世纪90年代起，该问题成为一个重要的议题。可持续消费包括价值观和行为方式的双重改变，采取的措施有生产绿色消费品、提供绿色市场供应链、形成绿色政府和施行工业“绿色采购”等。可持续消费还需要考虑日益增多的环境足迹，有时还需考虑贸易惯例。对快速发展的发展中大国，如中国，还需要供应足够的资源、增加工业经营和能源利用方面的生态效益。

（二）理论视角和政策关系

从新古典经济学的观点看，环境和社会问题处于从属于经济问题。目前，关于环境与社会关系的普遍观点认为，当社会总财富达到一定水平后，就会具有一定的经济手段、成熟的政治体系和制度以及技术能力来应对环境挑战，正如“环境库兹涅茨曲线”①中描绘的。这种观点有两个假设：(1)经济增长可以增加收入、消除贫困，从而改善环境和社会；(2)市场机制是可持续增长的最佳催化剂。但这两种假设往往会导致一些错误的诠释和结论。事实上，虽然比较富裕的国家在污染环境问题上具有更多的经验，但没有证据可以表明收入与环境保护之间存在必然的联系。

然而，即使在主流经济学中，人们对市场的局限性也是公认的。市场存在某些缺陷，有时还会失灵。这些局限性适用于以下多种情况：如环境服务和公共资源等“非市场”商品和服务；大气污染等外部效应或公有物(或公害)；污水系统、

① Selden T M, Song D. Environmental quality and development: is there a Kuznets curve for air pollution emissions? . Journal of Environmental Economics and management, 1994, 27(2): 147-162.

排水系统、公共卫生、公共交通或能源供应等自然垄断产品和服务。碳排放总量控制和交易体系试图用市场的手段解决环境问题，但这种做法的结果好坏参半。同时，在环境服务中引入定价等市场机制的做法，往往会形成新的不公平。

制度经济学和生态经济学以及其他社会科学（如政治经济学和政治生态学），从不同的研究视角提供了大量的替代性方法。这些方法对经济、环境与社会问题之间更为复杂的关系进行了分析——包括市场嵌入社会的方法，这反映了更广泛的制度安排、社会和权力关系以及不同的价值观和优先级别。例如，制度经济学家阐述了与公共财产资源相关的集体行为问题；政治生态学家则关注环境如何影响或限制经济发展及造成环境退化的结构性（如性别）不平等问题。其他社会学科从社会、伦理、文化和哲学观点出发，对自然和环境赋予了不同的价值认识，并且对风险的认知和对合理性的解释也各不相同。这些多样化观点为认识并解决各种环境与社会相关的分歧和潜在冲突提供了更多思路，也有助于建立一个联系社会科学、自然科学与政策制定的一体化框架。

当今世界各国正在努力建立全球共识，为国际政策和实践建立一个主要框架，即可持续发展框架。1987 年发布《布伦特兰报告》后，可持续发展的主要观点便是经济、社会和环境协同发展，实现“三赢”。在环境限度内为个人和社会创造利益，并保证现在和未来都拥有充足的环境资源和服务。事实上，到目前为止，经济、社会和环境领域仍未得到平等的对待，即便是当前流行的“绿色经济”理念，也主要是强调环境和经济的关系，在社会方面仍然相对较弱。各方面的批评促使人们不断努力重新构筑三者之间的关系。其中一种方法是一种内嵌式的方法，即经济是社会的一部分，而社会和经济都在环境承载力和生态系统功能范围内。这种观念表述见图 3-1。

虽然存在理论和概念上的辩论和挑战，但在人们的共同努力下，仍然形成了大量的分析工具和政策创新，旨在更加系统地解决被忽略的环境和社会关系。如以下三个实例：

1. 资本

世界银行环保工作人员提出一种方法，即通过扩张和收缩资本存量（自然、社会、人力、建造或制造的资本，有些还包括金融资本）来验证经济、社会与环境之间的关系②。此方法证明不同类型的资本可相互影响，例如：自然资源可用

① 乐施会圆环图 http：//www. oxfam. org/sites/www. oxfam. org/files/dp-a-safe-and-just-space-for-humanity-130212-en. pdf.

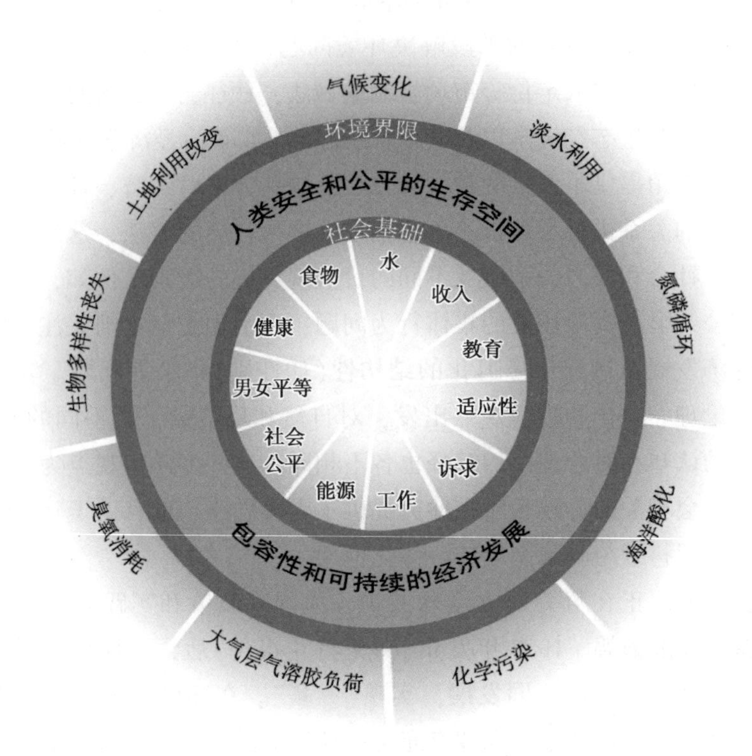

图 3-1　地球承受极限和社会资源利用及环境影响最小化之间的平衡关系①

于投资教育和医疗保障，因此增加了人力资本，而有些人力资本可能会流向公共机构，进而增强社会资本。资本方法论有助于确定以生态产品和服务形式而呈现的必要自然资本、城市绿地的必需开放水平，以及可再生资源的水平，从而为当代人和后代人提供物质需求。

2. 风险评估、环境影响评估(EIA)和社会影响评估(SIA)

这些是决定项目和政策存在的环境和社会问题的重要工具。但是这些工具多为单独使用；最常使用的是 EIA。EIA 和 SIA 一起考虑能更透彻地了解环境保护和社会发展间的关系，并减少结果的不一致性。风险评估能为社会和环境问题提供更细致的定量评估。但是，客观地检测问题存在的风险和概率以及可能产生的影响，需要由独立机构根据完善的科学知识、实际行动和有效监督进行透明的评估。还需要公开信息，确保信息、检测机构和公众传播的可信度。有必要使所有

① http：//www. forumforthefuture. org/project/five-capitals/overview.

受到结果或可能受到结果影响的人都参与到决策制定过程。

3. 环境和区域或城市发展规划

这是一体化解决发展规划中各种社会和环境需求的重要方法。该规划必须综合各种信息，进行大量复杂的分析，解决各种关键问题，比如建立绿色运输系统、绿色停车场和其他开放空间以及解决自然灾害风险并将加剧环境和社会问题的土地利用冲突降到最低程度。

可持续发展的研究方法能够综合反映出各种观念、假设之间的差异，还反应出了基本价值观或传统观念的变更，其中包括市场和国家的相对作用；效率与公平的相对权重；多种物质和非物质资源（如环境或文化）的评价方法；当代和下一代福祉的平衡；“弱永续性”和“强永续性”的不同路径选择等。中国目前注重生态文明，这正反映出思想重点从收入和 GDP 的增长开始转向非物质的商品和服务。在全世界范围内，尤其是 2012 年召开的联合国可持续发展大会情况看，围绕绿色经济的争论也发生了类似的变化。

关于环境与社会的政策框架和重点领域随着时间和空间变化也在发生改变。从时间上看，我们看到工作重点从 20 世纪 60 年代到 80 年代早期（绿色革命时）的技术方法，转到 20 世纪 70 年代到 90 年代盛行的以社区为基础的资源管理方法，再转变到近期的绿色经济方法。最新的方法则是融入环境保护和食品安全因素的综合方法。从空间上看，分清不同地方或不同层级在环境与社会发展上的责权是十分困难的，特别是在一些跨界问题上，这使得应对措施变得更加复杂。

（三）机遇和挑战

尽管在现实中还存在种种挑战，但世界各国用来系统解决环境和社会发展问题的政策创新仍不断涌现。这些政策有可能增加社会公平、降低风险、更有效地实现环境目标。例如，将环境和社会目标纳入长期发展规划和影响评估中，按照目前流域保护中采用的生态补偿方法，建立以环保为目标的社会政策，制定促进教育与培训和绿色就业的政策，以及增强向公众发布环境信息措施，促进公众参与到环境评估中并提高监督等。

当前，世界各国的决策者们越来越意识到，他们对环境与社会关系的重要性理解不到位，对于社会政策在实现环境目标，并将其转为政策和实践的潜力方面的理解也很有限。这种状况亟需转变。目前，最主要的问题包括：环境变化或恶化对人、社区和社会群体健康的影响；人类行为对环境的影响；经济增长、各种不公平现象和资源分配对环境和社会的影响；制定缓解紧张关系和控制潜在风险的各种机制。

人们对社会政策在实现环境目标方面的潜力的认识还不够深刻。社会政策中包括许多公共行动，可以控制生活资料风险、保护人们不受意外事故的伤害（如疾病和收入损失）。人们意识的提高以及公众的参与同样非常重要。社会政策在满足可持续发展要求而进行的各种转变中发挥着重要的作用，如减少由资源获取不公平引起的福利赤字；促进绿色就业和技术转变；创造激励机制，改变消费者行为；促进机构间的社会包容、合作和信任，反过来可降低社会紧张关系和冲突。

所有的变革过程都不可避免的会产生一些新问题，比如收益不平衡、新资源冲突、社会骚乱等，而且没有一蹴而就的解决方案。环境与发展之间的矛盾不会自然化解，也不会通过技术手段和市场得到解决；相反，市场往往会加剧已有的不公平分配现象或权力关系，而技术解决方案对社会和分配问题又不敏感。国际实践和理论清楚地告诉我们：对环境影响意识和担心增加而产生的社会矛盾不会自动消失——尤其是在资源利用压力加剧和城镇化过程中。如果环境保护方面的进展没有跟上这个步伐，人们对解决方案的关注和压力将演变成大的政治和社会问题。

换言之，要解决环境变化的社会问题，国家需要采取一致行动，降低环境对部分群体的负面影响，并形成适当的社会管理制度和公众参与机制。此外，还需要对污染型企业等行为主体采取强有力的治理措施。在中国，各级政府需要提高管理创新能力，为公众社会群体和公民行动提供更广大的空间，明确各种环境与社会权利和责任等。

三、环境保护与社会发展相协调的理论框架

我们面临的重要挑战是将研究重心从只针对某个部门的理论与实践视角，转移到创造统一政策的框架上。目前，世界各国正在努力就一系列国际政策与实践达成共识，该政策与实践为可持续发展提供了更加强有力的框架①。但是，上述框架应以在中国复杂的国情下发展各种机制为基础。

本课题组研究出一些模式，用以展示人类社会、自然环境、环境行为与环境

① 2015年后全球为实现可持续发展目标以及在全球范围内建立绿色经济与绿色经济发展所做的努力（UNEP、OECD等）。

治理四个关键因素之间的关系与相互影响。

图 3-2 模式的依据是：假设适宜的生产和生活行为有利于维持自然环境质量，就能够帮助找到解决环境问题与社会进步的方法，使得环境保护与社会发展更加协调。这里所描述的是一个理想的状态，事实上，无论是政府、企业还是个人，所谓行为主体都存在一些不文明或不合理的环境行为，这也说明在现实生活中，环境与社会的紧张局势虽然是局部的，但问题还是普遍存在的。

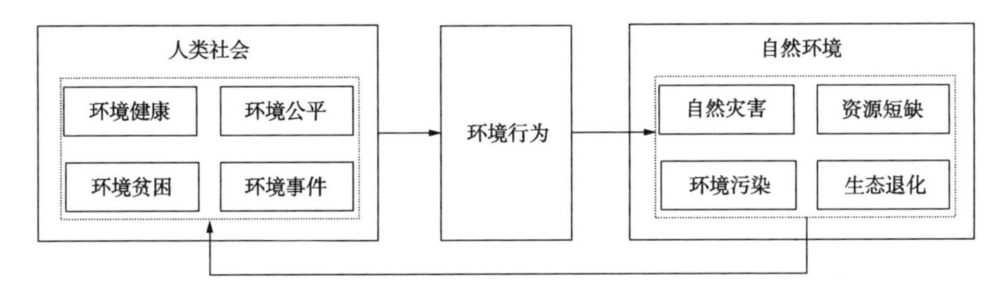

图 3-2　环境行为对环境和社会的影响模式

环境治理是影响环境与社会发展关系又一重要变量。一方面，针对不同的环境行为可以采取有针对性的治理模式和制度政策进行调节；另一方面，通过先进的技术手段可以对出现的环境问题进行治理，遏制环境恶化的趋势，改善环境质量。20 世纪中期之后，随着人类环境意识的觉醒，世界各国开始采取措施对环境问题进行治理，环境行为和环境治理开始成为影响环境与社会关系的双重变量（图 3-3）。

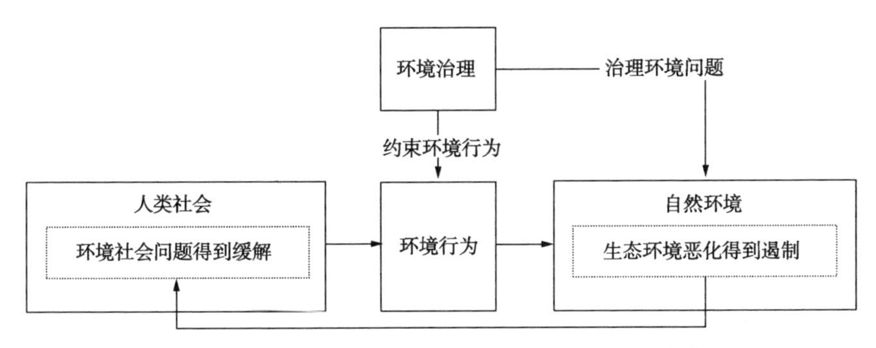

图 3-3　环境行为、环境治理对环境和社会的影响模式

20世纪80年代至今，世界各国逐渐形成了生态环境保护的环境价值观念，环境保护和可持续发展也逐步融入了各国政府的发展理念，人类开始从根本上转变经济社会发展模式，探索经济、社会和环境相协调的发展道路。这种理念在中国主要体现为科学发展观和生态文明的理念。在这种理念下，通过进一步创新环境公共治理模式，可以从根本上改善环境质量、促进社会和谐稳定，实现经济、社会和环境的可持续发展。环境行为、环境治理和环境价值观念三个维度相互作用，共同构成了环境与社会的最新理论关系框架（图3-4）。

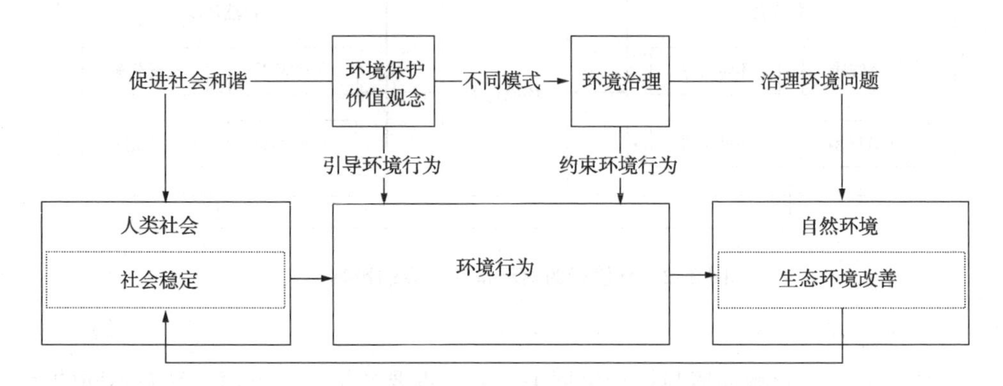

图3-4　考虑环境行为、公共环境治理与环境价值的模式

第四章　行动框架

为实现生态文明建设的长期愿景和目标，课题组提出了以下政策行动框架，建议有关部门通过实施这一框架，将社会发展和环境保护有机结合起来，使环境得到进一步改善、社会发展更加和谐，全面建设“美丽中国”。

一、基本原则

为制定将环境保护和社会发展联系起来的政策，课题组确定了五条指导原则。

• 多方参与性原则。实现可持续发展人人有责。经验表明，中国政府在过去的三十年来专注于经济发展，造成许多环境问题和社会问题。环境群体性事件呈现高发态势，有些甚至损害了政府的公信力。这其中的重要原因是没有为利益相关者和其他社会群体提供足够的参与机会。因此，应当明确不同参与者的作用和责任，进一步转变政府职能，建立能够使个人、企业等发挥积极作用的参与机制，以更好地促进经济、社会和环境协调发展。

• 长期与短期目标相结合的原则。课题组强调，环境保护和社会发展不但需要立即采取实际行动，还需要制定长期目标和计划，为子孙后代提供保障，建设“美丽中国”。因此，在政策制定过程中，不但要对长期目标进行清晰描述，还要制定短期目标，并设定相应的任务。

• 政策目标一致性原则。经济、社会、环境政策应相互依存、相互促进，而不应相互矛盾，在确定政策共同目标以及其各自目标时更是如此。

• 以法制为保障的原则。课题组强调法律法规的重要性。法律法规的设立是为了对经济、社会和环境目标以及要求提供支持，而不是为了满足既得利益群体、企业或个人的某些偏好和主张，这是保持社会长期稳定的重要保障。

• 公平正义原则。环境权益和责任应是公平分配和平等的，不同区域、群体都应拥有平等地享受良好环境的权利，同样，每个社会主体都应当履行保护环境的责任和义务。因此，在制定相应的政策时（例如绿色采购政策、可持续消费政策），不仅要考虑能够保证环境权益在不同群体间的公平分配，还要能够推动各个社会主体积极参与环境保护，履行环境义务。

二、2050 年目标/2020 年行动

课题组提出了环境保护与社会发展的 2050 年目标/2020 年行动框架。考虑到基础设施建设、空间格局调整以及融资所需的时间，应当立即采取行动，以逐步实现 2020 和 2050 年的目标①②③。基于对目标完成进度的倒推，不难看出某些政策应当尽快出台，尤其是在“十三五”规划之前，这样才能确保该目标有机会实现。第五章对需要采取的行动给出了具体建议。初步研究之后，还会进行多项有针对性的、以政策为导向的研究。以下是课题组提出的中国环境保护与社会发展的阶段性目标(图 4-1)：

图 4-1　环境保护与社会发展的 2050 年目标/2020 年行动框架

---

① WBCSD（2010），2050 年愿景，企业新议程，世界可持续发展工商理事会，日内瓦。

② PBL（2009），进入迈向 2050 年的正轨。欧盟辩论引文。PBL 荷兰环境评估局和斯德哥尔摩应变中心。PBL，Bilthoven，荷兰。

③ TIAS（2010），展望——回溯方法的对比研究，综合评估社会 http：//www.tias.uni-osnabrueck.de/backcasting/.

• 到 2015 年，确保实现“十二五”规划中的生态、环境和社会目标。人民生活持续改善，初步形成环境保护的主流价值观念和有利于环境保护的生产、生活方式，促进环境与社会协调发展的法治体系更加健全，管理制度和政策体系趋于完善，环境与社会发展的矛盾和突出问题得到初步解决。

• 到 2020 年，全面建成小康社会。国土空间格局和环境功能区划基本形成并发挥作用，资源节约和环境友好型经济结构和体系基本建立，资源利用效率接近世界先进水平，单位 GDP 能耗大幅降低。随着主要污染物排放量的显著减少，整体环境质量得到明显提升。生态文明观念在整个社会的牢固树立。有利于环境与社会协调发展的法治体系、政策体系、社会风险防控体系和环境保护公共管理服务体系初步完善。

• 到 2030 年，更多环境污染问题得到解决。环境质量全面达到目标要求，群众健康需要将得到满足。生态系统稳定健康并能提供稳定的服务功能，生物多样性得到有效保护。科学合理完整的国土空间布局和环境功能区划形成，经济和产业结构达到生态文明建设的要求。资源利用率达到世界先进水平。生态文明理念进一步普及，形成保护环境的主流价值观和低碳环保的生产方式、生活方式，科学完善的环境公共治理体系基本形成，绿色繁荣的和谐社会初步建立，美丽中国初步呈现。

• 到 2050 年，环境保护和经济社会发展相协调将成为一条准则，人与自然的和谐相处将会达到一个更加合理的水平。将建成生态文明良好的“美丽中国”。大多数人将住在宜居的城市中，大部分生态恶化景观都得到恢复。为减轻气候变化的影响，适应和缓解气候变化的措施得到有力实施。能源利用模式将彻底改变，通过采用高效节能的工业运输系统和实践，对矿物燃料的依赖大大降低。

三、政策领域和行动

为应对现在和将来的各种挑战，寻找社会发展和环境政策之间的平衡至关重要。这需要中国社会不同行动主体的参与来完成，如下所述：

在环境意识维度，要在主流价值观念领域采取政策行动，努力建立与生态文明相符的价值观和行为准则。通过完善法律法规、开展宣传教育、制定和实施政策措施，将环境权利明确为公民的基本权利，使环境保护成为全社会的主流价值观念，让保护环境成为全社会的共同责任和基本义务。

在环境行为维度，要在公众、企业和社会组织三个领域采取政策行动。要采取激励政策鼓励公众积极参与环境保护，通过宣传和教育手段培养公众形成环境保护的生活习惯和行为准则，在全社会建立绿色和可持续的消费模式；通过进一步完善和落实环境保护法律法规、标准，健全环境保护的经济政策和激励手段，培养企业的环保理念，推动企业履行环保社会责任；通过引导性政策，促进公益性环保组织发展，充分发挥行业协会、社区的参与作用，形成环保公益组织广泛参与环境保护的局面。

在环境公共治理维度，应着力于改善立法、社会和环境风险控制，并扩大环境公共服务的分布范围和覆盖面。在法律领域，要立法保护公众对环境保护的知情权、参与权和监督权，进一步完善信息公开、环境听证、环境公益诉讼、环境损害赔偿等制度，严格环境执法，建立独立的环境仲裁和专业审判机构，推动形成较为完善的环境保护法律法规体系和执法体系；在政策领域，要把环境保护放在与经济、社会发展同等重要的位置，加快研究制定环境保护的社会政策，建立健全重大政策的环境社会评估机制和独立的环境政策评价机制，形成环境保护与经济社会发展相协调的政策体系；在社会风险控制领域，要探索重大项目环保社会风险评估机制，健全环保社会风险民意沟通和利益诉求机制，强化社会风险化解工作机制，构建突发环境事件的应急处置机制，探索建立政府、企业和公众定期沟通、平等对话、协商解决的机制；在环境公共服务领域，要建立环境基本公共服务投入的长效机制，积极引导社会资本进入环境公共服务领域，不断提升政府环境信息的公开性和透明度，创新环境保护的社会管理机制。

图 4-1 也概括了未来 35 年内各方参与者在 8 项政策领域内所采取的关键行动。第一阶段是从现在到“十二五”规划期结束。在这期间，重点是建设基础设施和支持体系。此阶段的主要工作是对主要的环境和社会问题进行充分调查、健全法制、在有条件的地区和领域进行环境和社会政策的试点。第二阶段是到 2020 年，即整个“十三五”规划期间，重点是完善制度。这个阶段的主要任务是在确保经济、社会和环境三个领域发展强度下，实现三个领域的平行、协调发展。第三个阶段是到 2030 年。此阶段主要着力于完成中期目标，即环境污染得到全面解决，环境质量基本满足群众身体健康需要，生态系统稳定健康，其服务功能得到明显恢复；第四个阶段是到 2050 年。这个阶段主要着力于远景目标，即实现环境保护与经济、社会发展相协调，人与自然相和谐，全面进入生态文明时代，建成美丽中国。

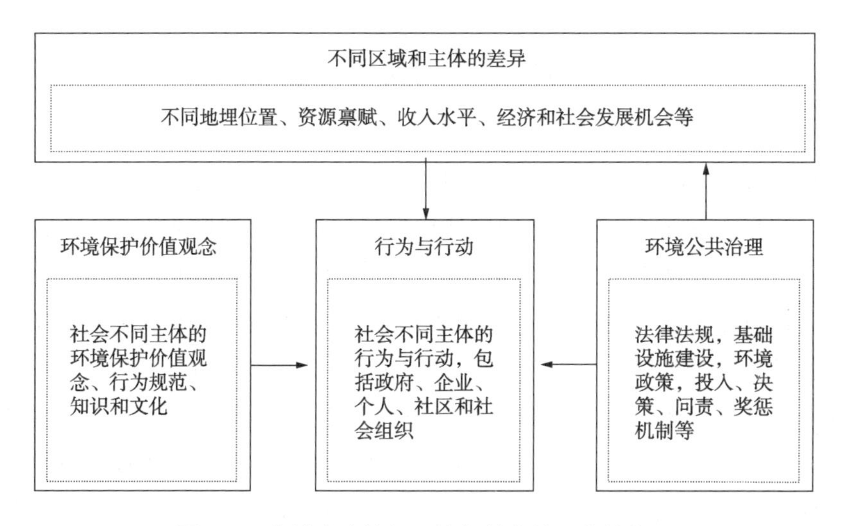

图 4-2　各维度在协调环境与社会关系中的作用

第五章 政策建议

生态文明是一项宏伟的愿景。实现该目标的关键在于通过明确、协调和一致的政策、平衡经济增长、环境保护和社会发展的目标。到目前为止，中国和其他国家对环境保护和社会发展的关系关注较少。本报告中第四章的2050年愿景/2020年行动框架提出了解决这一问题的路线图。该框架将21世纪中期生态文明和美丽中国的长远愿景与近期需要施行的政策和采取的行动有机结合了起来。

考虑到当前面临的突出问题，课题组在若干关键事项上提出了一些政策建议。这些建议仅仅是给出了政策的方向，还需要进行具体的政策设计和制定实施细则。所有任何近期的行动都应作为战略转型的一部分，且应与整体愿景保持一致。例如，环保措施应考虑其社会影响，而社会发展举措应确定并解决其对环境的影响。

课题组注意到，本报告中有一些政策和实践方法已在之前的报告中提出过，但有些内容重复提出是因为需要对其格外关注，并加大其实施的力度。

经过充分论证，课题组提出以下六个方面的政策建议：

一、制定2050年愿景和阶段性目标

课题组的首要建议是，进一步制定指导短、中、长期行动的2050年愿景/2020年行动框架，真正实现中国经济、社会和环境层面的协调。课题组建议按照以下三个维度研究政策框架(图4-1)。第一，环境意识维度。从提升和建立生态文明的主流价值出发提出政策建议。第二，环境行为维度。从公众、企业、社会组织三个行为主体分别提出政策建议。第三，环境公共治理维度。分别从增强法律保障、建立独立的环境政策、提高社会风险控制和提高环境公共服务四个方面提出政策建议。

二、形成生态文明的社会规范和价值观

社会政策的制定以价值观和社会规范为基础。课题组认为，与中国生态文明

有关的社会价值和规范，是制定未来环境保护和社会发展政策和开展实践活动的基础。因此，应当在全社会形成生态文明的社会主流价值，使广大公民认识到，环境权利是公民的基本权利，良好的环境是一项公民的基本福利，而保护环境也是公民的基本义务。政府在促进全社会形成生态文明价值观中负有重要责任。建议采取如下政策措施：

1. 制定教育和培训计划。一是加强干部培训。在各级党校、行政学院和其他培训中心开设环境与社会方面综合课程。二是将环境基础知识和可持续发展理论纳入学历教育，包括九年制义务教育和大学本科教育，宣传倡导相关规范和行为，突出操作性、趣味性。三是通过职业教育和继续教育系统，对各种新生就业人口和已就业人群给予帮助，如为涌向城市的农民工提供培训。

2. 支持理论和政策研究。围绕经济、政治、文化、社会、生态文明"五位一体"的总体建设目标，进一步丰富与中国传统道德和文化理念相契合的生态文明价值体系，使之成为维护社会公序良俗、推动可持续发展的重要基础。

3. 广泛传播生态文明价值观。运用新闻媒体、互联网等传播载体，开展形式多样的社会宣传活动；鼓励社会组织开展各种实践活动，褒奖先进人物，建立教育基地，推广各具特色的文艺作品和出版作品。

三、鼓励所有社会主体发挥作用

为适应日益多样化和多元化的社会需求，我们强调发挥各个社会行为主体的作用，同时在政府、企业、社会部门和民众之间建立一种合作互动的良性关系。建议采取如下政策措施：

1. 鼓励健康、可持续的生活方式。促进形成适度、公平和以人为本的生活方式，强调物质以及精神和文化层面消费的质量而不是数量。通过宣传和教育，培养公众形成可持续的生活习惯和行为准则。特别是发挥社会组织、企业家、以及公众人物在引领健康生活方式的示范带头作用。

2. 公众参与决策过程。通过公开环境信息、立法保证公众环境权益等手段(如第三章所述)来保护和强化公众的知情权。中国大规模的城镇化为此提供了独特的机遇，例如可以尝试创新型、参与式的城市规划方法。

3. 促进企业履行环境社会责任。通过进一步完善和落实环境保护法律法规、标准，健全环境保护的经济政策和激励手段，培养企业的环保理念，推动企业履行环保社会责任。推动企业参与超出环保义务之外活动，如通过开展公益事业与社区共同建设环保设施。通过独立监管和公众参与，对企业环境行为进行评级和信用评价，将企业环境信用评价作为审核企业发债、上市、银行贷款等资格的重要依据①。

4. 支持环保社会组织的进一步发展。社会组织作为独立的评估者和监察者，有助于保护公民权利、提高环境和社会意识、展开调查研究、促成社区活动、保护生物多样性，并且为政策制定者建言献策。这些社会组织超出了当前中国正式注册的范围。因此，有必要考虑改变社会组织注册相关政策，放松其开展环境、社会领域相关活动的限制。很明显，促进发挥公益性环保组织的作用，需要创造条件解决他们面临的注册难、经费难、社会参与难的问题。具体措施包括，一是积极鼓励和引导城乡社区参与环境保护，例如，发挥他们在宣传动员等方面的作用；二是鼓励社会组织广泛参与重大项目的环境影响评估、社会风险评估，推进重大项目立项做到公平、公正、透明；三是鼓励、引导社会组织有序参与提供公共服务。政府购买公共服务中，将环保组织纳入招标范围，这将有利于密切政府和社会组织的关系，弥补政府提供公共服务的不足。

四、加强环境公共治理

实现生态文明愿景的关键是政府制定一套清晰完善的法律法规体系和采取行之有效的政策行动。中国政府应当同时完善环境政策和环境社会发展政策，并保持其一致性。环境和公共健康领域的国际经验表明，实现这些环境与社会目标需要拥有足够的政治意愿②。同时，根据中国复杂的生态和环境问题，有必要建立等同于经济、社会政策的强有力的综合环保政策。建议采取如下政策措施：

1. 从“十三五”规划开始，中国政府将每五年的规划改为“国民经济、社会发展与环境保护规划”。在五年规划中，环境政策与经济和社会并列成为同等重要的内容。

---

① 课题组欧洲考察期间讨论的见解。见 Schijf，B. 和 Boven，G. van，：发展实践中的战略环境评估。近期实践回顾。2012. OECD/DAC 报告 9789264166745（PDF）；9789264166738（印刷体）DOI ：10.1787/9789264166745-en. OECD，巴黎。也见“观点与经验 nr 11，2012，荷兰环境评估委员会”，下载自 http：//www.eia.nl/en/publications.

② 尤其，世界卫生组织在全球案例的基础上，指出实现有效合作通常需要国家上层领导人的强烈支持，同时暗示，中国环境和环境相关问题引起的担忧正好为制定相关政策策略提供了一个千载难逢的机会。很明显，总理为代表的责任制应落实到各级政府。社会事务和卫生部出版物 2013：9. ISBN 978-952-00-3406-1（印刷）ISBN 978-952-00-3407-8（在线发布）URN http：//urn.fi/URN：ISBN：978-952-00-3406-1.

2. 中国各级政府在每年“两会”上所提交的“国民经济与社会发展报告”也相应地改为“国民经济、社会发展与环境保护报告”。

3. 建立重大政策的环境社会评估机制。建立环评结果的追溯机制和责任制，环评单位和个人要对环评报告负责，加大环评违法的处罚力度。譬如，欧盟委员会的事前政策影响评估是实现政策一致性的重要手段，可以作为参考借鉴。

4. 完善政绩考核和政府绩效评价体系。改革政绩考核方法，逐年提高生态环境、社会发展等方面指标在评价体系中的权重，促进地方政府主动在生态环保上加大投入。

五、建立健全环境社会风险评估、沟通、化解机制

课题组建议，凡涉及公民环境权益的重大决策、重大政策、重大项目、重大改革，均纳入环境社会风险评估。政府应该建立一套全面的环境和社会风险评估方法。建议采取如下政策措施：

1. 实行“前置审批”制度。对具有社会影响的重大项目、涉及公众环保权益的政策和改革进行“前置审批”，包括进行程序合法性评估、政策合理性评估、方案可行性评估、诉求可控性评估等。

2. 建立征求和吸纳民意的规范程序。在重大项目决策前，通过座谈会、听证会、社会公示等多种形式，邀请人大、政协、行业协会以及社会各界代表对社会风险评估报告进行审评，以获得民众的理解、信任和支持。

3. 建立环境社会影响问责制。对履行评估程序不严格、造成“评估失灵”的干部严肃处理，对不重视社会风险评估结果的决策者严格问责。

4. 构建突发环境事件的应急机制。制定完备性强、可操作的应急预案，明确各级相应机制的启动条件、启动时间、对应人员及装备等。

5. 提高环境信息的公开性和透明度。在应对环境事件过程中，发布及时、准确和实际的信息，以避免误导、失实报道、猜测和谣言。应该充分利用新媒体平台，例如网络、微博等。

六、提高环境基本公共服务水平

这里强调环境公共服务是为了体现政府有能力实现改善和保护环境、满足公民健康和福利期望的目标。在需要基础设施规划和决策的城乡层面快速城镇化和

显著变革的环境下，提供环境公共服务的政策一致性显得尤为重要[①]。基本公共服务是由政府主导提供的，旨在保障全体公民生存和发展基本需求的公共服务。人类的基本需要包括水源清洁、空气清新、土地肥沃的宜居环境。此外，制度安排、标准和法律等无形服务也逐渐纳入其中[②]。具体行动如下：

1. 制定适当的协调机制，保证环境基本公共服务均等化。合理确定环境基本公共服务的范围和标准，如配备污水处理、垃圾处置等设施；保障公众清洁水权、清洁空气权及宁静权等；环境应急响应机制；环境信息服务，如保障公众环境知情权和环境监督权。

2. 通过购买服务提高环境基本公共服务的水平。例如，调动社会组织开展环保监测、评估和提高环保意识的宣教活动。

3. 逐步提高环境基本公共服务在财政支出中所占比重。建立多种资金渠道，完善中央转移支付和跨区域转移支付的机制，为各地实现环境与社会政策目标提供资金保障。

4. 建立生态补偿机制。加大对重点生态功能区的均衡性转移支付力度，研究设立国家生态补偿专项资金。鼓励、引导和探索下游地区对上游地区、开发地区对保护地区、生态受益地区对生态保护地区的生态补偿。使保护生态环境也可以增加地方收入，造福当地群众。

课题组开展的是初步的、框架性的研究。我们建议下一步开展多个有关复杂重点问题的战略研究，重点包括以下三个课题：

1. 生活方式和行为。国外经验表明，促进生活和行为方式的改变是一项艰巨的长期目标。目前，中国政府在影响生活方式和行为上的直接作用较小，而社会组织和企业家等组织在引领趋势方面更富影响力。因此，应重点考虑设计最有效的举措，避免各种负面影响。

2. 社会发展和环境保护的法律基础。急需探索如何解决与环境有关的社会问题，例如，由污染引起的抗议或“群体性事件”，部分原因是由于在缺乏司法渠道或者司法系统不适于解决这类问题的情况下，公众缺乏表达诉求的其他渠道。有效的法律机制应具备诚信、权威和长期一致性的特征。为此，需要评估的关键方面包括：表达诉求的司法与行政渠道的平衡；实际法律修订过程中的参与制度，包括预期变化的公众通告制度、舆论处理和回应制度以及法律听证制度，

---

① 参见国合会课题组报告：中国环境保护与社会发展研究报告，127-134，218-243.

② 世界银行(2012年)中国2030年。构建现代、和谐和创新的高收入社会。会议版。世界银行和中国国务院开发研究中心。世界银行，华盛顿。

等等。

3. 环境保护和社会发展相协调所需的资金来源。中国高度集中的财政体系为建立和优化环境保护和社会发展的积极关系带来了机遇和挑战。即使有强制的政策，也往往无法及时为各级政府落实相关工作提供适当的资金。因此，建议对如何落实政府愿景所需的资金展开研究。

中国经济社会发展面临很多重大任务，如经济结构调整、公共治理创新，这既是重大的挑战，又是促进环境保护和社会发展相协调的难得机遇。深入研究环境保护与社会发展之间的关系，将有助于制定有效的政策。中国政府在这些方面的成就将吸引许多其他国家和国际组织的关注。

Pandect

1 INTRODUCTION

Sustainable development is broadly understood as a process that must consider simultaneously economic, environmental and social factors. This requires a systemic approach to policy, with an understanding of complex linkages, synergies and trade-offs among these three policy domains. However it has proven to be difficult to conceptualise or implement—in China or elsewhere.

Since the start of its reform process, China has prioritised economic growth development which has led to enormous pressure on the environment in the form of air, water and soil pollution, resource over-exploitation and environmental degradation. Harmful impacts are most directly experienced in areas with soil depletion and deforestation, air and water pollution, water scarcity, and industrial environmental accidents that directly affect public safety, health and livelihoods. All citizens are exposed to risks associated with climate change[①], ecological damage that affects the stability of water basins and coastal zones, and contamination of soil that affects food safety. Population growth combined with increasing incomes, and changing consumption patterns (such as the inclusion of more meat in the daily diet), further exacerbate pressures on an already constrained natural resource base.

Ecological and biodiversity degradation is felt less directly by all, but it is visible in deforestation and desertification and loss of species. Loss of intangible benefits such as the beauty of China's landscapes, and the loss or threat of extinction of iconic species such as pandas and river dolphins, are also of significance to China: indeed, their protection and survival provide hope and inspiration for the creation of a 'Beautiful China'.

It is not surprising then that while economic growth has generated impressive improvements in the living standards of the Chinese people, it has also led to rising ine-

① Intergovernmental Panel on Climate IPCC Working Group I assessment report, Climate Change 2013: the Physical Science Basis concludes that human influence on the climate system is clear, and this is evident in most regions of the globe.

qualities and conflicts related to the environment. Current environmental conditions have created serious concerns for public health, and also contribute to the unjust distribution of resources and consequent living standards across China. An uneven pattern of economic and social development, across regions and social groups maps in different ways onto environmental inequalities, while social and economic inequalities are exacerbated by environmental problems in specific geographic contexts. Poor rural populations are more likely to be located in ecologically rich regions but fragile regions are vulnerable to environmental degradation; they may depend on such environments for their survival, yet further degrade the environment in the pursuit of viable livelihoods. The poor, whether rural or urban, have few if any choices about where to live, so are more likely to suffer from poor air and water quality, which in turn negatively affects their health and productivity. In cities, living conditions for both poorer people and the emerging middle class may not allow for a 'moderately well-off standard of living' even where income levels are above poverty level. At the same time, an educated and financially well-off middle class now has higher expectations and demands for acceptable environmental conditions.

Such factors have led to an intensification of social conflict around environmental concerns. Unrest may be associated with changing land use, siting of industrial activities, and acute incidents such as oil or chemical contamination, or chronic pollution problems such as air quality—problems that disproportionately affect the poor and tend to reinforce pre-existing inequalities or deprivation. Environmental issues are often at the core of protests against large projects, particularly when there is uncertainty about the magnitude of negative social and environmental impacts. Sometimes these concerns may be more perceived than real but nonetheless they need to be addressed. Limitations in the existing information disclosure system, leading to a lack of accurate, freely accessible information, sets the stage for the spread of rumours or inaccurate information, which in turn can aggravate tensions.

China recognised the importance of environmental protection at a relatively early stage in its reform. The Second National Conference on Environmental Protection in 1983 explicitly emphasized the need for a coordinated advancement of both economic development and environmental protection. With the increasing severity of environmental problems, however, the government more recently prioritised a balanced emphasis on economic growth and environmental protection in its 11th (2006-2010) and 12th (2011-2015) Five-Year Plans. Some positive results have been achieved in industrial restructuring, energy conservation and emission intensity reductions, while coordinated efforts in both environmental protection and economic development have been implemented through a series of laws and regulations, programs, and policies. Despite these efforts, major problems persist, with the health impacts of environmental damage in particular attracting increasing public attention.

Consequently attention is now being directed at trying to better understand relationships between the three dimensions of environment, economy and society. This includes

- The relationship between environment and society. Critical aspects include the relationship between the environment and population, poverty and inequality, health and wellbeing, consumption, disaster prevention and mitigation, the provision of basic services, the improvement of peoples' living environment necessary for good health and livelihoods, and the role of public participation in environmental governance.
- The relationship between environment and economy. This covers areas such as links between the environment and economic growth, sustainable agriculture and rural development; industrial growth and environmental pollution prevention and control; sustainable development in the transportation and communications sectors; and sustainable energy production and consumption.
- The relationship between resource utilization and ecosystems. This covers the protection and sustainable use of natural resources; protection of the air, water and soil ecological systems; biodiversity conservation; sustainable utilization of the oceans and seas; desertification control, protection of the atmosphere, and environmentally harmless solid waste management.

For the past 25 years, the notion of sustainable development involving a mutually reinforcing relationship between development of the environment, the economy and society has been represented in a simple diagram such as Figure 1-1. In practice, achieving synergies and managing trade-offs between economic, social and environmental policies and actions necessary for sustainable development has been difficult—as recently reflected in the 2012 Rio Sustainable Development Conference, and has led to efforts to define a set of global objectives for sustainable development. ①

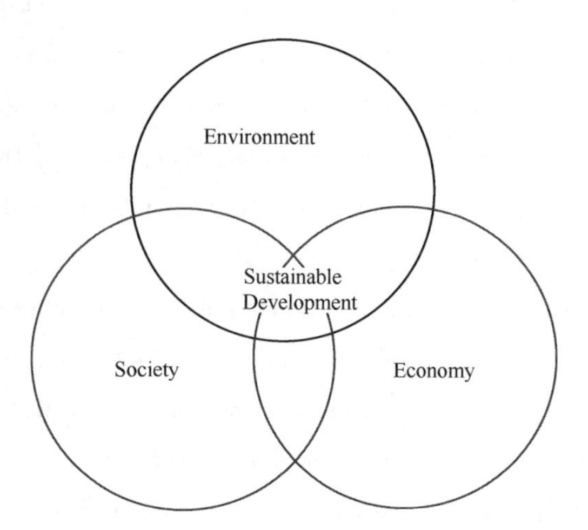

Figure 1-1 A view of sustainable development

Within China, the concept of 'Ecological Civilization', introduced in 2007, provides a vision for a harmonious society, sharing the fruits of development, and safeguarding social justice and equity. ② At the 18th Communist Party Congress in November 2012, the concept of Ecological Civilization was incorporated into the Party's constitution, and became a fifth element added to the existing four pillars of development policy-economic, political, social and cultural. ③ As a result, concerns over the relationship between environmental protection and social development have reached the highest political level, creating an urgent need to deepen the understanding of environmental-social relationships and identify priority fields for action.

This Executive Summary Report of the CCICED Task Force on Environmental Protection and Social Development examines the critical linkages between the environment and social development in China, with a view to suggesting a preliminary framework that can guide policy and practice both in the short and the long term. It also proposes some specific policies that could mutually support both environmental protection and social development. By comparison to the extensive work done by

① http://sustainabledevelopment.un.org/index.php?menu=1300.

② See the CCICED 2008 Issues Paper *Environment and Development for a Harmonious Society*. 26 pp.

③ http://china.org.cn/china/18th_cpc_congress/2012-11/15/content_27118842.htm.

CCICED over the past two decades on the relationship between environment and economy, the current Task Force's effort is ground-breaking in its focus on social development, but therefore can be expected to yield only tentative conclusions and recommendations at this stage.

Three important questions (Box 1-1) have been at the centre of the Task Force's work. Our Executive Summary Report draws upon information reported by the Chinese members of the Task Force in a longer report,[①] and the opinions and expertise of both Chinese and international team members based on a series of Task Force meetings and field visits held from August 2012 to September 2013. The Task Force also focused on urbanization as an issue that illustrates both the linkages among the three dimensions of sustainable development and the opportunities that exist for developing policies and practices that are mutually reinforcing. More complete referencing to support observations and conclusions drawn in this Executive Summary Report can be found in the longer report.

Box 1-1 Key questions to be addressed in this report

1. What is our understanding of the current and future relationship between environmental protection and social development in China?
2. What are the most important opportunities for policy and interventions that would address simultaneously and positively the twin objectives of environmental protection and social development, while minimising unintended consequences?
3. In a rapidly changing global context, how can China combine short-term actions, mid-term objectives and long-term visions to achieve social justice and sustainable development?

Chapter 2 focuses on China's environmental and social achievements and challenges, identifying critical environment, health and social issues and reflecting on the linkages among them. Chapter 3 draws upon international experience in environmental and social development, summarizes domestic and international research findings, and con-

① CCICED Task Force Report, *Report on Environmental Protection and Social Development in China*. In Chinese.

structs a conceptual framework based on notions of values, behaviour, and public governance. Chapter 4 illustrates how this framework could be operationalised by specifying objectives for action over various time horizons—by 2015, 2020, 2030, and 2050. Finally, Chapter 5 concludes by proposing some specific policy recommendations. Some terms used in this Report are explained in Box 1-2.

Box 1-2 Key terminology

Social development is both a process of change leading to the desirable objectives or outcomes decided by a society, and the outcomes or measurable achievements of those objectives. Definitions tend to include material, social and cultural achievements (such as good health and education); access to the goods and services necessary for decent living; a sense of security; and the ability to be part of a community through social and cultural recognition, participation and political representation. Social development is shaped by institutions and actors (such as households, communities, civil society organisations, the media, private or market enterprises, or the state). A core element of any social or 'people-centred' development is participated by all people in decision-making processes that affect their lives; along with mechanisms of accountability, redress and access to justice.

Environmental protection refers to activities, strategies and policy instruments aimed at safeguarding and prudently using the environmental resources that people and societies depend on for their livelihoods and well-being. The disruption of ecosystems, or specific environmental impacts such as pollution or climate change by, for example, economic activities can affect present and future human livelihoods, as well as health and wellbeing. Environmental protection behaviours are influenced by factors such as legislation, individual and group ethics, and education. Increasing understanding of the complex and inter-dependent relationships between living and non-living parts of the environment are seen to require more collaborative policy and action across government departments or between stakeholders to improve information and understanding, manage trade-offs, create synergies and improve policies and implementation.

A public service is provided by a government either directly or through the government financing or subsidising private or social organisation delivery. These services are those which society believes should be available to all people in order to live decent lives, regardless of their income. Examples vary across countries but tend to include energy provision, water, civilian and military security, environmental protection, waste management, education, social security and social services.

A public good is an economic definition of a product that one person can use or consume without reducing its availability to others (non-rivalrous), but from which no one can be excluded access (non-excludable). Examples include sewage systems, public parks, or air. They therefore tend to be provided or protected by the public sector.

Environmental justice focuses on the fair distribution of environmental benefits and burdens as well as equal access to decision-making and recognition of community ways of life, local knowledge and cultural and power differences. There therefore tends to be an emphasis on equality. Wider political and economic inequalities are believed to result in higher levels of environmental harm. In other words, those who are relatively more powerful or wealthy gain benefits from economic activities that degrade the environment. The relatively poor tend to disproportionately bear the costs of such activities.

Environmental rights relate to such things as: ensuring human access to natural resources that enable survival, including land, shelter, food, water and air; the ability to enjoy natural landscapes; and securing environmental justice. They can also include non-human rights such as the survival of a particular species. Environmental rights tend to be seen as basic human rights since people's livelihoods, health, and even existence depend upon the quality of, and access to, the surrounding environment. Environmental rights also tend to include rights to information, participation, security and redress.

2 THE CURRENT CHINESE SITUATION

2.1 Introduction

China has made some progress in harmonizing economic, social, and environmental development as it continues its quest towards sustainable development. However, a number of daunting challenges and obstacles remain. It is not possible in this short report to discuss fully all aspects of either China's achievements or major problems and issues regarding the relationship between the environment and social development. There are many sources that do so in both Chinese and international literature. The main purpose of the chapter is to introduce some of the themes of significance to the relationship between social development and environmental protection, and to provide some examples of both progress and difficulties.

Broad public awareness and major expressions of concern about pollution and environmental change became important in China 20 to 25 years ago. However, from the beginning of history, there is evidence that when China was primarily an agricultural society, the Chinese people took measures to improve and manage their local environmental conditions. Traditional land use practices such as rice paddy terraces and multi-species agriculture and husbandry had positive results for the environment. During often devastating natural disasters, revolution and war, the Chinese people have demonstrated resilience and the capacity to live as a conserving society, with per capita domestic consumption rates, sometimes close to the baseline levels for survival. Thus, until recently, despite its size, China has had a low total ecological footprint in comparison to western industrial countries.

Chinese society was however poorly prepared for the speed of environmental degradation and rising levels of pollution that accompanied a natural resource-intensive process of economic development commencing in the 1980s. In rural areas, the direct impacts on health and livelihoods quickly became major concerns of the rural population. Economic and industrial development, along with creation of infrastructure such as railways, roads and pipelines, has seriously affected the environment and social structure even in isolated communities. With the emergence of an affluent urban middle class, citizens are today increasingly aware of or engaged in environmental issues and activities. Millions of people are expected to migrate to urban areas in the next decade, increasing pressure on the use of energy and natural resources and intensifying the demand for social services.① The changing climate, extreme weather events and natural disasters impact millions of lives each year.

Given the prioritisation of economic growth in the early reform period, insufficient effort or expenditure were dedicated to environmental protection or social development. This changed with the latest (11th and 12th) five year plans (2006-10 and 2011-15). These plans more clearly integrate environmental and social goals, considering both the impact of social and economic development on the environment and the contribution of environmental protection to equitable and sustainable development. The adoption by the Communist Party of *Ecological Civilization* as one of five pillars driving policy is a powerful signal that opens up new possibilities for a strong and symbiotic relationship between environment and social development.

This Chapter describes some of China's recent achievements in environmental and social development. It also raises some unresolved challenges and questions, including how to respond to strong expressions of public protests; how to enshrine in legislation and policy the concept of environmental and social justice; how to develop indicators that accurately and consistently describe the state of the natural and social environment; how to improve the flow of information and increase knowledge to encourage a more informed dialogue about environmental and social impacts; and how to better understand expectations and deficiencies in the respective social responsibilities of civil society, enterprises and all levels of government.

2.2 Accomplishments in Linking Environmental Protection and Social Development

The First National Conference on Environmental Protection in 1973 placed environmental protection on the national agenda, reflecting heightened environmental awareness

① China Human Development Report. 2013: *Sustainable and Liveable Cities: Toward Ecological Urbanisation*, Beijing: China Translation and Publishing Corporation.

on the part of the Chinese Government. Subsequently the Government's increased attention to environmental issues became evident in the introduction of laws and the promotion of such measures as cleaner production, environmental labelling, and corporate environmental information disclosure. The environmental awareness of some Chinese enterprises has improved to the point where corporate social responsibility (CSR) has been introduced, and where banks and other financial bodies are incorporating environmental criteria into their lending practices.

The level of public environmental awareness is on the rise,① as the impacts of environmental degradation are more directly felt, as public understanding of and attention to environmental issues improve, and as the public is more directly involved in activities promoting environmental benefits.

In a 1998 survey, environmental issues ranked fifth after social security, education, population, and employment as an area of public concern. However, in a 2008 poll, environmental pollution ranked third as a public concern.② More recently, media reports about air pollution have also played an active role in raising public awareness and participation in efforts to support amendment to *Air Quality Standards* regulations.

A rising number of non-governmental organisations (NGOs) have come to play an important role in environmental protection. Since the birth of China's first environmental NGO, Friends of Nature, in Beijing in March 1994, the numbers have expanded rapidly: by the end of 2012, a total of 7, 928 environmental NGOs had registered with the Ministry of Civil Affairs (MCA).③

Improving environmental behaviour on the part of various social actors

There are signs that public and social organizations and some larger enterprises are also acting to improve the environment. The involvement of these actors is facilitated by the development of environmental regulations, policies, and standards, social supervision and management, environmental information disclosure and environmental impact assessment of major projects. Public engagement is also increasing, as illustrated

---

① Guodong Yan, Jiancheng Kang, et al. 2010. *China Trends in Public Environmental Awareness.* China Population, Resources and Environment, 2010 No. 10.

② Data come from *Survey Reports on National Public Environmental Awareness* 2007.

③ See CCICED Task Force Report, *Report on Environmental Protection and Social Development in China.* 22pp-23pp, 253pp-254pp. In Chinese.

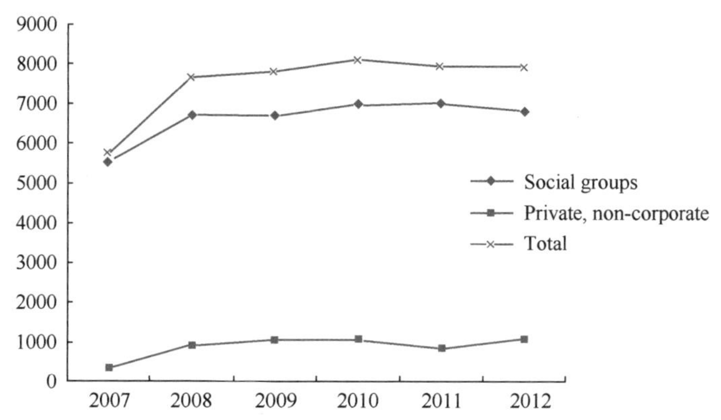

Figu e 2-1 Number of environmental NGOs registered in MCA, 2007-2012

in 2013 by demands for a better environment and for public involvement in national and local environmental decision-making triggered by issues such as air pollution in Beijing and the PX project in Kunming.

Corporate Social Responsibility

Amid growing attention from the media, the public and the government, Chinese businesses have become more active in implementing their corporate social responsibility. With the influence of two important priorities embodied in the 12th Five-Year Plan – improved livelihoods and energy saving and emission reduction – some enterprises have effectively developed business management plans that strategically focus on sustainability. The *Grant Thornton International Business Report* 2011① showed that Chinese mainland enterprises are increasingly aware of social responsibility, driven by such external factors as public opinion, tax incentives, and regulatory policies. That report also showed that Chinese enterprises placed the most emphasis on human resources and environmental protection. In their telephone survey of some medium and large companies in mainland China, 84% of respondents claimed active involvement in employee health and welfare; 75% stated that they improved products or services in order to mitigate adverse environmental and social impact; 69% and 63% of respondents reported efforts to conserve energy and reduce pollution emissions respectively; and 39% were said to have begun to calculate their own carbon footprint.

① See *Corporate Social Responsibility: the Power of Perception*. 24 pp in the IBR 2011 report. http://www.internationalbusinessreport.com/files/IBR_2011_CSR_Report_v2.pdf.

The Task Force visited the Elion Resources Group, a well-known example of one corporate group where social and environmental objectives are jointly considered in the restoration and sustainable development of China's seventh largest desert area.

Legislative and administrative framework for environmental protection

Environmental protection has grown in importance in national decision-making, especially since the 11th FYP period. In 2008, the State Environmental Protection Administration (SEPA) was upgraded to the Ministry of Environmental Protection (MEP), directly under the State Council. Systems for integrated management, pollution prevention and control and supervision and law enforcement have gradually improved. To date, the National People's Congress (NPC) has created 10 environmental laws and 30 resource protection laws. Among them, the *Criminal Law* dedicates a chapter to the "crime of destruction to environmental and resource protection" and the *Tort Liability Law* interprets "environmental pollution liability" in a special chapter. Local people's congresses and governments have developed more than 700 local environmental rules and regulations, and the departments of the State Council, have issued hundreds of environmental regulations, including 69 regulations formulated by the MEP. The SEPA issued *Interim Measures on Public Involvement in Environmental Impact Assessment* and introduced *Interim Measures on Environmental Information Disclosure (Trial)* in 2006 and 2007. More recently, the proposed amendment to the *Law on Environmental Protection* embodies environmental and social provisions, while the new ambient air quality standards released in February 2012, and the new *Atmospheric Pollution Prevention Action Plan* released in September 2013, demonstrate a shift in orientation to addressing environmental health hazards.

In theory, these efforts should provide some additional legal protection to the Chinese public in the expression of environmental demands. Furthermore, a number of new initiatives - in areas such as green credit, environment pollution liability insurance, power tariffs for desulphurization, improved coal-fired power generation, ecological compensation for watershed and mineral development, and ladder tariffs - have been implemented or are being tested in many parts of the country, and have some potential to bring social benefits. However, the real impact will only be seen when these initiatives are assessed, and when legislation, policies and practices are implemented effectively. Weak enforcement is a widespread concern.

2.3 Problems at the Intersection of Environmental Protection and Social Development

In spite of such progress, China currently faces significant challenges: environmental issues are already a major factor affecting social development (in areas such as health, livelihoods and equitable access to resources) and social stability, and may compromise future economic and social development.

Increasing mass incidents caused by environmental problems

Environmental petitions and complaints have increasedby an average of 29% annually since 1996 (Figure 2-2), focusing on such issues as food and water safety, persistent organic pollutants (POPs), hazardous chemicals, and hazardous waste. China witnessed 21, 985 "unexpected environmental events" (environmental incidents) between 1995 and 2010 in such areas as water pollution, air pollution, solid waste pollution, noise pollution, and earthquake hazards. The MEP has handled 927 environmental incident cases since 2005.

Examples of such disputes and mass incidents include the dispute on the Environmental Impact Assessment (EIA) for the Yunnan Nujiang hydropower development plan in 2004, the Xiamen PX project, the Beijing Liulitun waste incineration plant in 2007[①], the Liuyang cadmium pollution incident in Hunan in 2009, Oji Paper's wastewater discharge project near Qidong, and the Shifang molybdenum-copper project in Sichuan in 2012[②].

As a result, construction projects have been suspended or relocated due to strong public opposition and government-led projects such as the Liulitun incineration plant failed to proceed as scheduled because of the strong expression of public concern. Of even greater concern, danger to public health has sparked mass protests, which has even led to the occupation of government offices, judiciary organs, and companies involved, and there have been some cases of rioting. These negative incidents may be factors in social instability. On the other hand, they also highlight the need for greater attention to environmental impacts and scientific monitoring and better mechanisms for early public

---

① Wanxin Li, Jieyan Liu, Duoduo Li. 2012. *Getting Their Voices Heard: Three Cases of Public Participation in Environmental Protection in China*. Journal of Environmental Management, 2012, 98: 65pp-72pp.

② Gilbert, N. 2012. *Green Protests on the Rise in China*. Nature 488(7411): 261pp-262pp.

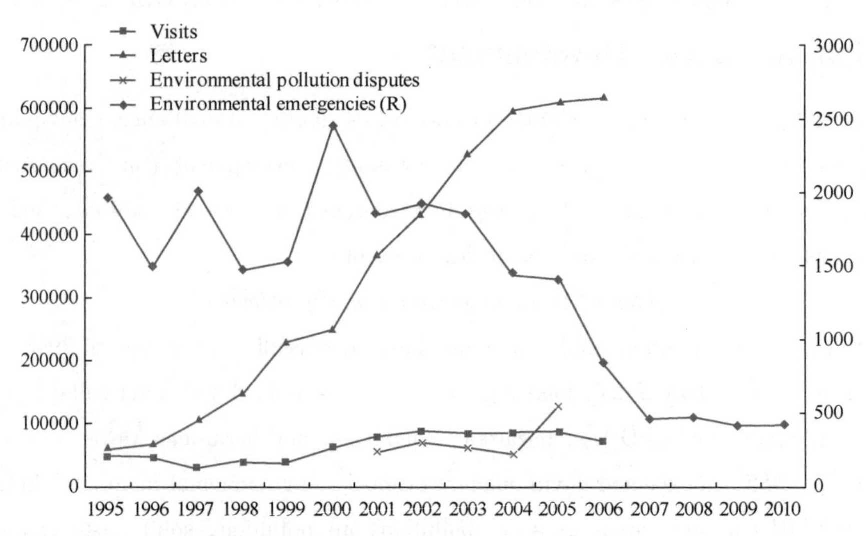

Figure 2-2 Number of environmental petitions, environmental pollution disputes, and environmental emergencies in China, 1995-2010①

input to planning of potentially controversial activities.

Public health hazards caused by environmental degradation

Surface water pollution affects the major economically developed and densely populated areas in China because of the concentration of industry in these areas. Farmland contamination is caused in part by mining and non-ferrous metal smelting. Many Chinese are directly exposed to environmental pollutants at levels much higher than international standards. Given the high concentrations and long duration in the environment of some pollutants, China's large and concentrated population base, complex and diverse channels of exposure, and historical accumulation of environmental pollution, health effects of environmental degradation are difficult to fully eliminate in the short term or perhaps even the medium term. The situation is made more complex by health hazards caused by new environmental pollutants, and the difficulty, in a situation where pollution has many causes, in identifying the main pollutants, pollution sources and health hazards.

The 2010 *Global Burden of Disease* studied by the World Health Organization

① Data source: Environmental Statistical Bulletin, 1995—2011.

(WHO) indicates higher stroke and heart disease mortality in China because of PM2. 5 pollution. Disease caused by outdoor air pollution grew by 33% during 1990-2010, and 20% of lung cancers in 2010 can be attributed to PM2. 5 pollution. ① A study in Xi'an indicated that, with every 100μg/m^3 increase in the PM2. 5 concentration, the total mortality, and the mortality for respiratory diseases, cardiovascular diseases, coronary heart disease, stroke, chronic obstructive pulmonary disease (COPD) would increase by 4. 08%, 8. 32%, 6. 18%, 8. 32%, 5. 13%, and 7. 25% respectively. ② In addition, environmental endocrine disruptors (EEDs), persistent organic pollutants (POPs) and new materials and chemical contaminants can be assumed to further complicate the picture of potential health impacts.

Environmental degradation and poverty

Environmental quality plays a decisive role in people's health, productivity or earning capacity, security, energy supply, and living conditions. In particular, environmental degradation exacerbates the vulnerability of the poor in rural areas who are dependent on land and other natural resources, and intensifies poverty in some regions. In turn, rural poverty may lead to overuse of very limited resources which accelerates ecological deterioration, giving rise to a vicious cycle. Yet the link between these impacts in China's development policies has not been fully recognized.

New social injustices brought about by environmental issues

Against the backdrop of deep-rooted regional, urban-rural, gender and ethnic inequalities in China, environmental and social injustices have become increasingly prominent. These inequalities are reflected in income, access to environmental resources and services, environmental damage, and environmental pressure, and health and social security. In the field of resources and the environment, injustice is manifested in the sharp differences evident in the possession and allocation of environmental resources, and in access to public environmental services among regions and groups. Existing studies related to education and health care call for the integration of public environmental services into the Government's social objectives, given that public environmental services serve as an important means to create a green economy, economic restructuring

① WHO. 2010. *Global Burden of Disease*. http://www.who.int/healthinfo/global_ burden_ disease/en/

② Ke Zhao, Junji Cao, Xiangmin Wen. Urban Residents' Mortality and PM2. 5 Pollution in Xi'an. Journal of Preventive Medicine Information, 2011, 27 (4): 257pp-261pp.

and the desired "service-oriented governmental transition". ① Moreover, the 12th *Five-Year Plan for Public Services* underlines environmental protection as an important aspect in the equalization of basic public services. Solid waste treatment and urban and rural drinking water safety are considered priorities in environmental protection, and access to public environmental services is also an important indicator of environmental justice.

However, studies reveal distinct provincial differences in the environmental performance of basic public services (see Figure 2-3). Most eastern provinces enjoy better public environmental services than do provinces in the central and western regions. Inequality also exists between economically developed areas and economically underdeveloped areas, and between urban and rural areas. It is evident that, if public environmental services are neglected in the acceleration of economic development in underdeveloped areas, environmental injustice will become more widespread and the relationship between environment and social development will deteriorate further.

Mounting pressure on resources and the environment with rapid urbanization

According to the national census, China's urban population grew from only eleven percent in 1949③ to thirty-six percent in 2000 and now to over fifty percent. ④ China's urbanization rate has accelerated and more than 10 million people flow into cities each year, creating huge challenges for local governments. By 2030, it is projected that the urbanization rate will reach 70% and 300 million people will have moved from the countryside to the cities (see Figure 2-4).

The core challenges are to consider how to reconcile the intensive utilization of resources and the carrying capacity of the environment, how to integrate the development of city centres and the surrounding hinterland with ecological protection, and how to coordinate urban and rural development and equitable access to public services among regions. The Wuhan Urban Circle and the Xiangtan Urban Agglomeration, for example,

---

① State Council. *National* 12th *Five-year plan* (2011-2015), 2011.

② Education, health care and transportation, environmental protection are all important components of the public service system. In China, environmental public service is performed by central and local governments for ensuring the basic supply of high-quality environmental products, such as environment infrastructure construction, environmental management, water resources protection and pollution treatment. Hongyou Lu, Guangping Yuan, Sixia Chen, et al. 2012. The Quantitative Measurement of Basic Public Service Efficiency of China's Provincial Environment. China Population, Resources and Environment, 2012, 22 (10): 48pp-54pp.

③ Third National Population Census 1982, 5th National Population Census 2000.

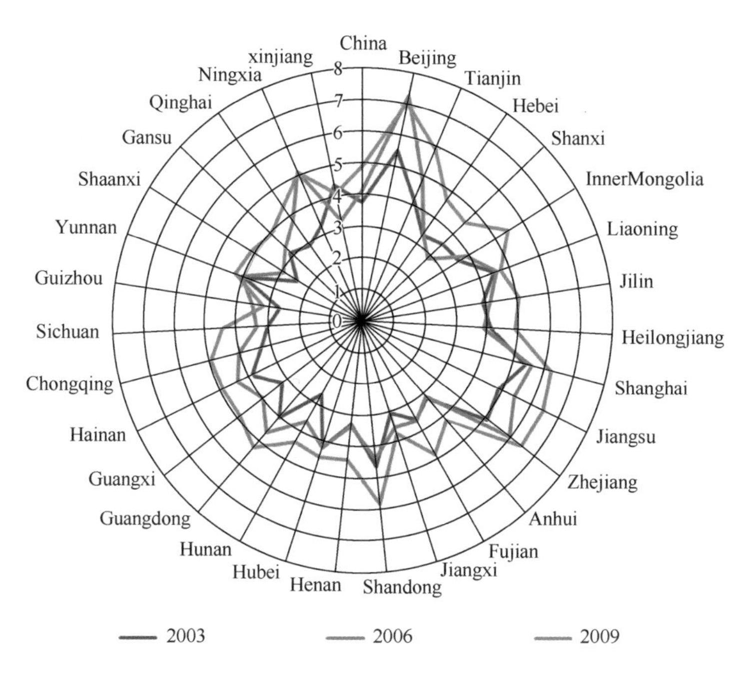

Figure 2-3 Assessment of basic public environmental services by province, 2003—2009①

were authorized by the State Council in 2007 and 2008 respectively as pilot projects in the construction of "resource-saving and environment-friendly society". Subsequently, several local governments have begun to explore new models of urbanization.

Urban land use efficiency is low. Land devoted to urban construction land grew by 6.04% annually during 2000-2010, much higher than the rate of urban population, 3.85%. As a result of the rapid expansion of urban space and construction, cultivated land is dwindling. Over 400 of 650 cities face water shortages, of which about 200 are serious. The spatial distribution of cities and towns does not match the carrying capacity of resources and the environment. Urban agglomerations are not designed appropriately. Mounting population pressure in some large cities intensifies the degree to which environmental carrying capacity is exceeded, while middle and small-sized cities, due to the disparity between cluster industry and population needs, have not fully exploited their potential. Uncoordinated urban spatial distribution and size structure push up economic,

① National Bureau of Statistics, China Statistical Yearbook 2012.

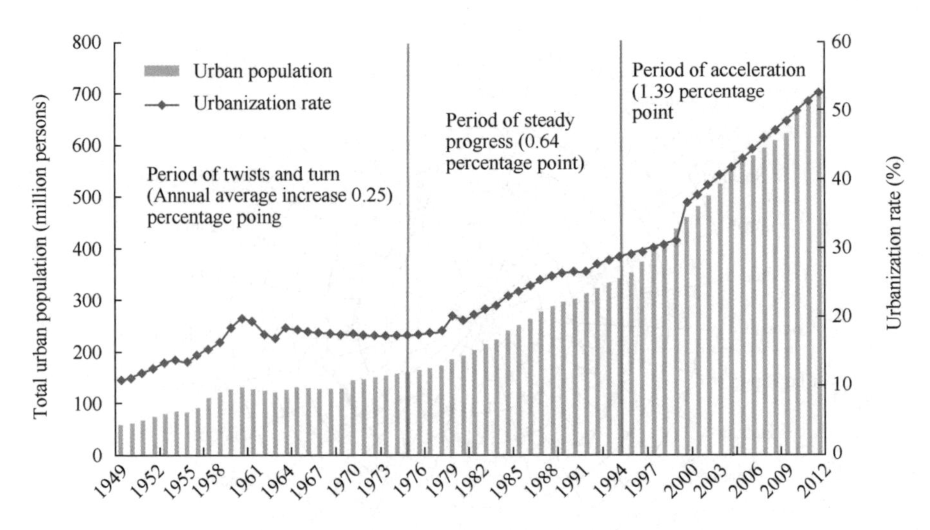

Figure 2-4 Stages of China's Urbanization

social and environmental costs. Further, urban construction frequently ignores the protection of existing natural ecosystems, and in most cities and towns, the native natural ecosystems are withering quickly. Due to the mounting costs of the construction of urban environmental infrastructure coupled with weak environmental protection, regional urban environmental problems are increasing. ①

Current models of urbanization will therefore face bottlenecks in water, available land for construction, energy (see Figure 2-5), and eco-environmental quality. Environmental issues arising from urbanization will become increasingly intertwined with social issues, creating a huge challenge for sustainable development. Forecasts for urbanisation suggest there will be a net increase of 1. 89 times in the demand for energy, 0. 88 times water, 2. 45 times construction land, and 1. 42 times, eco-environmental overload pressure. ②

2.4 Obstacles to Environmental and Social Development

① See CCICED Task Force Report, *Report on Environmental Protection and Social Development in China*. 218-243. In Chinese.

② Chuanglin Fang, Jiawen Fang. *The Analysis on the Resoures and Environmental issues of China's Urbanization*. China National Conditions and Strength, 2013, 4: 33-34.

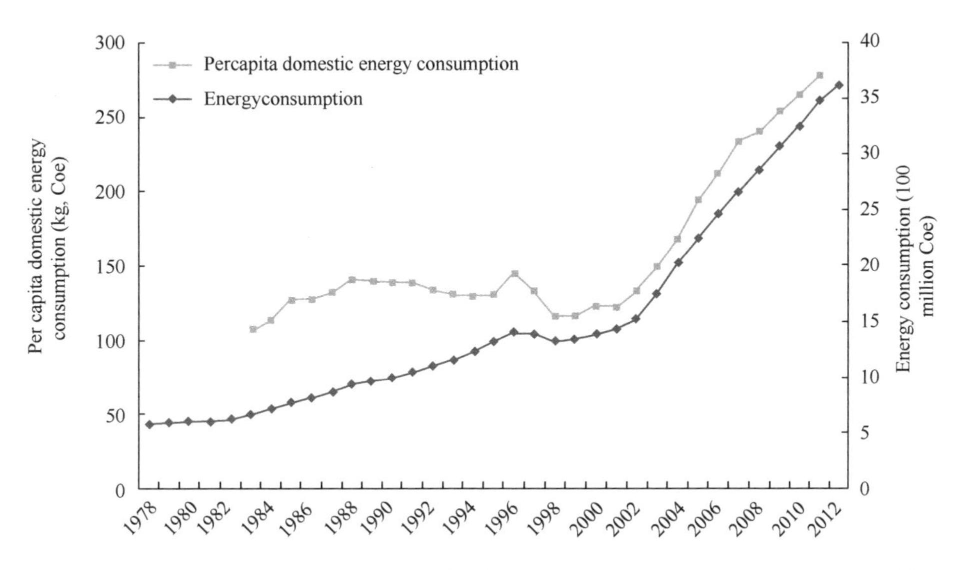

Figure 2-5 Energy consumption in China's rapid urbanization process 1978-2012 ①

Addressing the above problems at the intersection of environmental and social problems requires attention to a number of critical obstacles. Here we focus on three clusters of bottlenecks identified as significant obstacles to progress. These are: lack of knowledge; inadequate identification and fulfilment of appropriate roles of all actors in the system; and, deficiencies in governance.

Environmental perceptions and lack of information

At present in China there is inadequate research, knowledge and understanding about the inextricable relationships between the environmental and social dimensions of sustainability. Consequently those relationships, whether positive or negative, are not yet a well-developed focus of policy concern. The prioritisation of economic growth at all levels of government means that it has been difficult to design a development path that also meets social and environmental objectives. Frequent and serious environment pollution incidents in recent years, such as groundwater contamination, illegal dumping of hazardous toxic waste, and intentional concealment of pollution are all too evident.

The Chinese public is understandably concerned about local environmental issues that directly affect their daily lives, and in the absence of information or alternative

① Source: China Statistical Yearbook 2012

channels for redress increasingly resort to protest. Inadequate knowledge may also lead to government planning, such as spatial or urban planning, that may meet economic objectives but cause unintended impacts on the environment or fail to recognize the way in which the daily lives of people are affected.

Challenges also stem from the lack of public trust in government and enterprises. This is in part a result of poor quality or lack of access to information, for example, official environmental and social reports. The belief that environment-based mass incidents are compromising social stability is leading to government measures that affect the use of media, particularly regarding information sources that may prove to be false or open to varying interpretation. It is therefore increasingly difficult to reach a consensus on policies and measures for environmental protection and social development, or to achieve public acceptance of these policies, especially when the process requires compromise or negotiation.

Inadequate fulfilment of social responsibility by all actors in society

Sustainable development requires the fulfilment of roles and obligations by all actors. Businesses, the public and social organizations are not yet fulfilling their responsibilities in ways that would contribute to coordinated environmental and social development. Nor have they been adequately empowered or mobilized to do so. Raising the levels of personal and institutional commitment to practice a green lifestyle, and to engage fully in consistent support of sustainable, green development is difficult.

Although the Chinese leadership, in its drive for Ecological Civilization, has recognized the need to actively engage people to help improve policy design, delivery and implementation of sustainable development, the actual mechanisms for doing so are still lagging. The full range of activities that could systematically and consistently allow for monitoring, investigating and reporting on social and environmental impacts and changes have not yet been implemented. This lack of supportive systems and infrastructure also make it difficult to create a vibrant and active civil society around these concerns. Similarly, corporate social responsibility is in its infancy in China. There is also limited understanding about the role and limits of the market in meeting social and environmental objectives.

Deficiencies in the public governance system

Our preliminary, analysis has pointed to legislative, financial and structural deficiencies that limit the realization of positive impacts that would come from understanding and acting on social, environmental and economic objectives at the same time. For example, although the investment in environmental protection has increased, local government funding remains inadequate because of fiscal decentralization, and thus the quality and delivery of environmental and social public services among regions is uneven and may create social instability. Notwithstanding the considerable efforts and declarations of the Chinese Government, a green lifestyle has not been realized and the intensity of resources and energy use has not been optimized.

The rule of law underpinning environmental and social development is weak in its implementation and coverage. Increasing public awareness of environmental benefits and the growing incidence of transboundary damage highlights weaknesses in environmental compensation mechanisms and inadequate mediation capabilities. Increasing environmental violations and mass disputes have underlined inadequacies in the operability and enforcement of environmental regulations. Environmental petition and litigation procedures are lengthy and complex, not well understood or managed by officials or the courts. There is also poor access to, and limited availability of, appropriate redress for people and communities.

In terms of policy development, the current policies on social development and environmental protection and their implementation, as well as economic policies are formulated separately and implemented independently. This lack of integration weakens the likelihood of creating a harmonious and productive relationship between them. Not only are opportunities for synergy not identified and maximized but opportunities to deal with inevitable tensions before they become serious problems are missed. Furthermore, public and stakeholder involvement remains inadequate in policy design, formulation, implementation and evaluation.

This also means that incentives that promote negative behaviour patterns continue to dominate. For example, existing societal norms, pricing of goods and services and how policy outcomes are measured, influence what is deemed to be important and are not challenged. Local government officials are not fully trained, evaluated or rewarded on the basis of achieving environmental protection or social development goals. Reinforcing status norms among the public, such as automobile ownership, or wasteful eating patterns, leads to unsustainable consumption.

Finally, China has profound inequalities between different geographic areas, between rural and urban residents, between genders and ethnicities. These inequalities manifest themselves in income, in access to and benefits from environmental resources and services and in relative exposure to environmental harm and threats to health and social protection. Reducing these inequalities for present and future generations is more likely to be achieved with policies and actions that are based on an improved understanding of the relationship between the environment and society.

2.5 International Observations on Environmental Protection and Social Development in China

Three major multilateral organizations have previously addressed the relationship of environmental, economic, and social development in China and the need for policy harmonization among them. Brief highlights of these reports appear in Box 2-1. It is noteworthy that the OECD recommendations, presented two planning cycles ago, remain appropriate today, and indeed have become even more urgent.

Box 2-1 International studies on environment and social development in China

1. The Organization for Economic Cooperation and Development (OECD) stated in its *Environmental Performance Review of China* (2006)[①] that, to improve integrated environment and social development, improvements are needed in six areas: (1) increasing the proportion of the population with access to better environmental services (including safe drinking water, basic sanitation and power.); (2) accelerating the collection of environmental health and health-risk information; (3) improving the quality, frequency, scope, and reach of information disclosure on exposure to environmental health hazards; (4) improving channels for the general public's access to environmental information; (5) improving environmental education and dissemination; and (6) strengthening government cooperation and partnerships with enterprises, the public and NGOs to promote corporate social responsibility.
2. The World Bank indicates in *China* 2030[②] that China has the potential to become a modern, harmonious, creative, and high-income society, but reaching that goal requires a new development strategy. This strategy should include the implementation of structural reforms to strengthen the foundations of a market-based economy; the

① OECD. 2006. *Environmental Performance Review China.*

② World Bank. 2012. *China* 2030: *Building a Modern, Harmonious, Creative, High-income Society.*

acceleration of the pace of innovation and the creation of an effective and creative innovation system; seizing the opportunity to "go green" through a mix of market incentives, regulations, public investments, industrial policies, and institutional development; the promotion of social security for all by facilitating equal access to jobs, finance, quality social services, and portable social security; and ensuring the provision of adequate financing to local governments to enable them to meet their responsibilities.

3. The *China Human Development Report* 2013 of the United Nations Development Programme① conducted in-depth research on China's urbanization process from the perspective of human development and long-term sustainable development. The Report argues that human development should be the over-riding issue and primary benchmark in China's urbanization process and that strengthening governance in the social sphere is the key to future success. In the absence of strong and effective governance structures and mechanisms, it will be difficult to meet the complex challenges inherent in future urbanization.

① UNDP-China. 2013. *China Human Development Report* 2013: *Sustainable and Liveable Cities: Toward Ecological Civilization.*

3 PERSPECTIVES FROM INTERNATIONAL PRACTICE AND THEORY

3.1 Introduction

This chapter elaborates on international experience and perspectives concerning the relationship between environment and society. It examines: (1) the historical association of environmental protection and social development particularly since the industrial revolution; (2) international theoretical research on environment and social development, including alternative disciplinary perspectives; and (3) construction of a conceptual framework for the integration of environmental and social development building on the widely accepted definition of sustainable development. The chapter concludes with some implications for environmental protection and social policies in China.

3.2 Environmental Protection and Social Interaction since the Industrial Revolution

A look into environmental history reveals that the relationship between humans and the natural environment has undergone profound changes in the transition to a modern economy and lifestyle that began with the Industrial Revolution. Yet in that transition from agricultural society, when the dependence on water, land and biological diversity was absolute, to an urban and post-industrial society, the point is sometimes lost that people still have an absolute dependence on nature for their existence and well-being. However much societies may believe that it is possible to "have dominion over nature" or to substitute for the many natural goods and services provided by ecosystems, there are rude awakenings, sometimes in the form of "natural disasters" that often are the result of human action①.

Historical review

This short review (Box 3-1) examines key landmarks at the intersection of environmental, social and economic development, starting with the Industrial Revolution. This marked a critical transformation, with a large share of agricultural labour moving to urban areas, as well as changing production and consumption patterns. Industrialisation enhanced the capacity of humans to use and change their natural environment, while industrial activity led to the deterioration of ecological resources and environmental pollution. Throughout this process, environmental and social issues became increasingly closely linked. ②③

Box 3-1 Timeline of some events and actions affecting environment and society

Time	Major Events and Actions
1760s -	The first technological revolution occurred, in which the steam engine was widely used as a power machine. Social wealth expanded dramatically alongside the production machines. Social lifestyles changed with the development of a large industrial economy. Migration to cities and industrial towns created unsanitary living conditions and local pollution. Natural resources were harvested on a larger scale, often from other less developed countries around the world.
1820s -	The world's population, human activities, and environmental pressure mounted. Because the rapid population growth happened in Europe and large emigration to the Americas. Oceania and Africa, the distribution throughout the world changed as well. ④

① Some scientists believe human influence on the environment is so great that we have entered the Anthropocene epoch. http://www.anthropocene.info/en/home.

② A more detailed timeline of key environmental events and actions from the early 1960s to 2012 is available from IISD. (International Institute for Sustainable Development.) http://www.iisd.org/pdf/2012/sd_timeline_2012.pdf.

③ For regionally differentiated assessment worldwide, considering timelines and linkages between environment and social issues, see for example www.unep.org/GEO and Kok, M, et al.: Environment for Development —Policy Lessons from Global Environmental Assessments. Netherlands Environmental Assessment Agency. the Netherlands: Bilthoven; 2009 - www.pbl.nl.

④ Vries, Bert JM de: Sustainability Science. Cambridge University Press, 2013. Data as compiled in the History Database of the Global Environment (HYDE) http://themasites.pbl.nl/tridion/en/themasites/hyde/.

Continuwtion sheet

Time	Major Events and Actions
1870 - 1920	The second technological revolution was marked by the wide application of electric power, internal combustion engines and new means of transportation, new means of communication, and the birth of the chemical sector. The world moved into the "electric era" and increased use of oil-based energy production which fuelled development. Western countries took a number of measures and enacted a series of laws and regulations, such as the British Alkali Act and Rivers Act, Plant Management Regulations of Osaka, Japan, and early pollution prevention regulations of the United States and France. Conservation measures and national parks became popular. Water supply and sanitation was the focus of attention, especially in the new era of urban planning.
1900 -	Local social organizations were active in nature and landscape conservation and tried to achieve the access to and long-term ownership of natural and cultural heritage. Urban planning linked public housing promotion with environmental and social objectives by improving the indoor and outdoor living environment.
1920-1950	Air, soil and waters were subjected to on-going pollution with the formation and development of coal, metallurgy, and chemical industries, consumer and war-time industries thrived but without much pollution control or eco-efficiency, and post-war urbanization led to suburban development. The first great climax of pollution issues arrived, including the farmland water pollution in Ashio copper area in Japan, air pollution in Belgian Maas valley industrial zone, photochemical smog in Los Angeles and the Donora smog.
1950 -1970	The third technological revolution broke out, with significant inventions and breakthroughs in atomic energy, computer, aerospace engineering, and biological engineering. The Western powers competed for development after World War II, accelerating the industrialization and urbanization processes. Health problems increased dramatically as a result of industrial activities and private car use. A variety of air and soil pollution and food contamination incidents intensified, including the Minamata disease in Japan during 1953-1965, the Toyama Prefecture during 1955-1972, and the rice bran oil incident in 23 counties in Japan, including Aichi and Kyushu in 1968. The Western countries began to set up specialized agencies for environmental protection, and promulgated and developed a series of environmental regulations and standards to strengthen the rule of law, especially after the USA established the National Environmental Policy Act in 1969. The awareness at the international dimension of environmental change (cross-border air and water pollution, regional and global issues) was enhanced, for example through the International Joint Commission between Canada and the USA.

Continuwtionsheet

Time	Major Events and Actions
1970 - 1990	People became aware of and demanded action concerning such environmental issues as illegal logging and land reclamation, overfishing, stratospheric ozone depletion, chemical pollution, and climate change and demanded action. The first Earth Day (1970) involved thousands of organizations. Major watersheds in industrialized countries were gradually restored, with urban air pollution brought under control. New environmental departments and non-governmental departments began to use specialized, integrated and systematic means to address issues associated with public health, natural resources and landscape. Frequent major industrial accidents forced the adoption of more stringent laws and regulations and also increased voluntary action such as Responsible Care in the chemical industry. The United Nations Conference on Human Environment was held in 1972 in Stockholm and launched the International Human Dimensions Programme on Global Environmental Change (IHDP). The concept of sustainable development was raised by the IUCN and in 1987 by the World Commission on Environment and Development (Brundtland Commission) and gradually accepted globally and at community levels. UNEP was founded and began to play a coordinating and facilitating role through the United Nations and its agencies. Global environmental change was recognized in the early 1970s. International NGOs (such as the Club of Rome) joined in environmental action together with other international, national and local civil society organizations. A number of regional multilateral environmental agreements came into existence, such as the European *Convention on Long-range Transboundary Air Pollution*, and some protocols on desertification and chemicals.
1990 - 2010	Humans entered the era of globalization, and computer network technology, information technology, biotechnology, genetic engineering technology, and microelectronics integration technology were becoming highly integrated and industrialized. Ministerial Conference on Environment and Development held in 1991 in China adopted and announced the Beijing Declaration. In the same year, the CCICED was established in Beijing.

Continuwtionsheet

Time	Major Events and Actions
1990 - 2010	In1992, the United Nations Conference on Environment and Development convened in Rio de Janeiro, Brazil and adopted two programmatic documents, namely the Rio Declaration and Agenda 21, marking that sustainable development had been generally recognized by countries of varying ideologies about development. 2002 UN World Summit on Sustainable Development held in Johannesburg, brought attention to poverty eradication and environment links, and to the creation of the Millennium Development Goals. The systematic and integrated assessment and outlook of national, regional and global environment situation, human development and other issues were conducted by the UNDP, UNEP, WHO, and OECD among others. Following introduction of the world wide web, and various social media, communications related to environment and social issues expanded dramatically. A number of international environmental conventions were made, such as the Vienna Convention, and Montreal Protocol, and some consensus and principles on global environmental governance reached, such as "common but differentiated responsibilities". The 2009 Copenhagen Climate Change Conference discussed the global agreement on greenhouse gas emissions reduction by 2020.
2010s -	The 2012 UN Conference on Sustainable Development which took place in Rio focused on two topics: (1) role of a green economy in sustainable development and poverty eradication; (2) an institutional framework for sustainable development including the creation of Sustainable Development Goals.

The above review suggests that the relationship between societal issues and the environment have become complex over time and the dependency between them increasingly close. Environmental protection measures and policies are increasingly constrained and driven by social issues, and the reverse is also true. The following international statements are examples of the clear recognition of this complexity.

Economic and Social Development Links: We recognize that poverty eradication, changing consumption and production patterns and protecting and managing the natural resource base for economic and social development are overarching objectives of and essential requirements for sustainable development. ①

Green Jobs: …Coordinated global action and investments of about US $ 1.8 trillion to achieve a series of sustainable development objectives might lead to 13 million new green jobs per year until 2050. Considering that higher costs in energy supply would replace other jobs through lower consumption, net global job creation would be less, possibly substantially so…Under no conceivable assumptions will green jobs alone be an answer to the global employment challenge to create on the order of 63 million decent new jobs per year until 2050. ②

Liveable and Sustainable Cities: ... Denotes urban areas managed to provide for people's basic needs and comfort in the short and long term. Some indicators include sound urban planning and design, urban form, the availability of well-maintained public spaces, adequate and widely available services, the preservation of culture and tradition, the promotion of cultural services and infrastructure and cultural industries, clear sky and clean water, and efficient use of natural resources…③

Lessons from historical experience

This brief review (Box 3-1) reveals an accelerating pace of economic development and social change that can work both for and against environmental protection and social development, as well as the increasing complexity and dependency in this relationship. The timeline also reveals the influence of innovation and of disruptive technologies that can provide new solutions but also create new problems. It also highlights the important roles that cities have played in development, and the great need to make them liveable.

Even this cursory overview of changes that have been important in shaping development in various parts of the world yields significant conclusions that are important for China's future sustainable development planning and decision-making. Several key points are highlighted below.

- Coal-based and other resource-dependent industries have accelerated industrialization and urbanization, but in the process they have stimulated the rise of unsustainable

---

① 2002 *Johannesburg Statement on Sustainable Development*. http://www.un-documents.net/jburgdec.htm.

② Rio 2012 Issues Briefs *Green Jobs and Social Inclusion*. http://www.uncsd2012.org/content/documents/224Rio2012%20Issues%20Brief%207%20Green%20jobs%20and%20social%20inclusion.pdf.

③ UNDP China. China National Human Development Report 2013. *Sustainable and Liveable Cities: Toward Ecological Civilization*. http://www.undp.org/content/china/en/home/library/human_development/china-human-development-report-2013/.

lifestyles and consumption patterns, leading to serious environmental and social risks and challenges.

- For countries well on in the process of industrialization, government actions to address environmental and social issues can be dated back to the late 19th and first three decades of the 20th century. In general, these actions are driven by: the direct impact of industrialization on human health, but also, sometimes by the conservation of ecosystems; and increasing awareness of the relationship between poverty, ill-health and the environment through public and community health campaigns. The U. S and Japan, for example, took many appropriate measures to clean up public waterways, formulated laws on factory management, and carried out public health control measures and social policy initiatives to increase investment in basic public services. However, with a lack of effective policy instruments to address emerging public policy issues, these early actions were quite constrained.

- More complex public policy responses, pollution governance and environmental regulations were rolled out in the second half of the 20th century, with improvements in technology, public awareness of environmental pollution and human health, and the level of attention given to environmental issues. From the 1950s onwards, industrialized countries began the clean-up of contaminated waterways and smog abatement by introducing new environmental laws and organizations, and increasing government spending for environmental protection, (for example, to levels of 1% to 2% of GDP of the United States and Japan). Environmental movements emerged in the late 1960s, registering citizen concern for environmental degradation impacts on human well-being and the economic costs. Environmental issues gradually became the focus of global attention. In 1972, the first United Nations Conference on Human Environment was convened in Stockholm.

- The international background of discussions on the environment has undergone tremendous change as many traditional regional and local environmental issues have evolved into global issues, such as illegal logging, air pollution, climate change, and over-consumption. It is noteworthy that scientific and technological progress, as well as the institutional frameworks, governance mechanisms and social movements for environmental protection in developed countries generally do not mitigate the environmental impact on developing countries, while the discussions on economic development, human welfare and environmental rights become more heated in developing countries.

- Many long-term changes arising from human activities such as emissions, excessive natural resource use, biodiversity and habitat loss are now recognised to be irreversible. Scientific studies of the global impacts of human activity on the environment suggest that some ecological limits are being exceeded. This shifts the calculus of risk, requiring greater emphasis on precaution and preventive measures.

- Actions to protect the environment are intertwined with a variety of political, economic and social issues, such as the liberation movement of workers in Europe in the 1920s, the Western anti-authoritarian sentiments during the Vietnam War in the 1960s and 1970s, the turmoil of the centrally planned economy in Poland in the 1980s, and the minority (Kurdish) national issues in the large-scale water and mining projects in Turkey during the 1990s.

- Over the past four decades, social organizations and NGOs have played an important role in environmental policies and actions, as well as on other social issues. Civil society activities are now widespread in both rich and poorer nations and in the international community. Transnational networks play significant roles in shaping policy action, and are frequently at the leading edge of social and environmental matters.

- In some developing countries, including China, better-educated, wealthier middle-class citizens with higher environmental awareness have raised new demands from governments. Beyond simple health, livelihoods and short-term environmental issues, their demands incorporate higher aspirations, including participation in decision-making, transparent governance process, information disclosure, and better government attention to environmental issues. In this sense, economic prosperity and rising expectations press for new and better requirements to deal with contaminated products; but also that unsustainable consumption may lead to worsening of environmental problems. Meanwhile, in the context of increasingly quick and transparent information dissemination, the tensions between different interest groups (social organizations, businesses and governments) are more likely to spark public mass incidents.

- Green economic transformation and international partnerships have become a new international trend. In the short run, green economy policies may have similar goals to current policies in promoting economic growth and employment. However, over the longer term, investment should be reallocated to enhance social and environmental benefits. According to UNEP's *Green Economy Report*① an annual input of 2% of the world's GDP, i. e., USD 1.3 trillion at the current level, to ten major economic sectors from now to 2050 could be used to catalyze the transition to a low carbon green economy. Under the guidance of green economy policies at national and international levels, priority areas for investment would include agriculture, construction, energy, fisheries, forestry, manufacturing, tourism and transportation. While stimulating growth and creating jobs, an appropriate 'green development' strategy should be designed to reduce pressure on water and other critical resources, and contribute to the eradication of extreme poverty and the mitigation of climate change.

- The current global ecological and environmental protection crisis is resulting in a rethinking of environmental, social and economic policies, in order toidentify a new path of green development in the 21st century. Countries in the world recognize that the global ecological environment is also a commons of concern to all people. Increasing population and increasing per capita consumption have shifted the world's attention to climate change, planetary boundary limits, and the Earth's carrying capacity.② Through the UN Earth Summits of the past two decades and current negotiations over a set of Sustainable Development Goals, finding pathways to sustainable development has now become a priority for the international community. However in practice a failure to implement appropriate policies, the dominance of economic growth and disagreement over the allocation of responsibilities make this new path a profound challenge.

3.2 Theory and Practice at the Intersection of Environmental Protection and Social Development

No single theoretical or disciplinary approach or 'model' is adequate for illuminating or explaining the complex and multi-dimensional relationships between environmental and social issues, or the conditions under which they lead to conflict. The goal of an 'integrative' framework, that appropriately balances these different elements, remains elusive. The taskforce sought to develop a simple framework that might be of help in identifying some of the more important linkages that should be taken into account when

① http://www.unep.org/greeneconomy/greeneconomyreport/tabid/29846/default.aspx.

② Recent reports on these subjects have been produced by the Intergovernmental Panel on Climate Change (IPPC), the Stockholm Resilience Centre, and the WWF.

balancing environmental protection and social development needs, in order to identify 'win-win' solutions and minimise trade-offs and to provide guidance to policy makers. This section identifies some of the theories and practices that form the basis for such a task; a simple framework is proposed in the following section.

Key policy issues and research fields

Any society or organization in the process of transformation will face certain basic contradictions and problems. These include conflicts between economic, environmental and social objectives, over the reallocation of resources, and among vested interests. As a consequence, tensions among different social actors and groups will arise, sometimes exacerbated by long-standing inequities and uncertainties. Public policy and governance mechanisms and institutional arrangements must play a role in the resolution of such conflicts.

Some key areas of interaction between environment and social issues, potentially leading to tension or conflict, are seen in the following fields:

- Environment and poverty. Pressures from environmental degradation, water scarcity, and climate change pressures fall on vulnerable groups through a series of mechanisms. Environmental issues could therefore exacerbate social differentiation. Rural poor populations are often regarded as the managers of the natural resources on which they depend, but they may also be responsible for environmental degradation, usually because of the lack of alternative livelihoods. The urban poor population is likely to be subject to hazards from their living and working environment. In addition, poorer people may be more prone to natural disasters due to their geographical location or limited response capacity.

- Environment and population. Important progress has been made in reducing pollution and environmental degradation and improving the efficiency of resource use in large parts of the world. This improvement can be attributed to a significant extent to technological progress. A fear that population pressure would wipe out these gains has been mitigated by a slowdown in the rate of population growth. However, controversy exists over whether the needs of a still growing population can be met within environmental limits through technological advances alone, or whether more profound changes in consumption and lifestyles are required.

Another demographic trendis occurring in many countries, including China, with rapid population aging. This places burdens on government budgets and social policies, with high and rising costs in care. Elderly people are also more vulnerable to environmental problems such as urban air pollution, which implies that environmental-related mortality and morbidity (and associated costs) will rise as populations age. Relatedly, the World Health Organization (WHO), the World Meteorological Organization (WMO) and the Intergovernmental Panel on Climate Change (IPCC) have drawn attention to the multilayer effects of metropolitan agglomeration, population aging, and extreme climate change.

- Environment, migration and urbanization. The movement to cities causes many problems both in the countryside and in cities and suburban areas. For the coming 20 to 30 years such migration will be extremely important in countries throughout the world mostly in Asia and Africa.① However none will match the scale of urbanization in China. Thus urbanization has rightly become a matter of intense focus for China's leadership.②

- Environment and health. Major public health and environmental management activities are driven by the relationship between environment and health. Poor environmental conditions undermine the health and the capacity of populations to cope with disasters or shocks; while conversely ill-health increases the vulnerability to other shocks including environmental hazards. Environment-related public infrastructure and services, such as water, sanitation, and solid waste management, have thus been a critical mechanism for improving public health, and are generally provided by states through public health programs.

- Environment and employment. Better employment is a critical mechanism for solving livelihood and poverty concerns, but also requires improvements in workplace safety and the work environment. In this respect, the current thrust towards creating "green" jobs and providing skills training for viable and sustainable economic sectors, may contribute to both social and environmental gains. At the same time, improved eco-

① UN HABITAT estimates that world's urban population is likely to increase from current fifty per cent to seventy per cent by 2050 http://www.unhabitat.org/documents/GRHS09/FS1.pdf.

② Premier Li Keqiang has advocated for a new type of urbanization: "people's urbanization" which should be human-centered, ensure the prosperity of the people, and support China's growth. http://www.chinadaily.com.cn/china/2013npc/2013-03/18/content_16314958.htm.

efficiency and other innovations could help to meet environmental protection goals in the workplace in ways that would also improve the competitiveness and profitability of enterprises.

- Environment and social justice. In low-income areas, environmental and social tensions mainly relate to the conflicts arising from the use of resources (minerals, land, water) and forests, grasslands and other ecosystems to achieve basic livelihoods and well-being. Generally, poorer populations face higher health risks, and are more susceptible to industrial and workplace-related pollution. However, with respect to the rich and those living in more developed regions, environmental concerns tend to relate more to quality of life, or consumption and behaviour patterns, lifestyle expectations, and information needs. Environmental issues rise to become pressing political issues when the environmental economic costs become apparent (e. g., reduced productivity and rising health costs) or social conflicts and protests break out.

- Environment and sustainable consumption has been a major concern since the 1990s but progress on reducing overconsumption in richer countries haring been slow. Sustainable consumption involves a complex mix of values and behaviour changes, and depends upon enabling measures such as access to greener consumer products, green market supply chains, and green government and industry procurement practices. Sustainable consumption also needs to take into account growing environmental footprints and sometimes, trade practices. For large, rapidly developing countries such as China, particular dilemmas include gaining access to sufficient resources, while increasing eco-efficiency in their industrial operations and in energy use generally.

Theoretical perspectives and policy linkages

In mainstream neoclassical economics, environmental and social issues have generally been subordinated to economic ones. A widely held but disputed view is that once societies reach a certain level of aggregate affluence they have the financial means, sufficiently mature political structures and institutions, and technological attainment to respond to environmental challenges, as represented in the 'environmental Kuznets curve'①. While experience does point to more affluent countries effectively tackling

① Selden T M, Song D. Environmental quality and development: is there a Kuznets curve for air pollution emissions? Journal of Environmental Economics and management, 1994, 27(2): 147pp-162pp.

problems of pollution, there is no evidence for a deterministic relationship between income and environmental protection. Furthermore, this relationship hawing been misinterpreted in assuming that growth, by raising incomes and reducing poverty, will lead to better environmental (and social) outcomes; and that market mechanisms are the best facilitators of sustainable growth.

Even within mainstream economics, the limits of markets are acknowledged: markets are subject to imperfections or failures, or simply do not exist. These limits apply to many essentially 'non-market' goods and service such as environmental services and common property resources; to externalities and public goods (or 'bads') such as air pollution; cases of natural monopoly as in many environmental services such as sewage, drainage, public sanitation, public transport or energy supply. Cap and trade mechanisms that aim to combine markets with environmental limits have had mixed results. The application of such market mechanisms to environmental services (for example, through pricing policies) also has strong distributional impacts, tending to reinforce existing inequalities in the absence of strong redistributive measures. Often associated with mainstream economic approaches is a corresponding reliance on technological solutions to overcome environmental constraints. Social issues, in such approaches, tend to be relegated to a residual category with policies aimed at providing minimal assistance only to the most vulnerable.

A number of alternative approaches exist, coming from fields such as institutional and ecological economics or from other social sciences, such as political economy and political ecology. These tend to analyse the more complex links between economic, environment and social issues, including the ways in which markets are socially embedded and reflect broader institutional arrangements, social and power relationships, and diverse values and priorities. For example, institutional economists have shed light on collective action problems related to common property resources; political ecologists are also concerned with how the environment affects or constrains development, and the structural (including gendered) inequalities which are central to environmental degradation. Other social science disciplines point to a range of alternative social, ethical, cultural and philosophical perspectives, allowing for different values attached for example to nature and the environment, different perceptions of risk, and alternative interpretations of rationality. They thus provide greater scope for understanding and addressing sources of disagreement and possible conflict; and for developing more integrative frameworks linking social science with natural science and policy.

The main 'framework' around which efforts are currently being made to build consensus at a global level, and which informs international policy and practice, is that of 'sustainable development'. Following the 1987 Brundtland Report, the dominant view of sustainable development is of three inter-related domains or pillars, with presumed equality between them, which can be reconciled to create a 'triple win' scenario - delivering gains for individuals and societies within environmental constraints and ensuring adequate environmental resources and services now and in the future. In reality, these domains have not received equal treatment, and the social generally remains weakest, as illustrated by the current emphasis on 'green economy' solutions which focus principally on environmental and economic linkages. Multiple critiques have generated on-going efforts to reconceptualise the relationship. One approach is as a nested concept where the economy domain lies within the social domain, and should contribute to social goals, while both need to remain within the (shifting) boundaries of environmental carrying capacity and ecosystem functioning. One representation of this way of thinking can be found in Figure 3-1

Despite these theoretical and conceptual debates and challenges, a range of analytic tools and policy innovations have nonetheless been developed which aim to address more systematically the neglected environmental-social linkages. Examples include the following:

1) Capitals: An approach that has been promoted by World Bank environmental staff examines these relationships through consideration of expanding and contracting stocks of capitals: natural, social, human and built (or manufactured); some also include financial capital. ② This approach recognizes that there can be interplay among these types of capital, for example natural resources can be managed to invest in education and health care, and thus contribute human capital and also, perhaps, to institutions that enhance social capital. The capital approach is useful in defining necessary natural capital in the form of ecological goods and services, essential levels of open

① http://www.oxfam.org/sites/www.oxfam.org/files/dp-a-safe-and-just-space-for-humanity-130212-en.pdf.

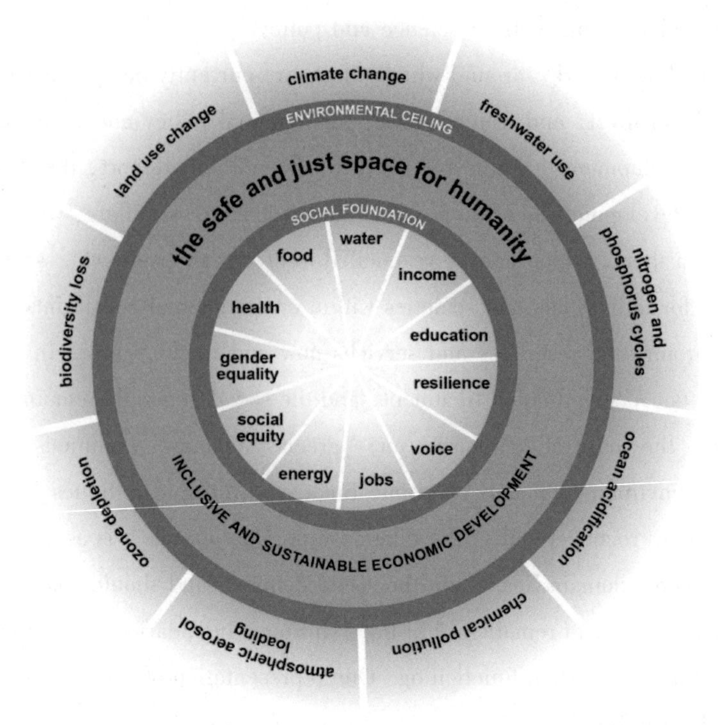

Figure 3-1 The Oxfam 'doughnut' showing the balance between planetary boundaries and a social minimum of resource use and environmental impact ①

green space in cities, and levels of renewable resources to supply the material needs of present and future generations.

2) Risk Assessments, Environmental Impact Assessments (EIA), and Social Impact Assessments (SIA) provide an important set of tools that help to shape decisions regarding environment and social concerns about projects and policies. However these tools tend to be used separately; most commonly an EIA is performed. Joint consideration of EIA and SIA would provide for a more thorough understanding of the relationship between environmental protection and social development and may lead to less conflict over results. Risk assessments allow for a more careful and quantitative assessment of both social and environmental concerns. An objective examination of the nature of risk and probability, and the possible impacts, however, requires that such assessments be

① See http://www.forumforthefuture.org/project/five-capitals/overview.

carried out in a transparent way by independent agents, and based on good scientific knowledge coupled with follow-up action and monitoring of effectiveness. It also requires openness of information flows and credibility in the information, institutions and processes among the wider public. Access to the decision-making process is also necessary for those affected or likely to be affected by the outcome.

3) Environment and Regional or Urban Development Planning have long played an important role as an integrative means of addressing a wide variety of social and environmental needs in development planning. The approach should be inclusive and adaptive both from an environmental and social perspective, and this is often a stumbling block. Also, the planning must draw upon a wide range of information with considerable sophistication in the analysis in order to address key concerns such as the creation of green transportation systems, parks and other open spaces, risks related to the siting of natural hazards, and to minimize conflicts of land use that raise environmental and social problems.

Approaches to sustainable development reflect varying perceptions, assumptions or preferences, and fundamental values or conventional wisdom. Among these are views about the relative roles of the market and the state; the relative weight given to efficiency versus equity; methods for the evaluation of various material and non-material resources (such as environmental or cultural resources); the balance sought between the well-being of current and future generations; and alternative choices between pathways towards 'weak sustainability' through incremental reform of current practices or the more transformative action needed for 'strong sustainability'. As far as China is concerned, the current focus on Ecological Civilization suggests a shift from the dominant focus on income and GDP growth to give more attention to non-material and ecological goods and services. On a worldwide scale, similar changes are observed in the discussions about the green economy, especially since the United Nations Conference on Sustainable Development in 2012.

Policy frameworks and priorities also change over time and space. From a temporal point of view, the focus of work changes, for example in the case of food products from the earliertechnical approach (as in the green revolution) of the 1960s to 1980s; the community-based resource management approach from the 1970s to 1990s; and then to the more recent emphasis on genetically modified crops and the green economy; and now further to a blend of all of these that takes into account sophisticated environmental protection and food safety factors. From a spatial point of view, regional and global differences can also be observed, further complicating responses given the difficulty of determining environment and social responsibilities, and distributing costs and benefits, in different places or at different levels. An added layer of complexity exists for cross-border issues, which may involve local boundaries or international borders, or small and isolated communities.

Current Opportunities

In spite of numerous challenges, innovative environmental and social policies are emerging to address systematically a range of interrelated environmental and social development challenges. Such policies have the potential of improving social outcomes, reducing risk and enhancing social justice while achieving environmental goals more effectively. For example, efforts could include the incorporation of environmental and social objectives jointly into long-term development planning and impact assessments, introduction of environmentally-targeted social policies as is currently done for some eco-compensation efforts for watershed protection, as well as the formulation of policies to promote education and training and green jobs. There is also a clear need to enhance environmental information release to the public beyond steps already taken, and to foster the participation of the public in assessments and improve oversight mechanisms.

Increasingly, decision-makers around the world are recognizing that their understanding of the importance of the environment and societal relationships, and the potential contribution of social policies to environmental goals into policies and practice is limited and needs to be transformed. At present, the issues of most concern include: the impact of environmental change or degradation on the livelihoods and health of populations, communities and social groups; the impact of human behaviour and consumption on the environment; the impact of such tertiary factors as economic growth, various inequities, and resource allocation on environmental and social outputs; and governance and participation, including the establishment of mechanisms to address tensions and manage potential or actual conflicts. In fact, understanding the social context helps to identify and analyze key factors in the environmental and social interaction, such as the role of different social actors and the formation of values, social equity and distribution, and social and public policies.

Among these, the potential use of social policies in achieving environmental objectives has not been fully explored, and in fact, may offer significant opportunities. Social policy encompasses a range of public actions designed to manage livelihood risks, protect people against contingencies (such as ill-health and loss of income) and invest in their capacities to contribute productively to the economy. It is also important in awareness-raising and public participation. Social policies thus have a significant role in the transformations required for sustainable development: by reducing well-being deficits associated with unequal resource access; incorporating environmental risks which disproportionately affect the poor; facilitating 'green' employment and skill transitions; creating incentives to change the behaviour of consumers; and fostering social inclusion, co-operation and trust in institutions, which can in turn reduce social tensions and threats of conflict.

A logical further step in the extension of social policies is thus to incorporate environmental objectives into the existing social policy system. Social policies can be designed to extend beyond the scope of protection and compensation mechanisms, to support a structural change towards sustainability of lifestyles, consumption and behaviour of individuals, businesses and governmental bodies, while ensuring the fairness of the results. These mechanisms may include the collective supply of social and environmental public goods, housing, energy and infrastructure investment for the poor, and low-carbon consumption incentives. In addition, the mechanisms should also cover the design and implementation process of public policies, the right of civil and social institutions to influence decisions, protection of the rights and interests of vulnerable groups, or relevant systems to supervise business and government and improve their accountability.

All change processes are inevitably accompanied by newproblems, such as unequal benefits, new resource conflicts and social unrest, and generally there are no easy solutions. Conflicts between the environment and development will not resolve themselves, or be resolved strictly through technical means and the market. On the contrary, the market tends to exacerbate the existing unfair distribution and power relations, while technological solutions tend to be insensitive to social and distributional issues. International practice and theories also clearly tell us that the social conflicts resulting from increasing environmental awareness and concerns over impacts will not melt away automatically, especially as resource use intensifies, and urbanisation proceeds. If environmental protection improvements do not keep pace, public concern and pressure for solutions will become a greater political as well as social problem.

In other words, to solve the social problems created in a changing environment, concerted and coordinated actions by governments are needed to reduce the negative environmental impact on some groups and to resolve conflicts. Such approaches need to be supplemented by appropriate governance, social management and participatory mechanisms. In addition, strong actions should cover polluting enterprises and local governments. The implication for China is that the government at all levels needs to be more innovative in the management of problems, as well as providing a wider space for civil society groups and citizen action, and clarifying environmental and social rights and responsibilities.

3.4 Explorations for a Suitable Framework Linking Environmental Improvement to Societal Action

An important challenge is to shift the focus from sector-specific theoretical perspectives and practices towards creation of an integrated policy framework. At the moment, countries around the world are making efforts to reach a consensus on an international set of policies and practices that provides a more robust framework for sustainable development.[①] However, such a framework will need to be grounded in the development of mechanisms that are relevant to, and implementable in, local situations in countries as complex as China. The contribution of this Task Force toward such an effort will indeed be modest due to limitations of time, and to some extent differing views among members.

The Task Force examined several simple word models to demonstrate the links and feedback loops between four key factors: human society, the natural (and in some cases also the built) environment, behaviour towards the environment, and environmental governance. These four factors are interactive.

The model in Figure 3-2 is based on the assumption that appropriate behaviour in

---

① See the efforts for a post-2015 set of global sustainable development goals. http://sustainabledevelopment.un.org/index.php?menu=1300; and also efforts to establish green economy and green growth experience throughout the world (UNEP and OECD among others).

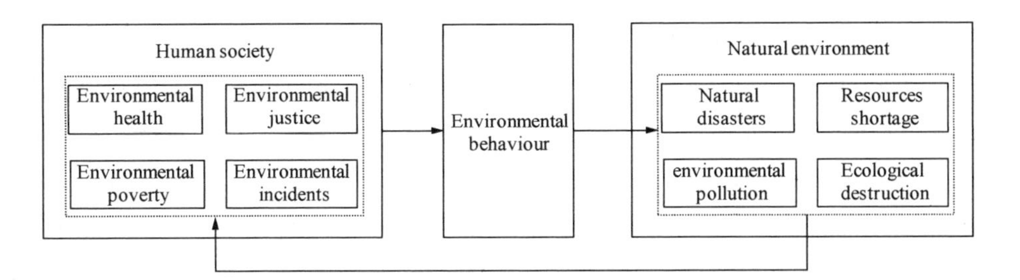

Figure 3-2 Model considering human behaviour towards the environment and the role of environmental governance

production and living is conducive to the quality of the natural environment, thus contributing to the solution of environmental problems and social progress, and thereby the coordinated environmental protection and social development. What is described here is an ideal, and points to the need for adjustment of uncivilized or irrational environmental behaviour—whether on the part of government, businesses and individuals. In reality, however, tensions between environment and society may be addressed at a local level, but remain problematic overall.

Figure 3-3 reveals in somewhat more detail what must be considered in both responding to environmental problems, and in satisfying society that problems are being properly addressed.

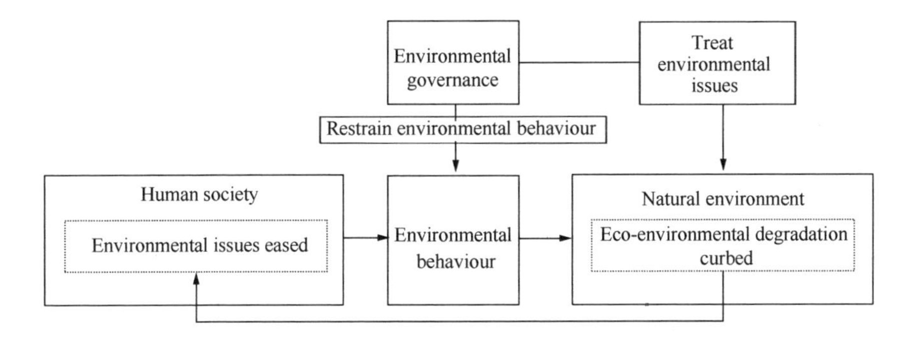

Figure 3-3 Theoretical model considering environmental behavioural variables

In Figure 3-4 a preliminary effort has been made by the Chinese members of the Task Force to put forward a conceptual framework suited to Chinese circumstances – one that will ensure progress towards achieving an ecological civilization. It incorporates concern for maintaining social stability while seeking environmental protection improvements, and for promoting social harmony and acting on environmental values. It assumes continued economic development and innovation in institutional, management and technological aspects, all related to the Scientific Outlook for Development. This model is likely to be quite different from models based on a western democratic society, and yet it will need to be robust in terms of improving both environmental protection and the social condition.

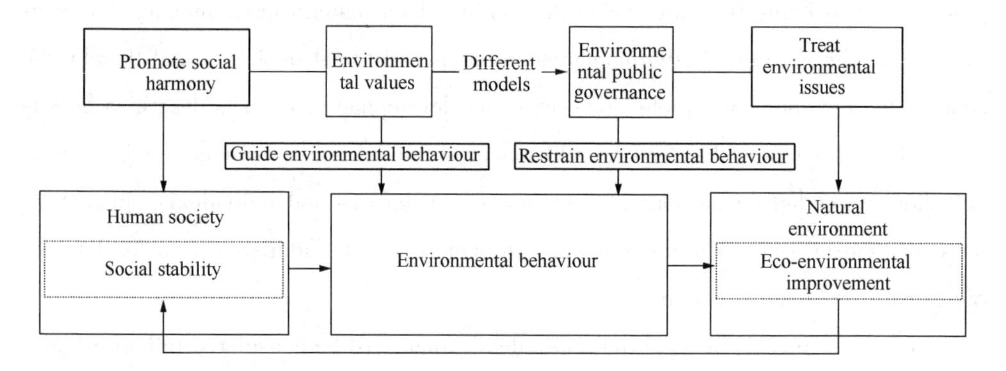

Figure 3-4 Model considering environmental behaviour, public environmental governance and environmental values

Importantly, its application to China will also have to consider some important dynamics over time that affects both environmental and social development. For example, the demographic changes already in motion, the growing importance of domestic consumption and the sheer pace of China's current changes. These are not shown in the simple version depicted below.

4 A FRAMEWORK FOR POLICY AND ACTION

To achieve a long-term vision or goal of Ecological Civilization the Task Force puts forward for consideration a framework for policy and action that integrates social development and environmental protection. The outcome should improve environmental and social harmony, and overall progress towards a "Beautiful China".

4.1 Basic Principles

The Task Force has identified five principles that can usefully guide the formulation of policies for linking environmental protection and social development.

- Multi-stakeholder participation. Attaining sustainable development is a shared responsibility. Experience has shown that the commitment of the Chinese government to economic development over the last three decades has generated many environmental and social problems and incidents have arisen, some of which undermine the credibility of the government. This can be attributed, in part, to inadequate opportunities or mechanisms for all actors and social groups to provide input into policy. It is therefore necessary to clarify the roles and responsibilities of different actors. Where necessary, it may be important to transform the functions of the Government, so as to more appropriately drive coordinated economic, social and environmental progress. Mechanisms to enable and recognize the positive contributions of businesses or corporations can also be created.
- Coherence between long-term and short-term visions and targets. The Task Force emphasizes that environmental protection andsocial development require immediate practical actions, but also a long-term vision and plan to safeguard future generations and build a "Beautiful China". Thus, the process of policy formulation should articulate clearly both a long-term vision and develop short-term targets and objectives to make progress toward reaching that vision. Achieving an Ecological Civilization is a long-term vision. To this end, the Task Force suggests setting clear objectives and tasks for

China's future development process associated with environmental protection and social development over short, medium and long-term timeframes.

- Policy coherence. Policies for economic, environmental and social development should be interdependent and mutually reinforcing, rather than conflicting or contradictory. For a country moving towards sustainable development, it is necessary to integrate and coordinate the development and implementation of economic, social and environmental policy objectives, even while recognizing the need for parallel and distinct means of accountability.

- A strong legal foundation. The Task Force highlights the importance of laws and regulations, designed to support the objectives and needs of economic, social and environmental development rather than being based on the preferences and propositions of vested interest groups, enterprises or individuals. This is an important guarantee for long-term social stability. It is particularly important to protect and safeguard the provision of public goods and services and to ensure that any framework considers the legal guarantees relative to other mechanisms such as market instruments. Laws and regulation should enable public access to information, and create a legal framework for robust and useful mechanisms for public supervision of development policies.

- Equity and justice. Environmental resources, rights and responsibilities should be distributed equitably. Access to a clean environment and an acceptable quality of life should be available to all. In order to fulfil their obligations to achieving environmental protection and social development individuals, organizations and enterprises must have the capacity, knowledge and means to behave responsibly. In formulating relevant policies in matters such as green procurement or sustainable consumption, not only equitable distribution of rights and interests among different groups should be secured, but also their obligations to participate in environmental protection and social development should be promoted.

4.2 Vision 2050/Action 2020

Adopting a Vision 2050/Action 2020 approach will require immediate actions in order to bring the 2030 and even the 2050 Visions within reach. That is due to the need to consider the time required to change and build infrastructure, spatial patterns and financial obligations.①,②,③ Through backcasting from a vision, it is clear that some decisions needed to be taken soon, especially for the 13th FYP, in order to ensure the chances of achieving the vision. Some of these actions have been identified and recommended in Chapter 5. Such backcasting is also important because it will give direction to a further program of targeted policy-oriented studies following this initial study.

To achieve policy coherence over various timeframes, a number of goals and objectives could be envisioned.

- By 2015, the ecological, environmental and social targets in the 12th FYP should have been achieved. People's lives will continue to improve and the main medium-term and longer-term goals for environmental protection, production patterns and lifestyles of quality will take initial shape. A more favourable and robust legal system for the coordination of environmental and social development will be established, while the management system and policy system will become better coordinated.
- By 2020, the aim is to have built a moderately prosperous *Xiaokang society*. Better spatial land patterns and environmental functional zoning will be in place. A resource-conserving and environment-friendly economic structure and system will basically have been built, although still in need of much more attention. Levels of efficiency of resource utilization should be closer to the most advanced levels in the world, while the energy consumption per unit of GDP will have been reduced substantially. As the total discharge of major pollutants decreases drastically, overall environmental quality will improve significantly. The concept of Ecological Civilization will be firmly rooted in the whole of society. Specific improvements will also have been made in the legal system, policy system, social risk prevention and control system, and public environmental management and service system.
- By 2030, environmental pollution problems will be much more fully resolved. Upon meeting environmental quality objectives fully, environmental public health needs will be met. Ecosystems will be stable and healthy with robust service functions and im-

---

① WBCSD (2010) *Vision* 2050. *The new agenda for business*. World Business Council for Sustainable Development, Geneva. English edition. Chinese edition available.

② PBL (2009) *Getting into the Right Lane for* 2050. *A primer for EU debate*. PBL Netherlands Environment Assessment Agency and Stockholm Resilience Centre. PBL, Bilthoven, The Netherlands.

③ TIAS (2010). *A comparative study of Visioning-Backcasting Initiatives*. The Integrated Assessment Society. http://www.tias.uni-osnabrueck.de/backcasting/.

proved biodiversity protection. The spatial land pattern and environmental function zoning will be fully established, while the economic and industrial structure will be able to meet the requirements of an Ecological Civilization. Resource efficiency likely will reach the world's most advanced level. With the further penetration of the concept of Ecological Civilization, the values of environmental protection and low-carbon and eco-friendly production and consumption patterns and lifestyles become dominant. A scientifically-based and sound public environmental governance system will have been put in place. A green, prosperous, harmonious society is at the inception, and a "Beautiful China" is being created. China will be widely regarded as having a highly functional Green Economy and a Green Development governance system fully in place.

- By 2050, the coordination between environmental protection and economic and social development will be the norm, and much more reasonable levels of harmony between people and nature will be realized. A "Beautiful China" with full ecological civilization will have been born. Most people will be housed in very liveable cities, but there will also be robust, ecologically-sound practices throughout the countryside and in China's ocean and coastal areas. Indeed, most ecologically degraded landscapes will be restored. Climate change adaptation and mitigation measures will be helping to lessen the impacts of climate change. Energy use patterns will be radically different from today, with much less dependence on fossil fuels and with eco-efficient industry, transportation systems and practices.

Previous successful studies have shown that there is greater likelihood of success when any analysis starts with the vision, then reasons back. This allows the identification of the critical steps that must be taken in time in order to keep the vision within reach. The longevity of infrastructure and capital stock is a key factor as illustrated by the urbanization cases investigated by the Task Force (Xixian New Area in China and Randstad area in The Netherlands). ①

4.3 Policy Fields and Actions

In response to current and future challenges, finding a symbiotic balance between

① See CCICED Task Force Report, *Report on Environmental Protection and Social Development in China*. 266pp-272pp. In Chinese.

social development and environmental policies will be essential. Finding those synergies will not happen by accident, thus an organized and disciplined set of actions and actors are described below.

All actions can be seen through the perspective of three functions: developing an*awareness* of appropriate values and norms in society, supporting appropriate *behaviour* of citizens, enterprises and other social organizations, and developing coordinated *governance systems*.

In terms of awareness, efforts should be made to establish values and norms compatible with an Ecological Civilization. By means of a combination of laws and regulations, dissemination and education, policies and measures, environmental rights should be identified explicitly as a right of citizens; and environmental protection and social development presented as a shared responsibility and a basic obligation of the whole society.

In terms of environmental behaviour, policy actions should be directed towards enabling and constraining the behaviour of public, government officials, enterprises and other social organizations. Incentive policies should be introduced to encourage public participation in environmental protection, and dissemination of information and education should cultivate environment-friendly habits and conduct and build sustainable consumption patterns throughout the whole society. While environmental laws, regulations and standards are to be further improved and implemented, economic policies and incentives should be put in place to cultivate among enterprises better incorporation of the concept of corporate social and environmental responsibility. With the development of guiding policies, the government should inclusively support public environmental organizations, industry associations and communities, and motivate a new pattern of broad participation in environmental protection and social development.

Various key roles are shown below in Figure 4-1.

In terms of the system of public governance, efforts should be made to improve legislation, social and environmental risk management, and the distribution and coverage of public services. In further developing the legal system, it is necessary to protect by legislation the public's right to know about, participate in and supervise or monitor environmental protection activities. To this end, improvement is needed with respect to information disclosure, environmental hearings, environmental public interest litigation, and environmental damage compensation systems. Where appropriate laws and regulations exist, attention should now turn to effective implementation. As already mentioned, environmental protection should be given equal importance alongside economic and social development.

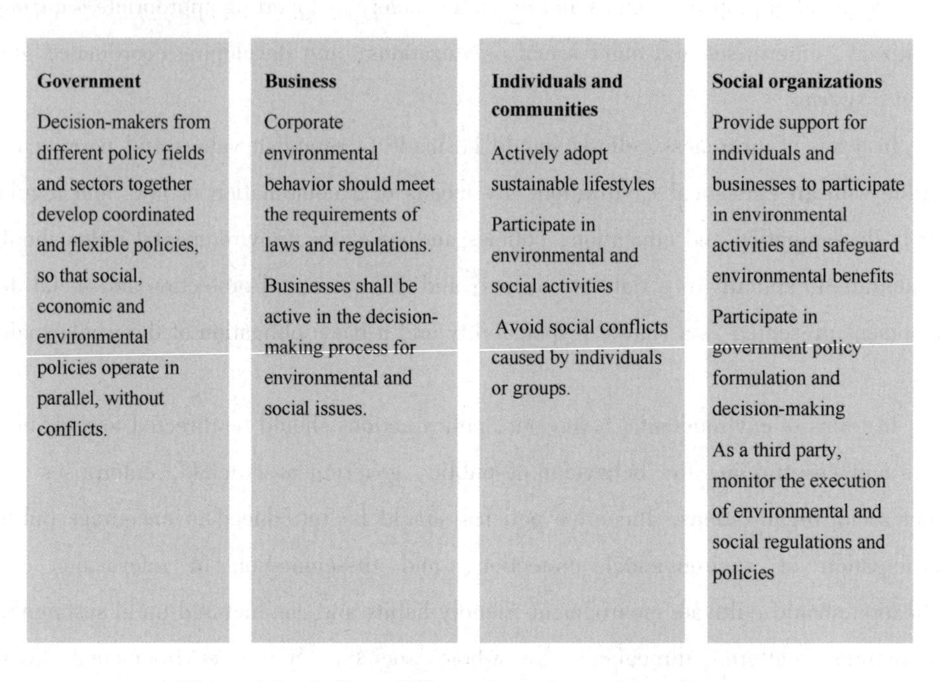

Figure 4-1 Actions of the various players in governance

Efforts are needed to accelerate the formulation of social policies for environmental protection, and to establish a sound assessment mechanism for major social policies. Consideration should be given to creating an independent mechanism for environmental and social policy evaluation that could be related to environmental impact assessment as is done in some other countries. Inrespect of risk control, it is important to set up a social risk assessment mechanism for major environmental projects and improve public communication and appeal mechanisms. An improved emergency response mechanism should be put in place to cope with sudden environmental accidents. Communication, dialogue, and consultation between the government, businesses and the public should take place on a regular basis. With respect to public services, it is important to build trust and social capital by continuing to improve government openness and transparency.

The following framework (Figure 4-2) summarizes the most important coordinated actions to be carried out in eight policy fields over the next 35 years by various actors in order to achieve a vision of Ecological Civilization and a "Beautiful China" with harmony between humans and nature.

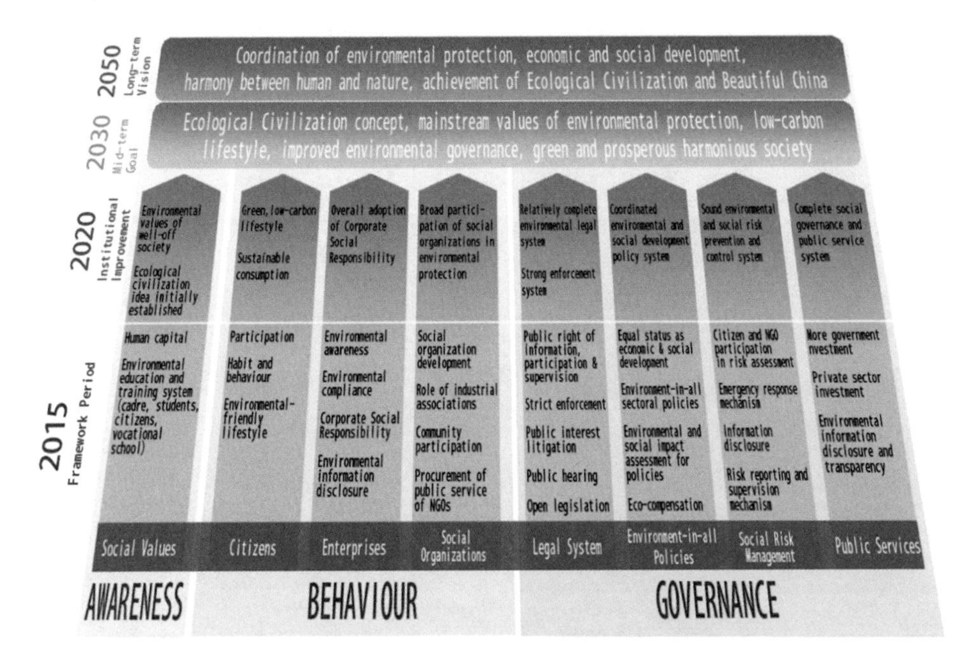

Figure 4-2 Policy and action framework

Stage One begins now and continues to the end of the 12th five-year plan. The focus during this period will be on building the appropriate infrastructure and support systems. Its main task will be to conduct thorough investigations of major environmental and social problems and policies, to establish a sound legal system, and to perform trial environmental and social policies in areas and fields where the conditions are right. Stage Two which is projected to coincide with the 13th five-year plan period until 2020 will focus on institutional improvement. The main task for this stage is to complete the development of parallel yet coordinated economic, social and environmental goals in a manner that ensures strength in all three elements. Stage Three will occur during the decade ending in 2030. This stage is committed to the completion of medium-term targets, that is, comprehensively solving environmental pollution problems, taking into account the contributions of appropriate social development so that environmental quality basically satisfies the health demands of the public, with stable and healthy ecosystems and restored ecosystem service functions. Stage Four, planned to conclude in 2050 should see the attainment of China's long-term goals, namely, securing the balance between environmental protection and economic and social development, establishing harmony between people and nature so as to realize an Ecological Civilization and a "Beautiful China".

5 RECOMMENDATIONS AND CONCLUSION

5.1 Introduction

Developing an Ecological Civilization is an ambitious vision. Central to achieving it is the ability to strike a good balance among the objectives of economic growth, environmental protection and social development through coherent, coordinated and consistent policies. Thus far, in China and elsewhere, relatively little focus has been put specifically on the relationship between environmental protection and social development. The Vision 2050/Action 2020 framework presented in Chapter 4 is one way of addressing this deficiency. In particular, the framework would help to connect the long-term vision of an Ecological Civilization and a "Beautiful China" by the middle of the 21st Century with policy decisions and actions that are necessary in the near term.

Considering these near term issues, it was apparent to the Task Force that for some of these there is already sufficient evidence to recommend immediate action. These recommendations are described only in intent - the actual details of design and implementation were not the mandate of the Task Force. The Task Force underlines that any short-term initiatives should be considered part of a strategic shift and so they should be consistent with the overall vision. For example, environmental protection initiatives should carefully consider social impacts and any social development initiatives should identify and address their impact on the environment.

The Task Force also recognizes that there are relevant policies and practices that have already been proposed, for example in earlier CCICED reports. Not all are repeated here. However a number are included since they deserve greater attention and strengthened implementation.

5.2 Recommendations

Recommendation 1. Elaborate a 2050 vision of coordinated environment and development and develop a phased plan of policy and actions that will be essential in achieving that vision. (Vision 2050/Action 2020)

The overarching recommendation of the Task Force is to further develop a Vision 2050/Action 2020 framework that will guide actions over the short, medium and longer time frame in a manner that will genuinely coordinate the social, economic and environmental aspects of development in China. See Figure 4-3.

In particular, the proposed framework will be a tool to identify, among the many important challenges and opportunities, those near term policy steps that are decisive in determining whether the long-term vision can be reached ('*back casting*'). The Task Force recommends that the contents of this framework will be elaborated based on specific follow-up studies in a Chinese context. The next recommendations provide an initial list of issues to be addressed in these studies.

In addition to charting key steps over time, the proposed framework serves to highlight the various societal actors that need to be involved - not only the government. We have expressed this by clustering the following recommendations as addressing three dimensions. The first dimension is *awareness* aimed at establishing and enhancing norms associated with environmental protection and social development. The second dimension is *behaviours* - in particular, the behaviour of the general public, businesses and social organizations. The third dimension is *public governance*. See Chapter 4 for a graphical representation of the framework and a brief discussion of these three dimensions.

Recommendation 2. Promote social norms and values related to ecological civilization ('*Awareness*' *dimension*)

The development of social policy begins with and builds on values and social norms. The Task Force acknowledges that social values and norms related to Ecological Civilization in China are the foundation for the development of future policies and practices in environmental protection and social development. Therefore, it should be a priority to advance the understanding, early on, that a sound environment is basic to the welfare of citizens. To that end, it is important to emphasize both the environmental rights and basic obligations of citizens. The government's role in transparently producing and disseminating information is particularly important. Specific actions could include:

Developing education and training plans such as: (i) Improving cadre training by developing or appropriately modernizing an integrated environmental-social curricula for the Party school system, colleges of administration and other training centres for cadres at all levels of government. (ii) Developing an educational initiative through China's vocational school system to ensure that groups that are socially disadvantaged such as the next wave of rural-urban migrants have the workplace skills to contribute to a sustainable modern urban environment. (iii) Investing in the future generation by incorporating basic environmental knowledge and sustainable development approaches within the school system and at universities.

Supporting conceptual and policy-oriented research on the development and implementation of the "five-in-one" system (economic, political, cultural, and social progress, and ecological civilization) emphasizing environmental values that are consistent with Chinese traditional moral and cultural philosophy.

Promoting values related to ecological civilization though extensive use of news media, internet and other communication channels, recognition of positive activities on the part of individuals and organizations and the promotion of distinctive literary and artistic works and publications.

Assessing and communicating actively the potential social risks of environment developments. The Task Force recommends that the Government establish a trustworthy mechanism to implement a comprehensive approach to ex ante environmental and social risk assessment based on principles of openness and transparency and meaningful public access. In other words, the approach should go beyond mere disclosure.

Recommendation 3. Encourage all in society to exercise their appropriate roles ('*Behaviour*' *dimension*)

To address increasingly diverse and pluralistic social demands, all individuals and organizations in society should be encouraged to play their respective and complementary roles in a positive and cooperative interaction with government and businesses. Achieving the vision of simultaneous social development and environmental protection in China can be greatly accelerated by connecting to the energy and flexibility of players other than the government. Specific actions could include:

Advocating healthy and sustainable life styles. It will be necessary to foster lifestyles that are healthy, resource-conscious and that consider quality rather than quantity in consumption and personal mobility. Advocacy and education shall be used to promote a sustainable lifestyle and behaviours, including through encouraging leaders of social organizations, entrepreneurs and other public figures to play a demonstration role by pursuing a healthy and sustainable lifestyle.

Public participation. Public participation in decisions that influence daily life, health, safety and enjoyment is important for coordinating environmental protection and social development. This engagement will be contingent upon the protection and enhancement of the public's right to know through disclosure of environmental information; the affirmation of environmental rights and interests of the public through the legal system; and the encouragement of citizen participation in development and environmental planning, as mentioned in Recommendation 3. China's large urbanization process offers a unique opportunity to make progress in this respect, for example through experiments with innovative, participatory planning.

Promoting acceptance by enterprises of their environmental and social responsibilities. The responsibility of enterprises to conserve resources in a socially and environmentally responsible manner should be encouraged, in a manner that mobilizes their creativity and innovation potential. This has the potential to dramatically reduce pollution and conserve resources, including energy, while strategically moving to longer-term business models. Obviously, this is contingent upon important improvements outside the scope of this Task Force, such as the establishment of sound economic policies putting in place true incentives for enterprises to move beyond environmental compliance to innovation. The Task Force also recommends development of an improved system of tracking corporate environmental impact assessments through independent oversight and public participation. [①] Similarly it is recommended to promote the active use by the financial industry

① Insights discussed during the study trip of the Task Force to Europe. See Schijf, B. and Boven, G. van, in: Strategic Environment Assessment in Development Practice. A Review of Recent Practice. 2012. OECD/DAC report 9789264166745 (PDF); 9789264166738 (print) DOI: 10.1787/9789264166745-en. OECD, Paris. Also in "Views and Experiences nr 11, 2012, Netherlands Commission for Environmental Assessment", downloadable from http://www.eia.nl/en/publications.

of environmental and social norms for evaluating loans, insurance and the potential worth of enterprises. These norms should include enhancement of environmental and social standards for access to formal qualifications of listed companies in China.

Supporting the further development of environmental and social organizations. Social organizations should be enabled to contribute by serving as independent assessors and supervisors of development activities, protecting citizens' rights, improving environmental and social awareness, conducting surveys, contributing to community activities, protecting nature and ecosystem services, and offering advice and suggestions for policy formulation. This would extend to a much wider range of social and business organizations than reflected in the current official registration in China. For example, trade associations have a potential role in environmental protection. It would be appropriate to consider policy change around registration of social organizations such as to easing restrictions on their ability to undertake activities across the environmental and social domains. Obviously, promoting the responsible partnership of non-profit environmental organizations is conditional upon conditions being created to overcome difficulties in registration, funds and social participation. Specific actions could include: (i) actively encouraging and guiding urban and rural communities to participate in environmental protection, for example, playing a role in publicity and mobilization, (ii) encouraging and enabling social organizations to actively participate in environmental impact assessment and social risk assessments of major projects that promote fair, impartial and transparent proposals as input to project planning, (iii) including environmental organizations in the bidding on government purchases of public services in order to establish a closer relationship between governments and social organizations, make up for the shortage of government's provisions of public services and enable social organizations to provide some public services.

Recommendation 4. Strengthen public governance ('*Governance*' *dimension*)

At the heart of realizing the vision of ecological civilization will be the development of a coherent and comprehensive set of legislative and policy actions by government, such as:

Establishing a highly functional environmental policy system. The Task Force recommends that the Chinese government sets out a strategy of simultaneously strengthening environment policy per se and also strengthening policy coherence on matters of environment and social development through the full breadth of its policies and institutions. International experiences from the environment and public health domains strongly suggest that a 'whole of government' approach is required to pursue these societal objectives with sufficient political power and at a credible scale.① At the same time, in view of the complex ecological and environmental problems in China, it is necessary to build a strong comprehensive policy field of environmental protection on an equal footing with economic and social policies. Specific actions could include:

1) From the 13th FYP, the five-year plan of the Chinese government should be listed as the National Economic, Social and Environmental Development Plan, so that environmental policy and the associated planning will truly become a significant item in parallel with economic and social policies. Meanwhile, the National Economic and Social Development Report submitted by the Chinese governments at the National People's Congress and the Chinese Political Consultative Conference (NPC & CPPCC) would then also have been changed to the National Economic, Social and Environmental Development Report accordingly.

2) To support this point, the Government, represented by the Premier in order to underline the whole-of-government approach, should submit to the National People's Congress an annual report with equal emphasis on the economy, society and environment. The report should list the achievements made by the Government with respect to economic, social development and environmental protection for all Chinese citizens in three clear and separate sections. In this way the Government will demonstrate responsibility for environmental protection in China, and clarify the relationship between environmental protection and social development through the report. To be a fair and credible assessment, the report should cover achievements in the past year based on ob-

---

① In particular, the World Health Organization, on the basis of worldwide examples, points out that strong support from the top policy level is always required to achieve effective coordination. It also suggests, by implication, that the period of increasing concern about China's environment and environment-related unrest provides a not-to-be-missed opportunity to embark on such a policy strategy. Obviously, the accountability that would be demonstrated by the Premier should find its way to other layers of government as well, but the Task Force focuses at the Premier to provide the strongest possible example. See (i) report of study trip Geneva and The Netherlands to be included in Long Report of the Task Force; (ii) Leppo, K. et al. eds. (2013) Health in All Policies. Seizing opportunities, implementing policies. Helsinki 2013. Publications of the Ministry of Social Affairs and Health 2013: 9. ISBN 978-952-00-3406-1 (printed) ISBN 978-952-00-3407-8 (online publication) URN http://urn.fi/URN: ISBN: 978-952-00-3406-1.

jectives that were set, using quantitative and qualitative indicators and measurements of success, as well as an assessment of the future significance of current developments and actions.

3) An environmental and social assessment mechanism should be established for major policies. An EIA traceability and accountability mechanism should be put in place to force EIA units and individuals to take responsibility for the assessments, and increase penalties for violations. Thus, *ex ante* policy impact assessments in the style of the European Commission would be a key instrument in pursuing policy coherence.

4) The environmental performance evaluation and government performance evaluation system should be improved, encouraging local governments to increase investments in environmental protection, by setting up a scientific evaluation system placing greater weight on environmental public services provided. The weight and therefore number of ecological environment and social development indicators should be gradually increased.

Recommendation 5. *Establishing a sound mechanism to assess, communicate, and mitigate the social risks of environment protection*

The Task Force recommends that the Government put in place a comprehensive approach to environmental and social risk assessment. To be convincing, the approach should be based on principles of openness and transparency and meaningful public access. Fundamental to achieving an effective and trusted risk management approach would be the systematization of information. Specific actions could include:

1) Establishment of a "pre-approval" system for major projects with environmental and social implications as well as policies and reforms involving public environmental interests to consider procedure legality, policy reasonability, program feasibility, and appeal rights.

2) To win the understanding, trust and support of the public, solicitation and incorporation of their opinion should be undertaken in advance of decisions on major projects through seminars, public hearings and public notices. In particular National People's Congress and Chinese People's Political Consultative Conference, industry associations, and community or social organization representatives should be invited to review the social risk assessment reports.

3) Cadres who fail to strictly follow the assessment process should be seriously punished in cases of "evaluation failure" and policymakers who do not attach importance to

risk assessment should be held responsible for this failure.

4) Building a more robust environmental emergency response mechanism should be given priority. Complete and operational contingency plans should be developed that clarify the conditions and timing when the response mechanism should be launched, as well as the personnel and equipment needed.

5) The provision of timely, and accurate information during environmental incidents is important, to avoid misleading and untrue reports, speculations and rumours. Full advantage should also be taken of new media platforms such as micro blogging to ensure more widespread and accurate knowledge of such incidents.

Recommendation 6. *Improving the level of public environmental services*

Public services regarding the environment are prioritized here as an opportunity to demonstrate that the government can meet the objectives of improving and protecting the environment and meeting the expectations of citizens regarding their health and well-being. Policy coherence in delivering environment-related public services is particularly important during a time of rapid urbanization and significant change at the urban-rural interface that requires infrastructure planning and decision. ① Basic public service is provided by the Government to meet the essential needs of all citizens for survival and development. Basic human needs include clean water, unpolluted air, and productive land. In addition to basic services, increasing attention is required for intangible services like institutional arrangements, standards and laws. ② Actions could include:

1) Setting up appropriate coordination mechanisms to ensure access throughout China to public services. Development of appropriate scope and standards for basic environmental public services, such as placement of sewage treatment and garbage disposal facilities, clean water, clean air and tranquillity, environmental emergency response mechanisms, environmental information services, the public right to know and to supervise environmental actions.

2) Consideration of outsourcing certain public services. For example, social organ-

---

① See CCICED Task Force Report, *Report on Environmental Protection and Social Development in China.* 127pp-134pp, 218pp-243pp. In Chinese.

② World Bank (2012) China 2030. Building a Modern, Harmonious, and Creative High-Income Society. Conference edition. World Bank and Development Research Center of the State Council, the People's Republic of China. World Bank, Washington DC.

izations can be mobilized for environmental monitoring and assessment and carry out "advocacy work" for improving environmental awareness

3) Gradually improving the proportion of spending on basic public environmental services. Measures should be taken to encourage multi-sourced financial mechanisms, better possibilities for cross-regional transfer, and better incentives for private investment. In this way, local governments can obtain adequate funding for their social and environmental policies in line with differing regional needs.

4) The formation of a more complete ecological compensation mechanism for different functional zones should be pursued, so that ecological environmental protection can increase local revenues and benefit the local people living in ecologically significant areas.

5.3 Final Considerations

The work of this Task Force was preliminary. Obviously elaborating the proposed framework will be a major undertaking. The Task Force recommends that several strategic studies on complex priority issues also be commissioned. These studies would involve more comprehensive analysis of previous international work and engagement with various societal actors. Three priority topics are:

Lifestyle and behaviour. A strategic study might be commissioned to explore the best alliances and an efficient package of government measures that would promote the shifts in lifestyle and behaviour that are necessary to achieve environmentally and socially sustainable outcomes. Experience abroad suggests that this is a long-term goal which is difficult to achieve. It also suggests that the direct role of government in influencing lifestyle and behaviour can only be modest, by current Chinese standards. Others, in particular social organizations and entrepreneurs, could be more influential in setting trends. Therefore, significant thought would have to be given to designing the necessary initiatives to be most effective and avoid unintended side effects. The study should advise on combinations of initiatives in education and communication policies with various 'harder' incentives such as changes to the fiscal system; taxes; and resource pricing. The current CCICED Task Force on Consumption may provide guidance on this matter.

Legal underpinning for coordinated social development and environmental protection. It is a matter of urgency to explore how to address the sense of injustice in relation to environmental issues and unfair exposure to pollution. For example, protest or 'mass incidents' occur in part due to a lack of alternative channels - often where the judicial system would not necessarily be appropriate or available as an option. An effective legal regime is built upon characteristics of integrity, authority and long-term consistency. The legal basis for environment policy as well as the system for expressing complaints are undergoing major revisions at the time this report is being written. It is therefore recommended to evaluate as soon as possible what the current changes mean in the light of the framework of Vision 2050/Action in four stages. At this point in time, key aspects for the evaluation seem to include: the balance between judicial and administrative channels to express complaints; the system of participation during the actual revision of the law, including the system of public notification of intended changes, the system of handling and responding to opinions expressed and the system of legislative hearings; and last but not least, the impact of the changes on feelings and opinions of injustice.

Financial resources required for implementing the twin mandates of environment protection and social development. The highly decentralised fiscal system in China creates challenges and opportunities to build and optimize the positive relationship between environmental protection and social development. It creates structures and incentives that undermine the government's capacity to implement its vision. Even when there is a policy statement of coordination in mandates, there is not always a financial allocation mechanism that provides an appropriate level and timely flow of financial resources to various levels of government to fully implement the mandate. Moreover, in current large urbanization projects, it happens that late central interventions, for example in land allocation, put a strain on the projects finances and therefore on the envisaged long-term combination of environmental, social and economic objectives.[①] Thus, a strategic issue is to understand these challenges on the basis of objective information and good analysis. The recommended study can show where the real commitments are to the implementation of the government's vision.

---

① See CCICED Task Force Report, *Report on Environmental Protection and Social Development in China.* 218pp-243pp, 266pp-272pp. In Chinese.

5.4 Conclusion

China is facing enormous challenges at present, including economic restructuring, and rapid innovation in policy and governance. These also provide rare opportunities to explore the relationship between environmental protection and social development. Understanding this relationship will allow China to better develop effective policies that will avoid unintended consequences and maximize the potential for successful outcomes. The achievements made by the Chinese government in these areas will attract attention from many other nations and international organizations.

ACKNOWLEDGMENTS

Financial support for the research was provided by the China Council for International Cooperation on Environment and Development (CCICED), who also convened the Task Force. The support from CCICED has enabled discussion, communication, and surveys among experts from China and internationally, which have served as the essential foundation of the research work. Special thanks to the Chief Advisers of the CCICED Advisory Committee, Professor Shen Guofang and Dr. Arthur Hanson, the Leader of Chief Advisers' Support Team Dr. Ren Yong, and the CCICED Secretariat and support office in providing information and organization and coordination support. The Task Force extends cordial thanks to the CCICED.

As part of the research process, Task Force meetings were held in Beijing, Xi'an and the Netherlands. A study tour was conducted and the Task Force thanks the hosts in the Netherlands and Switzerland for organizing meeting and visits to sites, relevant research units and government departments.

This Report was submitted by the Task Force on
Environmental Protection and Social Development

下篇 分论

第一章　中国环境保护与社会发展的现状

一、引言

环境与社会发展紧密联系，相互影响，互为因果。良好的环境是社会发展的重要条件，也是衡量社会文明程度的重要标准，同时也是社会民众的基本权利。保护环境，使社会全体成员共享环境保护的成果，才能增强整个社会的可持续发展能力。社会发展是环境保护的重要基础和外部条件，一个法治、公平、正义、诚信、有序的社会，可以维护和促进环境的改善与可持续性。反之，环境污染或者生态退化将直接冲击长期经济增长潜力，对人群健康和生态系统健康带来长期影响，并冲击社会关系与结构，进而引发严重的社会问题。而紊乱或者扭曲的社会制度、结构和关系会引致和加剧环境的破坏，如果社会关系和结构得不到及时调整，那么就会形成环境与社会关系的恶性循环。

经过 30 多年的经济快速增长，随着工业化和城镇化进程不断加快，中国的环境形势非常严峻。进入 21 世纪以来，环境事故与环境风险处于集聚期和高发期，环境污染和生态退化对社会发展的影响日益凸显，已经成为引致当前中国一些社会问题的重要因素和导火索，如因环境利益冲突引发的社会冲突、环境利益分配的不均衡和不公平、环境正义或者公平受到侵害、人群健康损害、环境引发的贫困问题、环境问题带来的社会成本和代价增加、贫困群体的基本公共服务水平降低乃至各类社会冲突加剧等问题。从目前的状态和趋势看，中国由环境问题引发的社会危机，有可能超越环境问题对经济的影响，成为影响中国社会关系和结构以及和谐发展的首要因素。

中国改革开放以来，在中国可持续发展的三维支柱中，经济是最为关注的中心维度，进入新世纪以来，随着科学发展观的提出和确立，对环境与经济关系的认识和理解、如何通过环境优化经济增长等得到不断加强，但是环境与社会的关系在整个中国可持续发展框架中一直较弱，环境与社会的相互作用与影响未得到很好的认识、理解以及应有的关注。中国政府在“十二五”规划纲要中明确提出要建设和谐社会，共享发展成果，维护社会公义和公平。环境保护与社会发展问题正是建设和谐社会的具体体现，体现了国家改善环境与社会关系的政治意愿。同时，社会公众的环境意识也日益增强，参与环境保护与社会发展进程的热情不断高涨，前不久社会公众热议的“$PM_{2.5}$”问题以及后续的相关政策和标准出台反映出国家环境公共治理模式转变的动态，可以说环境与社会管理转型具备了充分的社会条件和需求。

一直以来，中国在环境保护管理和政策方面，“自上而下”的行政手段以及基于市场的经济政策工具采用较多，“自下而上”的社会公众参与的治理手段相对较少，社会公众参与国家与社区环境治理的渠道和作用远未发掘，成为国家环境公共治理的短板。中国政府在“十二五”规划纲要中明确提出要坚持多方参与、共同治理，统筹兼顾、动态协调的原则，完善社会管理格局，创新社会管理机制，形成社会管理和服务合力。这为未来中国如何处理环境与社会关系，如何运用社会公众手段参与环境公共治理指明了发展方向。解决中国的环境保护与社会发展问题，促进环境保护与社会发展的和谐，就是要依靠社会力量和公众的参与，改进环境与社会管理的模式、方式与机制，推动环境与社会管理的转型。

在中国环境污染与生态退化对社会发展的影响日益凸显，国家落实科学发展观、促进环境与发展战略转型和建设生态文明的新形势下，开展中国环境保护和社会发展研究具有重大的现实意义和深远影响。实现国家环境保护与社会发展的和谐互动不仅能促进环境的改善与可持续，而且是实现社会稳定和国家长治久安的重要基础，是当前国家高层关注的重大战略问题。

顺应中国环境与发展的大潮，第四届国合会围绕环境与发展战略转型的主线，重点对环境与经济、能源与资源效率、生态系统管理、发展方式绿色转型等开展了系统研究。第五届国合会正处在我国发展转型的关键时期，改善民生、保障社会公众权益、加强社会管理和创新将是未来一段时期的国家战略重点。因此，在新时期以及新的历史条件和背景下，开展环境保护与社会发展研究，认真审视和认识当前及未来中国环境与社会关系的状况、存在问题、发展趋势和相关影响，梳理国家实现环境保护与社会发展和谐互动的基本思路正当其时。

本课题的总体目标是就如下两个方面向中国政府提出政策建议：一是国家促进环境保护与社会发展的总体思路和制度框架；二是国家促进环境保护与社会发展的政策建议。

本课题研究不是关于环境保护与社会发展的纯粹理论研究。而是在环境保护与社会发展的理论研究基础上，重点就环境保护与社会发展问题进行较为宏观的实证分析，同时就其中涉及的核心紧迫问题开展具体研究，提出促进环境与社会和谐发展的基本思路和相关政策。与此相关的环境与社会公义、环境与健康、妇女在环境保护中的参与等议题可进行概括性阐述，但不作专门具体的研究。

二、现状与问题识别

（一）环境问题引发群体性事件增多

近年来，中国的环境保护工作取得了积极进展，但部分地区的环境污染情况依然十分严重，严重威胁到社会公众的身体健康和经济社会的稳定发展。在大气污染方面，70%左右的城市不能达到新的环境空气质量标准，2013 年 1 月出现的长时间大范围雾霾天气，影响 17 个省（区、市），约占国土面积 1/4，受影响人口达 6 亿。在水污染方面，20%左右的国控断面水质依然为劣Ⅴ类，基本丧失水体功能；一半城市市区地下水污染严重，57%的地下水监测点位水质较差甚至极差。此外，中国还正处于社会转型和环境敏感、环境风险高发与环境意识升级共存叠加的时期，长期积累的环境矛盾正集中显现，环境风险继续增加，损害群众健康的环境问题比较突出，$PM_{2.5}$、饮用水安全、血铅事件和化学品污染问题，引起群众广泛关注①。

根据环境保护部的统计数据②，1995—2010 年间中国共发生突发性水污染、大气污染、固废污染、噪声污染及振动危害以及其他类型突发性环境事件（环境事故）21985 起，从 1996 年的 1966 件下降到 2010 年的 420 件，整体上呈现出下降的趋势。这说明中国长期以来的环境保护工作取得了较好的效果，突发环境事件的预防、监测、控制和管理有所加强。然而，如图 1-1 所示，随着突发环境事件数量的下降，环保部门的信访数量却呈现出明显上升趋势，与环境信访的急剧增加并行的是环境污染纠纷的凸现。2001—2005 年全国发生污染纠纷分别为 5.6

---

① 环境保护部．深入贯彻党的十八大精神 大力推进生态文明建设 努力开创环保工作新局面——周生贤部长在 2013 年全国环境保护工作会议上的讲话［EB/OL］. 2013. http：//www. gov. cn/gzdt/2013-02/04/content_ 2326581. htm.

② 环境保护部．全国环境统计公报 1995—2010。

万、7.1万、6.2万、5.1万和12.8万件①。这说明，随着公众环境保护意识的不断提高，环境污染和环境健康问题开始越来越受到社会公众的关注，并且环境污染问题已经不仅仅局限在突发环境事故上，而是已经发展成为涉及公众切身利益的社会问题。

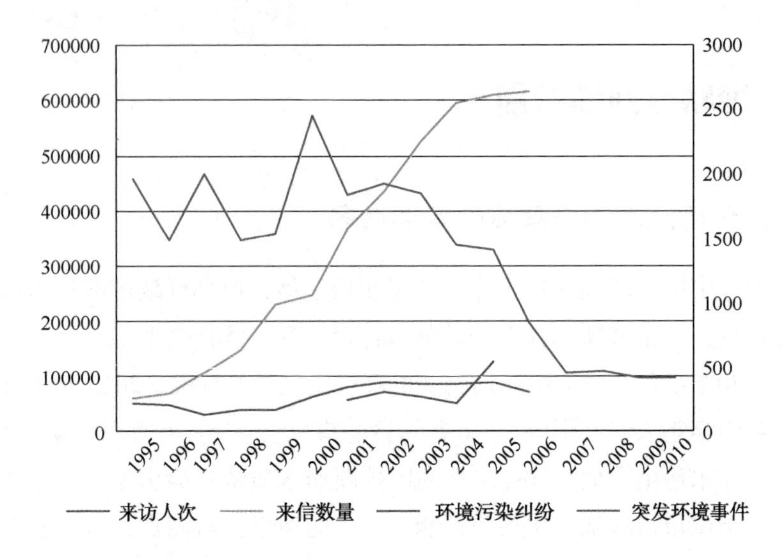

图1-1 1995—2010年中国环境信访人数、环境污染纠纷和突发环境事件示意图

群体性事件指由人民内部矛盾引发的、众多人员参与的危害公共安全、扰乱社会秩序的事件。过去的群体性事件的形成可由多种原因引发，如农村征地、城市拆迁、国企改制、基层选举、公民维权、腐败现象、安全事故等。然而近年来，环境问题却逐渐成为引发群体性事件的主因之一，并且环境群体事件比普通环保纠纷的影响更大、危害更强。环保部的统计显示，2005年以来，环保部直接接报处置的突发环境事件共927起，重特大事件72起，其中2011年重大事件比上年同期增长120%，特别是重金属和危险化学品引发的突发环境事件呈高发态势，这些突发环境事件进一步激化了社会矛盾、引发了群体事件，导致近年来中国的环境群体事件一直保持了年均29%的增速②。2004年的云南怒江水电建设

① 张玉林．政经一体化开发机制与中国农村的环境冲突［J］．探索与争鸣，2006，（5）：26-28.

② 环保部原总工程师、中国环境科学学会副理事长杨朝飞在十一届全国人大常委会第二十九讲专题讲座上的讲座材料《中国环境法律制度和环境保护中的若干问题》。

规划环境影响评价争议案、2007 年的厦门 PX 项目事件和北京六里屯垃圾焚烧厂事件[1]、2009 年的湖南浏阳镉污染事件、2012 年的启东王子制纸大型达标水拍海工程事件和四川什邡钼铜项目事件[2]等因环境问题导致的争议或群体事件便是其中的典型案例。

上述事件一些是建设项目因民众激烈反对而停建或迁址，一些垃圾焚烧厂之类的政府主导项目因为群众对于污染的强烈担忧而无法按原计划进行，更有一些是由于居民的身体健康受到伤害，继而引发群体性抗议，出现了围堵政府、派出所、相关公司及打砸行为。环境群体事件由于涉及利益群体广泛，负面效应容易扩散，一地的事件可能引发多地的效仿。环境问题所导致的利益分配不公平、环境贫困和环境健康等问题及其诱发的环境事件已经成为引发社会危机的重要因素。

（二）环境问题引发社会问题的本质

环境问题即是自然界出现的问题，也是一种社会问题，它不仅仅是人类活动作用于周围环境所引起的环境质量变化，还包括这种变化对人类的生产、生活和健康造成的影响。一方面，人类在改造自然环境和创建社会环境的过程中，自然环境仍以其固有的自然规律变化着。另一方面，社会环境也受自然环境的制约，也以其固有的规律运动着。人类与环境不断地相互影响和作用产生了环境问题，与环境问题产生所并发的，必定是又是各种社会问题的进一步出现。环境问题的凸显和由环境问题引发的一系列社会问题，给自然科学和社会科学提出了新的问题和挑战[3]。

在中国，随着环境问题的不断恶化，人们对环境问题的认识日益深化，环境问题已经从社会生活的边缘问题逐步上升为中心问题。在这种背景下，环境问题引发社会问题的核心本质归根结底在于：环境问题产生的原因、表现、影响和后果的整个过程中，都贯穿着社会不平等或不公平，进而带来了人们在利益分配、受害和责任分担上的不平等，从而表现出环境健康、环境贫困、环境群体事件或环境冲突等社会问题。

---

① Wanxin Li, Jieyan Liu, Duoduo Li. Getting their voices heard: three cases of public participation in environmental protection in china [J]. Journal of Environmental Management, 2012, 98: 65-72.

② Glibert, N. Green protests on the rise in China [J]. Nature, 2012, 488 (7411): 261-262.

③ 李友梅，刘春燕．环境问题的社会学探索 [J]. 上海大学学报（社会科学版），2003，10（1）：29-34.

从区域经济发展的角度来看，作为生存和发展的必要物质条件，资源环境基础始终是中国区域持续发展的根本支撑。改革开放以来，中国经济发展取得举世瞩目的成就，但区域经济社会的快速发展加大了对生态系统的压力，付出了高昂的生态环境代价，在发展的过程中还存在着巨大的环境不公平问题。首先，由于区域、流域资源禀赋的差异和区域发展战略的导向，东中西部的经济发展和环境保护水平显著不均衡①；其次，区域政府在权衡经济发展和环境保护的过程中，大都倾向于发展经济而把环境污染外部性转嫁给其他地区，或倾向于生态环境保护成本的外部化，跨区域环境污染纠纷猛增②；第三，经济发展的高水平区域对环境资源的跨区占用能力较大，低水平区域特别是中低水平区域的环境资源“被掠夺”处境仍然没有改观的迹象③；第四，长期以来对城市发展的重视和对农村发展的忽视，导致城市发展吸纳和支配了大量的环境资源，农村环境不断恶化，城乡发展不协调进一步加剧。

从社会群体的角度来看，好的生存环境对人类的生存和发展是必不可少的。人类社会发展到一定阶段，不同群体对资源的占有是不同的，进而决定了资源利用和支配的不同。改革开放以来，中国社会地位较高的社会群体占有了大部分社会资源和环境资源。这一小部分社会群体利用拥有的环境资源不断扩大生产，追求单纯的经济效益，而拥有少量社会资源的社会地位较低的群体只能在有限的环境空间内生产和生活，并且要承担小部分社会群体的生产活动造成的环境污染和损害。这种以大多数人的利益为代价的生产和生活方式，不仅加大了社会群体的距离，还激化了群体冲突，更加深了群体分化，这种群体分化的后果必定是“强者”越强、“弱者”更弱，进而导致社会发展走向两个极端：及其富有的群体或者及其贫穷的群体。

从环境保护的角度来看，理论上，污染者应当承担造成环境污染的责任。然而，长期以来中国主要依赖行政手段处理环境问题，由于环境保护的法制不健全，环境保护的权责关系不明晰，污染者往往可以逃避污染责任，进一步导致了少数人在环境问题上受益，而大多数人成为环境污染的受害者。与此同时，保护环境者由于得不到有效的激励和补偿，也进一步降低了社会公众参与保护环境的

---

① 杨兴宪，刘毅，牛树海等．我国区域发展中的生态环境特征分析［J］．长江流域资源与环境，2006，15（2）：264-268.

② 谢慧明，沈满洪．生态经济化制度和区域发展协调性——一个基于长江三角洲地区生态经济化制度安排的经验研究［J］．浙江社会科学，2011（8）：12-18.

③ 杨振．中国区域发展与生态压力时空差异分析［J］．中国人口·资源与环境，2011，21（4）：50-54.

积极性。

（三）环境问题引发社会问题的表现

1. 环境与健康

环境问题带来的首要社会问题便是对人类健康的影响。与发达国家不同，中国的环境与健康问题表现为环境污染浓度高，暴露人口众多，暴露时间长，暴露途经复杂多样，历史累积污染问题的健康影响难以短期消除，城乡差异显著等特点，导致传统型的环境健康问题没有得到解决、各种新型环境污染物的健康危害和风险呈现日趋严重的局面。

根据环境保护部历年发布的《中国环境状况公报》与《环境统计公报》，中国环境污染物的排放随着经济的快速发展，总体的环境污染物排放量总体呈持续增加态势。其中废水排放总量由 1989 年的 354 亿吨上升到 2010 年的 617. 3 亿吨，增加了 74. 4%。工业二氧化硫排放量由 1991 年的 1622 万吨逐年增加，近三年稳定在 2200 万吨左右，增加了 13. 8%。工业固体废弃物的排放呈现稳步降低，但贮存量从 1995 年起保持在 2. 2 亿吨到 3 亿吨之间，较 1989 年有明显的增加。由于中国的工业布局多集中在人口密集的城市与地区，地表水的污染遍及中国主要经济发达与人口密集区域，以及采矿业和有色金属冶炼导致的农田污染，导致中国的人群直接暴露于环境污染物，并且这些人群长时间以来对污染物的摄入量和暴露水平要远远高于发达国家的水平。同时，由于复合性污染重，导致很难确定污染区内的主要污染物、污染源及其健康危害，加大了人群健康损害效应的调查和干预的难度。

近年来，大气污染已被发现与呼吸系统及心血管系统多种疾病的发病及死亡存在相关关系。有关研究发现，当高血压、糖尿病等慢性病患者暴露于受到污染的大气时，其呼吸系统及心血管系统疾病的患病率及死亡率会更高[①]。在公认的大气污染物中，颗粒物（包括可吸入颗粒物 PM_{10}，细颗粒物 $PM_{2.5}$）与人群健康效应终点的流行病学联系最为密切。世界卫生组织曾经估计，在全球疾病负担（GBD）中，由大气污染而造成的失能调整生命年（DALY）损失占全部的 0. 5%[②]。《2010 年全球疾病负担评估》数据显示，在中国，由于 $PM_{2.5}$ 污染程度

---

① SCHWARTA J. Assessing confounding, effect modification, an thresholds in the association between ambient particles and daily deaths [J]. Environ Health Perpect, 2000, 108 (6): 563-568.

② 阚海东，陈秉衡．我国部分城市大气污染对健康的研究 10 年回顾［J］．中国预防医学杂志，2002，36（1）：59-61.

上升及中风、心脏病导致的死亡率有所上升，1990—2010 年，由室外空气污染导致的疾病负担增长了 33%，2010 年，中国 20%的肺癌由 $PM_{2.5}$引起①。赵珂等（2011）针对西安市大气 $PM_{2.5}$污染与城区居民死亡率的关系研究还显示，$PM_{2.5}$浓度每升高 100μg/m^3，总死亡、呼吸系统疾病、心血管疾病、冠心病、中风、慢性阻塞性肺病（COPD）的死亡率将分别增加 4.08%、8.32%、6.18%、8.32%、5.13%、7.25%②。

表 1-1 $PM_{2.5}$每升高 100 μg/m^{3}各种疾病死亡率增加的百分比（单因素分析）

污染物	总死亡	呼吸系统疾病	心血管疾病	冠心病	中风	COPD
$PM_{2.5}$	4.08	8.32	6.18	8.32	5.13	7.25

此外，新型环境污染物的健康损害问题的出现导致问题更加复杂化。例如，环境内分泌干扰物（EEDs）③、持久性有机污染物（POPs）以及新材料和新化学污染物的健康问题正在成为全球环境保护的热点，其中对人类最明显的危害是使生殖机能下降，影响人类生育体系，导致男女比例出现变化。这些不确定的因素潜伏在环境与发展之中，将会影响社会稳定。

随着社会的发展、进步，及全民环保意识和维权意识的提高，国际、国内社会对我国环境与健康问题的关注度日益增高。环境权益与健康权益关系人民群众的基本生存和切身经济利益，人们的环境参与权及健康权是社会公平的主要组成部分。环境与健康问题如果处理不好，很容易就会成为影响社会稳定的“导火索”，导致社会问题。

2. 环境与贫困

贫困是经济、社会、文化等落后的总称，是由低收入造成的基本物质、基本

① 世界卫生组织.2010 年全球疾病负担评估.

② 赵珂，曹军骥，文湘闽.西安市大气 $PM_{2.5}$污染与城区居民死亡率的关系.预防医学情报杂志，2011，27（4）：257-261.

③ 环境内分泌干扰物（Environmental Endocrine Distuping Chemicals，EEDCs），主要包括天然激素、人工合成的激素化合物和具有内分泌活性或抗内分泌活性的化合物。已被证实或疑似为 EEDCs 达数百种之多，包括邻苯二甲酸酯类（PAEs）、多氯联苯类（PCBs）、农药、烷基酚类、双酚化合物类、植物和真菌激素、金属类、人工合成避孕药等，其中农药及其代谢物占 60%以上，且以杀虫剂居多，杀菌剂、部分除草剂也有此作用。表明，环境内分泌干扰物对处于发育阶段的各类动物有明显的干扰效应。EEDs 可导致人类生殖障碍、出生缺陷、发育异常、代谢紊乱、内分泌相关肿瘤增加、神经系统病变等。

服务相对缺乏或绝对缺乏，以及缺少发展机会和手段的一种状况。它的一个基本特征是人们的生活水平达不到一种社会可接受的最低标准①。绝对贫困又称生存贫困，是指在一定的社会生产方式下，个人或家庭靠自己的劳动所得不能维持个人或家庭的最低基本生存需要，生命的延续受到威胁。环境与贫困之间往往存在直接联系，一系列的研究均证明贫困与环境之间有着紧密的相关性。

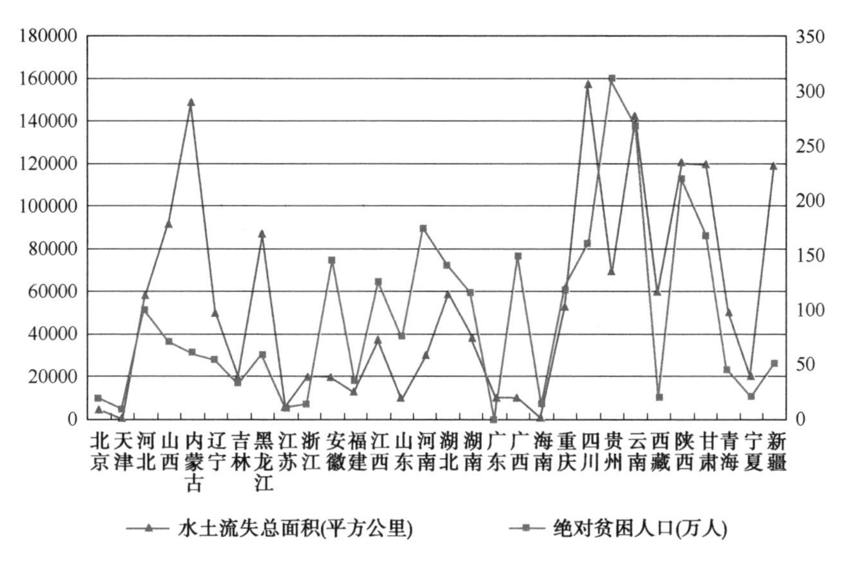

图 1-2　中国水土流失面积与绝对贫困人口分布关系示意图②

第宝锋等（2006）根据 2002 年的全国水土流失数据及 2004 年绝对贫困人口统计数据的分析表明，水土流失与贫困两者之间存在显著的相关性和耦合性，即水土流失面积自西部向东部有逐渐减少的趋势，同贫困人口的分布具有一致性，如图 1-2 所示。2004 年，水土流失较为严重的云南、贵州、山西、甘肃和四川 5 省区的绝对贫困人口占全国的 43%。大多数农村贫困人口生活在水土流失地区，居住在自然资源贫乏、缺少农用耕地、农业生产条件低下、自然灾害频繁、生态环境脆弱的区域。水土流失限制了对有限资源的有效利用，成为生态恶化和贫困的根源，而同时进一步的贫困又加速了水土流失和生态恶化，形成了恶性循环。

① 陈玉光，崔斌．深化农村经济体制改革与当代中国农村区域性贫困问题研究［J］．开发研究，1995（4）：30-31．

② 第宝锋，宁堆虎，鲁胜力等．中国水土流失与贫困的关系分析［J］．水土保持通报，2006，26（3）：67-72.

王雪妮等（2011）基于牛津大学水贫困指数和联合国开发计划署人类贫困指数比较了中国省际间水贫困和经济贫困水平的关系，研究显示中国大部分地区的水贫困与经济贫困存在较高程度的耦合度，即中国水资源短缺、水环境恶劣的地区往往与经济贫困并存①。

环境质量的好坏对人们的健康水平、收入能力、安全、能量供给及生活条件等方面都起到决定性的作用。特别是在农村地区，贫困人口对自然资源依赖性强，而环境退化又使得这种依赖性表现的更加脆弱。在一些地区，环境问题往往与贫困问题并存，环境问题的恶化往往会进一步加剧地区贫困，而解决贫困的前提条件则是解决贫困的根源，即这些地区的环境问题。

3. 环境公正

环境问题引发的群体性事件，一些学者认为其本质上是公众在追求环境公正。随着公众环保意识的不断增强，他们开始要求尽量回避环境风险，或至少与获益者公平分配、共同承担这些风险。然而，中国当前的环境不公正现象非常突出，主要体现在不同区域、群体间占有和支配环境资源的显著差异，以及不同区域、群体间享有环境公共服务的显著差异。

自然资源、能源和环境容量资源等要素是自然资本的重要组成部分，不同人群对自然资本占有的状况对远期的以货币表征的收入水平具有重要影响，是衡量环境公正或社会公平性的重要指标。相关的数据统计和研究表明，上世纪 80 年代以来，中国居民的能源消耗整体处于显著上升趋势，并且东部地区省份的增速普遍高于中西部地区。如图 1-3 所示，北京、上海、浙江、江苏、广东等省份的居民能耗增速和消耗量要明显高于其他省份，西部地区的云南、贵州、青海、宁夏等地区则一直处于较低水平。总体上来看，受益于西部大开发、振兴东北、中部崛起等国家发展战略和政策的引导，东中西部的这种差异正在逐渐缩小，但城乡间的差距却在不断扩大，区域间占有和支配环境资源的差距仍十分明显。

在已有主要强调教育与医疗等公共服务均等化的议题中，将推进环境基本公共服务发展纳入政府的社会目标，是发展绿色经济、实现经济结构转型以及向“服务型政府”转变的重要手段。中国“十二五公共服务规划纲要”中，环境保护已经成为基本公共服务均等化的重要方面，“各地（县）都应具备污水、垃圾

① 王雪妮，孙才志，邹玮等．中国水贫困与经济贫困空间耦合关系研究［J］．中国软科学，2011，(12)：180-192.

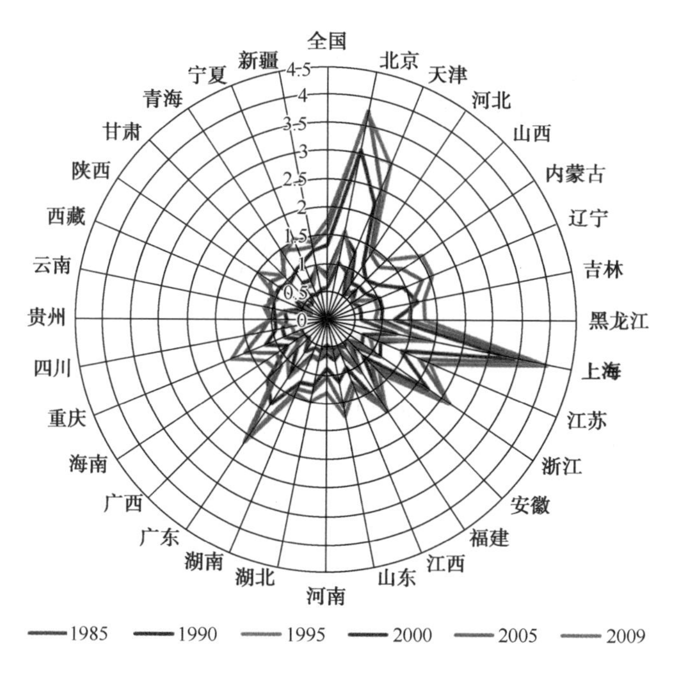

图 1-3　1985—2009 年中国各省份居民人均能源消耗趋势（单位：吨标准煤/tce）①

无害化处理能力以及环境监测评估能力”以及“保障城乡饮用水水源地安全”等作为环境保护的重点纳入评估目标。不同地区、不同群体的公众能否享有公平的环境公共服务，同样衡量环境公正的重要指标。卢洪友等（2012）构建了2003—2009 年中国环境基本公共服务绩效评估指数，指数中包括环境服务占用资源比重、城市污水处理能力等 15 项基本指标，对中国不同省份的环境公共服务效应进行了实证检验，如图 1-4 所示。研究表明，中国的环境基本公共服务绩效存在显著的省际差异，东部地区大部分省份环境基本公共服务的综合绩效水平显著高于中部省份和西部省份，经济发达地区与经济落后地区、城市地区与农村地区的公众所享受的环境基本公共服务明显是不均等的。很显然，如果在加快落后区域经济发展的过程中，忽视了这些地区环境公共服务水平的提升，将会进一步导致区域、群体间的环境不公正，以及环境与社会关系的进一步恶化。

① 余嘉玲，张世秋，谢旭轩．中国居民能源消耗趋势和差异分析［J］. 中国人口·资源与环境，2011，21（12）：351-355.

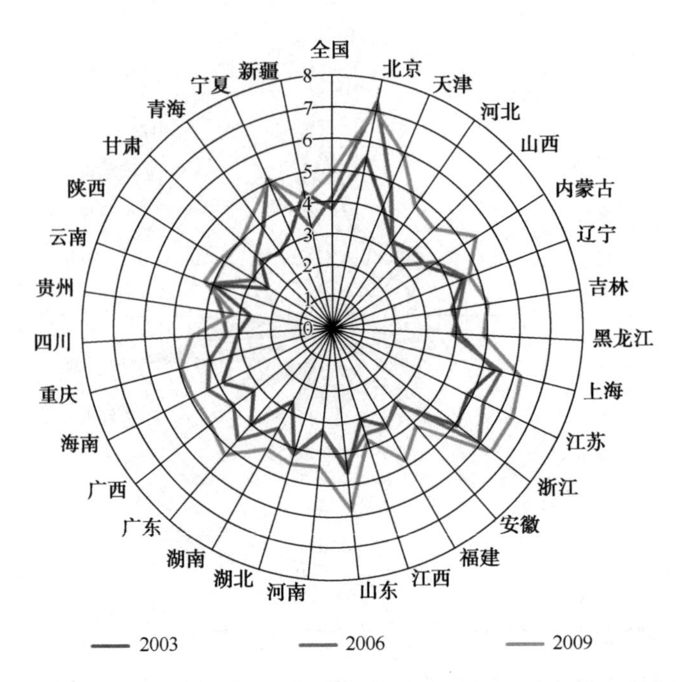

图 1-4　2003—2009 年中国各省份环境基本公共服务评估示意图①

4. 环境与城镇化问题

改革开放以来，中国的城镇化进程快速发展，从 1990 年的 26.41%快速增长到 2010 年 49.95%，年均提高大于 1 个百分点。目前，中国城镇化率已超过 50%②。相关预测显示，2015 年中国的城镇化将达到 52%，到 2020 年达到 60%左右③，2050 年将达到 77%④。城镇化推动了中国的经济增长，同一行业企业大量聚集，形成了“地方化”经济，并从知识与技术溢出、劳动力市场集中、激烈竞争中获益；而城市中不同行业企业大量聚集形成的“城镇化经济”则减少了交易成本。城镇化还推动了中国的技术创新与进步，提升了生产力，改善了人

① 卢洪友，袁光平，陈思霞等．中国居民能源消耗趋势和差异分析［J］．中国人口·资源与环境，2012，22（10）：48-54.

② 代表委员热议城市化：不能让农民“被城市化”．新华网［EB/OL］. http：//news.xinhuanet.com/society/2012-03/13/c_ 122827201.htm

③ 蒋洪强，张静，王金南等．中国快速城镇化的边际环境污染效应变化实证分析［J］．生态环境学报，2012，21（2）：293-297.

④ Asia Development Bank（ADB）. Key Indicators for Asia and the Pacific 2012.

们的生活水平[①]。

表 1-2　城镇化水平与变化（实际值与预测值）

地区	城镇化水平（%）			变化百分点数（%）	
	2000	2010	2050	2000—2010	2010—2050
欧洲	70.8	72.7	82.2	1.9	9.5
拉丁美洲和加勒比地区	75.5	78.8	86.6	3.4	7.8
北美	79.1	82.0	88.6	2.9	6.6
非洲	35.6	39.2	57.7	3.6	18.5
亚洲	35.5	42.5	62.9	7.0	20.4
中国	35.9	49.2	77.3	13.3	28.1
印度	27.7	30.9	51.7	3.3	20.8

中国向城镇化迈进也面临诸多难题，城市开发规模庞大、建设速度过快，带来的日益增长的能源需求和日益紧缺的土地资源、环境污染和气候变化，以及不断加剧的社会矛盾等诸多问题，这其中最显著的影响就是环境问题。城市基础设施、社会服务设施和居民住房保障的建设维护，日常生产生活需要大量的原材料和能源消耗，并产生大量的污染物排放，这些都将使资源与环境的压力不断加剧，城市的可持续发展面临巨大挑战。

首先，快速城镇化产生了大量污染物，造成严重的环境污染。相关测算表明，中国城镇化每增长 1 个百分点带来的城镇生活污水排放量增加 11 亿吨，城镇生活 COD 产生量和排放量增加分别为 79.6 和 3 万吨，城镇生活氨氮产生量和排放量增加分别为 6.7 和 1 万吨，城镇生活氮氧化物排放量增加 19.5 万吨，其中，由于机动车增长造成的氮氧化物排放量增加 8.6 万吨。城镇生活 CO_2 排放量增加 2525 万吨，城镇生活垃圾产生量增加 527 万吨，城镇化率的提高与主要污染物排放量的增加总体线性趋势呈上升状态。

其次，城镇环境基础建设滞后于快速城镇化。城市能否提供清洁用水与卫生设施，并妥善处理城市垃圾是衡量一座城市“绿色”程度的重要决定因素。然而，在中国快速城镇化的进程中，一方面是农村城镇化的基础环境设施建设滞后

① 中国—东盟环境保护合作中心．亚洲绿色城镇化的挑战与政策选择——对中国城镇化可持续发展的启示．亚太环境观察与研究，2013，(2).

于城镇化的速度，另一方面是中心城市环境基础设施建设滞后于城市的快速扩张，这无疑增加了快速城镇化过程中的环境隐患。

第三，快速城镇化造成了土地资源短缺等问题。城镇是农村经济发展到一定阶段的固有产物。在农村城镇化建设初期，一些地区对小城镇的功能、性质和定位的认识还不够清楚，盲目向外扩张，有的甚至放弃已经形成的原有集镇，重新征地建设新城镇。这样就使得城镇建设用地的集约化程度降低，造成土地资源的浪费，使人与环境之间的矛盾变得更为突出。而现有城市的进一步扩张也将加剧人与土地资源利用之间的矛盾。

此外，与快速工业化并行的应当是工业化的快速发展，城镇化率与工业化率（工业增加值占国内生产总值的比重）的比值应当被视为衡量城镇化程度是否适度的一个标尺。当城市的人数量超过工业化所能吸纳的程度时，就会造成“过度城镇化”。虽然短期内快速的城镇化可以拉动经济增长，然而，“过度城镇化”将不可避免地导致该地区出现诸如收入分配严重不公和两极分化、贫困发生率和失业率居高不下、住房紧张与贫民窟问题突出、医疗和教育资源不足等社会问题。这些社会问题在不同程度上又反作用于经济和环境系统，带来巨大的负面影响。因此，中国城镇化的扩张对地方政府和中央政府都是巨大的挑战，必须探索科学合理的治理模式，解决城镇化带来的能源、土地、水资源需求压力以及环境变化问题。

5. 环境运动与环境意识

公众环境意识于20世纪五六十年代在西方国家萌芽，并在世界各国迅速蔓延。中国公众的环境意识要晚于西方国家，相关研究显示，进入21世纪以来，中国公众环境意识的总体水平呈现上升趋势①，公众对环境问题的重视程度、公众的环保态度以及争取环境权益的意识不断增强。1998年的有关数据显示，56.7的公众认为中国环境污染状况“非常严重”和“比较严重”，远高于认为“不太严重”和“没有问题”的22.8%。但对面临的环境问题重视程度较低，排在社会治安、教育、人口、就业之后，位居第5。而2008年有关数据显示，2008年，公众对环境污染的关注度仅次于物价问题、食品安全之后，位列第3名，继2006年、2007年后连续第3次进前“三甲”，环境已成为公众最关心的问题之一②。与此同时，公众争取环境权益的意识也在不断加强，并逐步发展成为环境

① 闫国东，康建成等．中国公众环境意识的变化趋势，中国人口．资源与环境，2010，(10)．

② 数据引自自1998年《全国公众环境意识调查报告》。

运动。

一般来说，环境运动指的是由一定的支持者、参与者和组织构成，通过采取各种策略来保护地区、国家和全球环境的社会运动[①]。国际上，环境运动的思想和组织根源可以追溯到十九世纪后期开始的自然保护运动，根据环境运动的进展阶段及所包含的议题、机构、策略和战术的差异，大致可分为三个阶段：19世纪中后期的自然保护运动阶段、20世纪上半叶的环境保护主义阶段和始于20世纪60年代后期的生态保护主义阶段[②]。西方环境运动初期的议题是应对地方层次上可见的环境问题，从20世纪80年代以后，西方环境运动的主要目标转向跨界的或者全球性的环境难题[③]。相比较西方国家较为漫长的环境运动发展史，中国的环境运动起步较晚，在20世纪90年代后，随着一批民间环保组织的成立，才陆续出现有组织的环境运动。

根据参与主体的不同，中国的环境运动可划分为政府参与型、民间组织主导型和群体自发型。政府参与型是指政府依据环境保护相关法律推行的环境保护行动，例如2005年1月，原国家环境保护总局依据《中华人民共和国环境影响评价法》，叫停了大型企业未经环保审批的违法开工项目，由此在全社会产生了巨大反响，此次环保行动被舆论界称为“环评风暴”。在这次“环评风暴”后，关于信息公开、公众知情权和听证会等越来越多地被大众所接受，公众的环保意识得到了很大的提高；民间组织主导型是指民间组织发起的环境运动，是在20世纪90年代以后中国逐渐出现的各种社会性运动中较有影响力的一种。1994年3月，中国第一个民间环保组织“自然之友”在北京成立。此后，一大批民间环保组织陆续成立，截至2012年底，全国在民政部登记的生态环境类民间组织数量达到7928个（图1-5）。中国民间组织发起的环境运动虽然与西方的环境运动有较大差异，但是其在提高公众环保意识、解决环境问题等方面仍然起到了重要作用；群体自发型是指小规模的公众群体性环保运动，其行为特点是社区居民为维护自身权益参与到居住周边环保活动。随着公民社会不断成熟和环境权益意识的不断提高，中国社会公众将更多参与到捍卫自身环境权益的活动中。

然而，上述环境运动均存在一定的问题。首先，政府主导的环境运动主要是依靠行政命令手段，针对污染企业采取自上而下的关停、处罚等行动。这类环境运动具有影响力大、见效快的特点，但往往没有持续性，对污染者产生的仅仅是

① 王芳．西方环境运动及主要环保团体的行动策略研究．华东理工大学学报，2003，(2)．
② 岳世平．当代西方环境运动述评．河南大学学报（社会科学版），2006，(1)．
③ 崔凤，邵丽．中国的环境运动：中西比较．绿叶，2008，(6)．

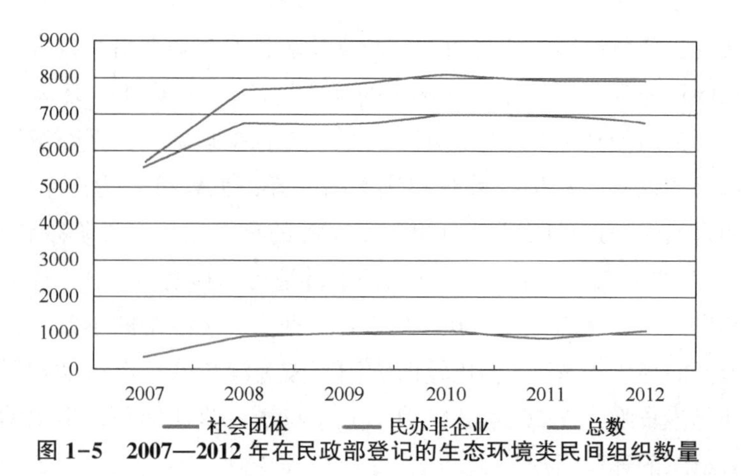

图 1-5　2007—2012 年在民政部登记的生态环境类民间组织数量

短期的威慑和影响；其次，中国具有政府背景的环保组织主要承担的是政府方面自上而下的衔接和沟通作用，其所承担的政府利益诉求相对较多。草根组织则多是由一两个环保倡导人发起，其组织规模较小、力量有限，关注的问题领域比较专一，活动范围领域较小。而国际组织的驻中国机构，其自身具有传达其组织理念，实现国际组织自身环境目标和利益的需求，在活动过程中，往往带有很强的国际组织烙印。总体来看，环保组织在中国还远未发挥其作用；第三，群体自发型环境运动虽然能促进环境问题的解决，但是如果参与的方式是无序的、非理性的，则很有可能引发环境对抗性群体事件。

随着公众环保意识的不断提高，中国环境运动也呈现出集中爆发的趋势，公众有序合理的参与环境运动可以起到影响政府决策、保护生态环境、维护公民环境权益的作用。但是，一旦环境运动成为非理性的、无序的环境群体性事件，则将对于社会经济发展及社会秩序的稳定带来负面的影响。一方面，非理性的、无序的环境运动将对社会经济发展带来损失。四川什邡事件中，宣布停建的“宏达钼铜冶炼项目”原定建成后年产钼 4 万吨、阴极铜 40 万吨，预计年销售收入将超过 500 亿元，利税超过 40 亿元，能解决当地约 3000 人就业，带动相关产业发展超过 400 亿元。而由于环评过程相关的信息公开、公众参与并没有到位，最后导致项目开工后遭遇众多民众的反对，项目无法执行。预期的经济效益无法实现，从一定程度上也阻碍了经济的发展。另一方面，非理性的、无序的环境运动也会影响了社会正常秩序，甚至影响政府执政的合法性。这些运动不仅充分揭露出中国环境保护与社会发展进程中存在的问题，还对维护国家和社会稳定造成了严重影响，对于中国政府执政的合法性带来巨大的挑战和威胁。从这个角度来说，环境与社会发展的矛盾比环境与经济发展的矛盾更加尖锐和深刻。

综上所述，环境问题已经在中国引发了显性的社会危机，环境与社会发展的不协调已经成为阻碍环境质量持续改善、实现社会稳定和国家长治久安的重要因素。其核心的本质是环境问题所带来的环境与经济利益的不协调、分配的不平衡等问题。长期以来，中国高度重视经济社会的协调发展，近年来，环境与经济的关系也得到了普遍关注。然而，环境与社会的关系，以及环境与社会发展的协调却长期被忽视。各级政府更强调运用“自上而下”的行政手段以及基于市场的经济政策工具解决环境问题，忽视了运用社会手段、动员社会力量参与解决环境问题的作用。社会公众参与国家与社区环境治理的渠道和作用远未发掘，成为国家环境公共治理的短板。因此，亟需研究和探索能够促进解决环境问题、化解社会矛盾和风险，同时又能保障环境民主和权益的环境公共治理模式，这对于未来中国处理环境与社会关系，运用社会公众手段参与环境公共治理，解决中国的环境与社会发展问题具有重要意义。

三、研究进展述评

（一）环境与社会理论研究发展历程

20 世纪 50 年代以来，全球工业化进程不断推进，人类开发自然的深度空前提高，资源短缺、环境污染等问题成为威胁人类生存和发展的根本性问题。在这种背景下，反对环境污染和生态破坏的生态运动崛起，成为一个有广泛群众基础的社会运动，国家开始参与环境管理，环境问题开始进入政治领域。环境与社会之间的关系成为学界关注的核心论题。

真正意义上的环境与社会关系研究，诞生于西方 20 世纪 60 年代末、70 年代初，环境社会学最初研究的重点，集中在具体环境问题的社会分析和政策研究，如全球环境变暖、空气污染、臭氧层空洞等环境问题的社会原因、产生机制和社会政策分析①。1971 年，环境社会学一词首次出现在塞缪尔·克劳斯纳（Samuel Klausner）的《自然环境中的人类》（On Man in His Environment）中。20 世纪 70 年代，美国成立了国家社会学会，使环境社会学得以制度化。到上个世纪 80 年代后期，90 年代前期，环境社会学不仅在美国蓬勃发展，而且在世界各国逐步

① 徐国玲．西方环境社会学研究的三种范式．

制度化并成立了国际社会学协会、环境和社会研究委员会等，为环境社会学全球间的发展提供了契机。进入 90 年代以后，环境社会学的研究扩展到世界各地，许多国家相继成立研究组织，学术研究非常活跃，形成许多不同的研究观点和研究方法。总体上说，环境社会学大致上经历了三个发展阶段，即 70 年代雄心勃勃的初创时期、80 年代彷徨的过渡时期和 80 年代末、90 年代以来柳暗花明的发展时期①。

经过 30 多年的发展，环境社会学逐步形成了特定的研究领域。20 世纪 70 年代，美国环境社会学家邓拉普（Dunlap）和卡顿（ Cottan）在其合著《环境社会学》中指出，环境社会学的基本研究领域为：（1）建筑环境；（2）环保团体、工业界及政府对环境问题的反映；（3）自然灾害与灾难；（4）（环境）社会影响评价；（5）能源及其他资源紧缺的影响；（6）资源配置与环境容量②。

1989 年，针对环境社会学的研究领域进行了一次调查。该调查报告中提到了以下研究领域：（1）环境研究的伦理标准、概念和方法；（2）对环境问题的描述；（3）环境污染的影响；（4）环境立法；（5）环境政策；（6）环境管理；（7）环境意识、行为、运动和环境犯罪；（8）环境教育；（9）环境与信息。

随着人类社会的发展、环境问题的不断演变，环境社会学的研究领域也在不断的深化。为了促进环境社会学的未来发展，一些学者提出了自己的建议或主张。邓拉普和卡顿认为环境社会学应当紧密围绕全球环境变迁（GEC）这一新的现象进行研究，以求自身的发展③。国内环境社会学研究学者洪大用认为，环境社会学的真正主题在于研究“环境问题的产生及其社会影响”④。这门学科的主要任务是具体分析环境问题产生的社会过程和社会原因，分析环境问题作为一种新的“社会事实”是如何影响现代社会的，分析现代社会对于环境问题的反映及其效果。这个观点的提出为环境社会学的未来研究工作指明了一个方向。

20 世纪 90 年代以来，国内学者开始关注环境社会学的研究，但大都是介绍国外环境社会学研究的发展。少数研究者欲结合中国的情况进行研究，但由于种种主观或客观条件方面的限制，没有对环境社会学学科体系做出系统完整的描述。目前，中国环境社会学还没有形成独立完整的学科体系，尚处于建立理论结

---

① 洪大用．西方环境社会学研究．社会学研究，1999，（2）.

② Dunlap, R. E.; Catton, W. R. Jr., 1979, " Environmental Sociology", Annual Review of Sociology, 5.

③ Dunlap, R. E.; Catton, W. R. Jr., 1994, "Struggling With Human Exemptionalism: The Rise, Decline and Rev italization of Environmental Socilolgy", American Sociologist, Sp ring.

④ 同①.

构框架阶段，现有研究多注重描述中国环境现状、管理现状与环境问题，社会学解释性分析不够，缺少社会共同属性的理论研究[①]。

当前，中国处于社会转型加速发展阶段，引发的各种环境问题愈演愈烈，中国环境社会学的理论建设已经进入到一个关键的发展时期。社会学家通过广泛的研究探索环境与社会之间的各种可能性，挖掘环境问题的社会层面原因和影响。环境社会学在中国起步晚却有着强劲的发展动力，随着环境问题由隐性问题逐步转化为显性问题，中国环境社会学将迎来一个崭新的发展时期。

（二）环境与社会理论研究主要范式

环境社会学经过30多年的发展，其经验和理论研究成果颇丰，各理论范式日趋多元。邓拉普等人在其所著的《环境社会学手册》的导言中提出，现在的环境社会学领域至少有如下九个明确的范式：人类生态学、政治经济学、社会建构论、批判实在论、生态现代化、风险社会理论、环境正义理论、行动者网络理论、政治生态学[②]。在西方环境社会学理论上影响比较大的学说，主要有新生态范式（New Ecological Paradigm）论、马克思主义政治经济学范式论、建构主义范式论三种观点。

目前国内的研究著作关于研究范式的划分并未达成共识。洪大用运用类型学方法将环境社会学分为环境学的环境社会学与社会学的环境社会学。吕涛认为社会学的研究对象是作为社会事实的“社会化了的环境变量”与社会的关系，并提出ESSP范式（Environment Socialization Society Paradigm）和SAEP范式（Society Action Environment Paradigm）。江莹将环境社会学研究范式概括为生态学范式、系统论范式、政治经济学范式、建构主义范式、社会转型范式、整合型范式6种[③]。这6种范式概括了环境社会学理论发展演变的基本范式。下面将对这几种研究范式的主要观点进行综述：

1. 新生态范式

1978年，卡顿和邓拉普在《美国社会学家》杂志第13卷上发表了题为《环境社会学：一个新范式》的文章，他们认为传统的社会学研究范式，全部停留在经济、政府等人类社会制度的语境之内，社会学的研究一直未能将社会和环境的

---

① 崔凤，秦佳荔．中国环境社会学研究综述．河海大学学报，2010，(4).

② Riley E. Dunlap，William Michelson. Handbook of Environmental Sociology [M]. Westport，CT：Greenwood Press，2001

③ 同①.

物质联系纳入视野。环境因素对社会事实的影响被忽略了。另一方面，“各种社会学理论尽管表面上分歧对立，但是都具有人类中心主义这一共同点”①。邓拉普和卡顿从生态学的角度提出了与传统范式不同的生态学范式。生态学范式与传统社会学范式不同之处在于，它强调了环境因素对于社会事实变化的作用②。

邓拉普和卡顿将传统社会学的“人类中心主义”范式概括为“人类豁免主义范式”（Human Exemptionalism Paradigm）（简称 HEP）。这种范式的核心假设是：（1）人类具有文化遗产和基因遗传，因此在地球中是独一无二的；（2）社会和文化（包括技术）因素是人类事件的主要决定性因素；（3）社会和文化环境是人类事件的关键内容，生物物理环境很大程度上与其无关；（4）文化积累意味着进化可以无限延续，使得所有社会问题最终都得以解决③。

这些理论假设在他们看来是错误的，忽视了资源稀缺与生态约束对人类进步造成的威胁。他们主张新生态范式（ New Ecological Paradigm，简称 NEP）来代替 HEP。他们认为 NEP 是以不同的假设为基础的：（1）虽然人类有突出的特征（ 文化、技术），但他们依然是包含在全球生态系统中的互相依赖的众多物种成员之一；（2）人类事件不仅受社会和文化因素影响，而且还受自然之网的复杂的因果联系和反馈的影响；（3）人类生存依赖于一个有限的生物物理环境，它对人类活动加上了潜在的限制；（4）如果超出了生态的承受极限，人类的创造力再强大都是暂时的，生态法则绝对不容违反④。

邓拉普和卡顿的生态学分析框架有两个思想来源 ，一是邓肯（Duncan, O. D. ）的“生态复合体”，一是帕克（Park，R. E. ）的“社会复合体”。⑤ 所谓“生态复合体”指的是人类、组织、技术和环境四者（简称 P、O、T 、E）之间的某种交叉依赖关系。其中，每一个要素与其他三个要素相互关联；任一因素的变化会引起另外因素的变化⑥。这种模型并未突出环境因素，为此，邓拉普和卡顿将环境的含义固定在自然环境或物理环境上，这样原来的模型也就变成了环境与其他要素（人类、组织、技术）之间关系的研究。

人类、组织、技术要素的结合就是帕克的“社会复合体”概念。邓拉普和卡顿借助这个概念将“生态复合体”中的组织要素细分为文化体系、社会体系

---

① 洪大用．西方环境社会学研究．

② 江莹．环境社会学研究范式评析．郑州大学学报，2005，(5)．

③ 徐国玲．西方环境社会学研究的三种范式．

④ 吕涛．环境社会学研究综述——对环境社会学学科定位问题的讨论．社会学研究，2004，(4)．

⑤ 同②.

⑥ 同①.

和人格体系。这样就形成了邓拉普和卡顿提出的有关环境与社会关系的新分析框架，即自然环境或物理环境与人口、技术、文化体系、社会体系和人格体系之间的关系。

1995年，邓拉普和卡顿在原有理论的基础上，又提出了“环境的三维竞争功能”（three competing functions of the environment）概念，认为环境对于人类具有三种功能：提供生活空间、提供生存资源和进行废物储存与转化①。他们认为，在特定区域内，这三种功能彼此相互竞争。今天，随着人口数量的增加和日益增多的人类活动，竞争比过去更加激烈，人类的需求也可能超出既定区域甚至地球生态系统的承载力。通过分析三种功能之间的冲突关系以及这些功能与关系的演变情况，可以解释当代环境问题的生态根源。

新生态范式超越了人类生态学只关注生存空间的局限，使主流社会学开始真正关注环境变量，环境社会学的地位得以真正确立。但是该模式并未摆脱传统社会学的深刻影响，其分析框架仍来自于传统社会学。

2. 系统论范式

系统论范式是吉尔贝托·加洛潘、巴勃罗·古特曼和埃克托尔·马莱塔等人提出的一种社会学系统研究方法。

加洛潘等人认为，应当采取系统论的视角研究环境与社会之间的关系。但是，与一般的系统论观点不同，加洛潘等人指出，社会—生态系统最好是看作一套因果轮回和有待提出的问题，而不是看作一套子系统。这样做，更具有适用性和灵活性，“能将要研究的变量或过程逐步组织起来，并且指导今后按不同的情况将系统分解成有关的子系统”。

他们认为，环境社会学应当主要研究影响自然生态系统的一系列人为活动以及影响社会系统的一系列自然产生的生态效应。这一范式与其他范式相比，不再片面强调技术性的主导作用，社会与生态系统的互动关系得到相当程度的关注。但是，该范式声称主体上具有宏观性，但却过于被动地滞留在生态和社会两大系统范围之内，而忽略了微观层面个体的能动影响，导致了一定成分的虚无消极因素产生，在实践中缺乏公共基础，难以具体落实②。

3. 政治经济学范式

该范式主要关注的问题是：环境衰退的社会根源是什么？究竟谁应该对环境

① Riley E. Dunlap，William Michelson. Handbook of Environmental Sociology [M]. Westport，CT：Greenwood Press，2001.

② 江莹. 环境社会学研究范式评析.

破坏负责？史奈伯格（Allan Schnaiberg）是其主要代表人物。他的理论是在广泛吸收马克思主义的政治经济学、新马克思主义和新韦伯主义的政治社会学等的有关观点和材料的基础上发展起来的。其理论内核是预设“社会与环境的辩证关系”以及提出“生产的传动机制（treadmill of production）”概念①。

社会与环境的辩证关系主要表现有三点：（1）社会的经济增长必然要求增加从环境中开采的资源；（2）开采资源的增加不可避免地造成环境问题；（3）这些环境问题为日后的经济增长预设潜在的限制②。

史奈伯格的“生产传动机制”，是指在现代资本主义社会中促进经济扩张的一种复杂的自我强化的机制，它在很大程度上是资本集中和集权的趋势日益发展的结果，同时也是资本主义国家与垄断经济部门关系发生变化的结果。他认为，人与生态系统关系中主要有两种传动：（1）向自然界的传动。企业应用更多的技术生产出更多的产品，获得了更多的利润，因此，就需要更多的原材料和自然资源，也导致了化学污染和更多废物的产生。（2）向社会的传动。在每一个生产的环节中，利润被分配到不同的企业，在他们的努力下形成了一种新的节省劳动力的技术，从而导致工人在生产过程中被驱除出去的结果，很多技术成熟的工人会失去工作机会③。

政治经济学范式认为环境破坏的原因在于人类建构的经济系统和政治系统的不平衡性。但是，社会主义国家的环境问题对于这种将环境破坏归咎于资本主义制度的观点提出了挑战。

4. 建构主义范式

这种理论范式关注的主要问题可以概括为两个方面。一是为什么有些环境问题早就存在，但只是到了某一时间才引起泛注意？二是为什么有些环境问题引起了广泛的注意，而有些则没有同样的效果④？该理论的代表人物是汉尼根（Jhon Hannigan）。他回避了对于环境问题的客观性以及客观原因的分析，而集中分析环境问题的社会建构过程，分析如何才能成功的建构环境问题⑤。

在汉尼根所著的《环境社会学》导论中指出，公众对于环境的关心并不直

---

① 洪大用．西方环境社会学研究．

② 徐国玲．西方环境社会学研究的三种范式．

③ Schnaiberg，A.，1980. The Environment：From Surplus TO Scarcity，New York，Oxford University Press.

④ 同②.

⑤ 同①.

接与环境的客观状况相关，而且，公众对于环境的关心程度在不同时期并不一定一致[①]。他指出，环境问题社会建构论的研究，应通过环境问题的聚集、呈现和竞争过程这样的分析工具来考察环境问题建构的过程。通过对全球环境问题的经验研究，他指出成功建构环境问题的必要因素，并且对环境风险和环境知识的建构进行了研究[②]。他认为，环境问题必须经由个人或组织的建构，现代社会中的两个重要社会设置——科学和大众媒体，在建构环境风险、环境知识、环境危机以及对于环境问题的解决办法方面，发挥着极其重要的作用。

汉尼根发现，成功地构建某种环境论题必须注意八个方面的因素：（1）某种环境问题必须有科学权威的支持与证实；（2）拥有科学普及者是重要的；（3）预期中的环境问题必须受到媒体的注意，正是媒体使得相关的呼吁变得真实而重要；（4）必须以非常醒目的符号和形象词汇修饰某些潜在的环境问题；（5）针对某一环境问题采取行动必须有可见的经济刺激；（6）应当有制度化的赞助者，他们可以确保环境问题建构的合法性与连续性。

汉尼根的建构思想在当今时代非常具有说服力，以往很多环境问题得到了充分重视，但是，其对环境本质客观性的忽视也使其具有一定的局限性。

5. 社会转型范式

国内学者洪大用在已有理论研究的基础上，提出了“社会转型范式”。他认为，国外环境社会学关于环境问题的各种解释范式虽然都从一定的角度解释了环境问题，但是不够全面，也无法解释中国当代的环境问题，只有深入研究社会转型的过程与趋势，才能寻求较为完善的环境保护对策[③]。

在《社会变迁与环境问题——当代中国环境问题的社会学阐释》一书中，洪大用采用了社会转型分析范式，他以中国为例，从社会学的角度分析当代中国环境问题的形成机制，评述环境保护的相应对策，并探讨了调整社会的发展目标，强调指出了优化社会结构的可能性和重要性。社会转型作为一个相对宏观的范畴，其发生作用必然会通过各种因素与主体。对于“社会转型”，他提出了三个方面变化：社会结构（工业化、城镇化、区域分化）、社会机制（建立市场经济体制、放权让利改革、二元控制体系）和社会价值观念（道德滑坡、消费主

---

① Hannigan, J. A., 1995, Environmental Sociology: A Social Constructionist Perspective. London and New York: Routledge.

② 赵万里，蔡萍．建构论视角下的环境与社会——西方环境社会学的发展走向评析．山西大学学报，2009,（1）.

③ 江莹．环境社会学研究范式评析．

义、行为短期化、社会流动加剧)。

该范式的研究基于以下两个假设：一是环境与社会是相互作用的；二是当代中国社会正处于社会转型加速期。从这两个预设出发，他推导出如下具体假设：(1) 转型社会的环境问题有其独自的特征；(2) 转型社会有其独特的造成环境问题的具体机制；(3) 当代中国环境问题在一定程度上是特定环境状况与特定社会过程交互作用的产物，即环境问题在一定程度上是经由转型期的特定社会过程建构的；(4) 应对环境问题的策略必须考虑到中国社会的特点；(5) 由于社会转型所加剧的环境问题，最终只能通过进一步的社会变革去解决①。

社会转型范式的主要结论是：需要通过组织创新，优化社会结构，促进社会的民主化以促进中国的环境保护。这一理论的提出在当前中国环境保护的形势下具有创新性的意义。长期以来中国“政府主导型”的环境保护模式存在局限性，随着公民社会的成长，需要转变政府职能，促进越来越多的“第三部门”和社会公众参与到环境保护工作中。

6. 整合型范式

整合性研究范式（Integrated Research Paradigm，简称 IRP）是美国社会学家 Hannah Brenkert，Julie L． Gailus，Aaron Johnson，Megan Murphy 在 2004 年提出来的，他们响应 Dietz 和 Rosa 在 2002 年的呼吁，希望在环境社会学领域里能够有一个具有广泛意义的理论框架来指导研究②。

他们将生物物理层面与社会层面放在一起研究，将环境社会学的研究范畴分为三部分：一是有关自然物质现象方面的生物物理子系统；一是有关整体化社会框架的宏观社会子系统；还有一个是具体到个体行动者自身心理框架的微观社会子系统。其目的是充分体现出生物物理子系统、宏观社会子系统和微观社会子系统三者之间的互动关系③。该范式将个体、社会、环境纳入一个“系统”研究，各子系统之间相对独立又相互联系，通过这些联系来处理系统自身复杂性带来的问题。就三个子系统的互动关系而言，存在九种关系组合，这九种关系折射出 IRP 所强调的三个子系统之间的多维度多层面的互动结果。

这种整合性的研究视角对于环境社会学理论研究来说，是一种积极的尝试，对于我们研究环境社会问题提供了一种多元化视野。

---

① 洪大用．社会变迁与环境问题——当代中国环境问题的社会学阐释．首都师范大学出版社 2001.

② Hannah Brenkert，Julie L. Gailus，Aaron Jonhson，Megan Murphy. （2004）. Integrated Research Paradigm：A Neorealist Model for Environmental Sociology. Institute of Behavioral Science. WorkingPaper.

③ 江莹，秦亚勋．整合性研究：环境社会学最新范式．江海学刊，2005，(3).

（三）环境与社会实证研究主要观点

环境与社会领域的理论研究经过三十多年的发展成果颇丰，而随着环境问题的日益加剧、环境运动的不断发展，其引发的社会冲突和社会现象集中爆发，关于环境与社会的实证经验研究成果也不断涌现。

根据目前已有的研究成果来看，环境社会的实证经验研究主要包括：环境公正、环境运动与环境意识、环境利益分配、环境与贫困、环境与性别、环境公共治理模式等。

1. 环境公正

20 世纪 80 年代以来，随着西方公众环境意识提高、环境运动高涨以及环境问题研究的深入，环境公正成为一个新的研究话题，引起了比较广泛的关注。环境公正是从社会学角度审视环境问题的重要切入点之一，涉及各种自然资源以及环境污染风险在社会成员之间的分配。

目前学术界对于环境公正还没有统一的定义。美国环保署（EPA）对环境公正给出的政策性定义是：在环境法律、法规和政策的制定、实施和执行等方面，全体国民，不论种族、肤色、国籍和财产状况差异，都应得到公平对待和有效参与环境决策①。洪大用认为，环境公正的核心在于在环境资源、机会的使用和环境风险的分配上，所有主体一律平等，享有同等的权利，负有同等的义务②。

环境公正研究目前仍然处于初级阶段。西方学者基于本国环境问题（包括环境事件、环境运动等）对环境公正进行了理论和方法探讨，并已经呈现了一些很有前途的理论模型和方法，但目前还没有形成成熟的理论和方法体系，学科建设也处于摸索发展阶段，根据相关学者的研究③，20 世纪 90 年代以来，出现了以下相关的理论模型：

一是基于地域性研究建构的理论模型，包括理性选择模型、社会政治模型、种族歧视模型、合作主义视角、“环境不公平的形成”视角；

二是基于全球视野建构的理论观点，其关注点是跨国界的自然资源剥夺和有害废物贸易。一方面，全球少数的富有国家人民消耗、浪费过多资源，并制造大量废弃物；另一方面，多数经济贫穷国家人民则缺乏资源并承受最多的环境危害。

---

① 美国国家环境保护署主页，http：/ / www. Epa g ov/ chinese/ pdfs/ EJ%20Bro chure_ CH I pdf

② 洪大用．环境公正研究的理论与方法述评．中国人民大学学报，2008，(6).

③ 同②.

环境公正的研究方法包括定量研究和定性研究。定量研究方法在环境公正研究中的使用晚于定性研究方法，进入20世纪90年代后，定量研究得到长足发展。定量研究的经典方法是“居住单位与遭受污染风险巧合分析法”（Analyzing Unit Hazard Coincidence）。首先，选择一个事先确定的地理单位（如县、地区代码、人口普查区域），再辨别其中哪些单位隐匿了环境风险，而哪些单位对风险作出了处置，确定一组恰当的比较单位（特别是那些不包括危险的单位），然后比较这组单位中的人口学特征。因为这种研究方法存在局限性，墨海等人在其基础上又提出了“距离分析法”，这种方法使用地理信息系统收集数据，并描绘出环境风险的精确位置，列出这些风险位置距附近居民点的距离。该方法能将特定距离内的所有单位（不仅仅是主要研究单位）的人口学特征与更远距离的单位的人口学特征进行对比，以发现是否存在环境风险不公正分布问题。

定性研究方法中，较受关注的是历史分析法和过程分析法。这两种方法用于阐释环境不公平的起因和建构过程，解释种族、阶级和公众的环境知识等因素如何随时间的推移而影响环境风险在不同人群中的分配，强调随时间的推移，人们各种观念、价值的变化对环境公正问题的不断重构。

2. 环境运动与环境意识

国内外学者对于环境运动与环境意识的研究成果很多，就国内研究而言，具有代表性的观点有：张玉林认为中国的环境运动包括3个部分：城市知识分子的环境启蒙、城市精英的环保行为、基层民众的申诉与抗议行动，且三者缺少联系与呼应，而基层民众的抗争主要是指农民在宏观的政治经济制度下污染与抗争的循环[①]。崔凤将中国的环境运动与西方的环境运动进行比较，认为中国的环境运动有民间组织、知识分子和政府发起3种，主要还处在以局部利益为出发点的“地方性思考、地方性行动”阶段[②]。在环境意识的研究方面，洪大用则提出了“新环境范式量表”（New Environmental Paradigm Scale）在中国的应用，对于量表本身的修正分析与环境关心性别差异分析等[③]。

3. 环境利益分配

自然环境与资源是人类赖以为生的基本因素，由于资源占有的不均与社会经济地位的差异，环境利益与环境负担在社会各行为主体间存在不公平的分配。部分群体享受环境利益而将环境负担转嫁给社会其他群体，这就会导致“环境不公

---

① 张玉林．中国的环境运动．绿叶，2009，(11)．

② 崔凤，邵丽．中国的环境运动：中西比较．绿叶，2008，(6)．

③ 洪大用．环境关心的测量：NEP量表在中国的应用评估．社会，2006，(5)．

平”。

从我国的社会结构现状来看，地区之间、城乡之间、群体之间环境利益存在不同的分配情形[①]。

（1）地区利益差异

就地区差异而言，东部沿海地区利用政策优势与对外交通便利，社会经济等各方面发展较内陆地区先进，这种先进以率先利用国家资源与造成环境污染为代价，污染的环境则由全国人民来承受其害。环境利益与环境负担不匹配。两者只有在地区之间较公平地分配，才能扭转由于环境利益差别而形成的地区发展畸形。

（2）城乡利益差异

城乡之间环境利益的差异更彰显环境利益分配不公，这是我国城乡二元体制社会治理下社会问题的集中凸显。在我国社会转型之际，城乡之间的环境不正义必须得到矫正，使农村也能享受到因环境资源带来的利益，而避免环境污染的更大承受。

（3）群体利益差异

群体之间的环境利益差别关系到个人权利的保障与自我实现。由于社会经济地位的不同，不同群体对环境利益与环境负担的分配不同，占有强势地位的群体更多的是环境利益的享受者，而不大可能受到环境污染的侵害，弱势群体往往是环境损害的承受者，并且他们没有便利的条件将损害填补。这种利益分配的不公平很可能导致环境群体性事件的集中爆发，不利于社会秩序的稳定，群体之间的环境利益与负担必须要在社会结构内得到合理的分配，才能最终对维护社会秩序起到积极作用。

解决环境利益分配不均的重要途径之一是环境司法的专门化，运用法律的约束合理分配各主体之间的环境利益，通过制度的作用塑造环境社会的公平正义，从而维护社会秩序的稳定。

4. 环境与贫困

环境贫困是环境恶化带来的一个重要问题。有学者指出，中国的“贫困——人口过度增长——环境退化”的恶性循环（PPE 怪圈），与中国农村的“社会发育程度低——农村经济结构单——农民文化素质低”（RAP 怪圈）相结合，环境

① 杜健勋，陈德敏．环境利益分配：环境法学的规范性关怀——环境利益分配与公民社会基础的环境法学辩证．时代法学，2010，(10).

问题不断恶化，贫困问题也持续严重[①]。环境问题本身导致的贫困也时有发生。在我国，环境贫困状况存在严重的地区差异，中西部和少数民族地区是环境贫困的重灾区，应对环境和改善环境能力严重缺乏，此外，城乡之间、城市区域之间也存在差异。对于城市来说，城镇化带来的环境压力，使城市变得脆弱不堪，粗放的经济增长方式、急剧增长的人口压力、空气质量的日渐下降、工业所造成的水体污染等等，都严重危害了城市的环境。有学者认为，我国的三大城市群：珠三角城市群、长三角城市群和京津唐城市群，已经形成三大城市污染群，主要污染类型为大气复合污染[②]。相对其他污染来讲，大气污染还是很“公平”的，每个人都必须承受，但是固体废物污染、水污染、噪声污染的受众则基本上都是城市中的贫困者。

环境贫困已经很明显的凸现，而且危害正在加剧。环境保障的实施虽然刻不容缓但却存在很多的困难和障碍，需要国家和社会每个人的共同努力。

5. 环境与性别

20 世纪 70 年代，在环境运动的助推下，西方女性主义者开始探索妇女与自然的关系，发展出生态女性主义理论，并在发展领域实践其理论主张。中国对妇女与环境互动的关注始于 1995 年世界妇女大会，在很大程度上受到西方生态女性主义理论及其实践的影响，但目前仍是一个尚待理论化的主题。“生态女性主义”这一概念由法国女性主义学者奥波尼（Eaubonne）于 1974 年首次提出，她以此号召女性领导一场生态运动，重新认识人和自然的关系[③]。

生态女性主义关心的核心问题是自然与妇女的联系，其宗旨在于揭示人类思想领域和社会结构中普遍存在的贬低女性与贬低自然之间的一种特殊关系，反对父权制的世界观和二元式思维方式统治下的对女性与自然的压迫，把反对性别压迫、追求妇女解放和解决生态危机一并当作自己的奋斗目标，倡导建立一种人与人、人与自然之间的新型关系。

其基本的理论观点包括三个方面：一是强调女性与自然的认同。生态女性主义认为女性对自然界有一种认同感，以一种具体的、爱的行动与自然界相联[④]。二是用“保护和养育”张扬女性原则。主张弘扬女性原则，尊重地球和妇女。

---

① 冯凯旋．环境问题中的环境贫困现象初探．赤峰学院学报，2009，(2)．

② 专家发言：三大城市群大气污染重．南方都市报，2008-06-13（AA08）．

③ 杨玉静．生态女性主义视角下的中国妇女与环境关系评析．妇女研究论丛，2010，(7)．

④ M. Kheel. Ecofeminism and Deep Ecology [A]. in Reweaving the World: The Emergence of Ecofeminism [C]. ed. By I. Diamond & G. F. Orenstein, Sierra Club Books, 1990.

三是批判“父权制”的文化及二元对立的思维方式。

围绕“是否将妇女与自然相联系”这一问题，生态女性主义者形成了不同的理论流派，主要有：精神的生态女性主义、社会主义的生态女性主义、文化/自然的生态女性主义以及社会/社会—建构主义的生态女性主义。其中后两者对中国的妇女与环境关系建构产生了一定影响。

6. 环境公共治理模式

20 世纪 90 年代，随着民间环保非政府组织纷纷成立，我国民间环保运动兴起，非政府组织、公共知识分子等非政府群体参与环境保护运动，对传统的政府公共管理模式带来新的发展机遇。在环境问题治理中，非政府组织、公众以及政府的互动关系成为各界探讨的焦点。

在我国政府职能转变，从威权政府向“有限政府”和“责任政府”转变的大趋势下，西方公共治理理论的引入引起了学界的热切关注。公共治理理论强调的是多中心治理结构，强调国家政府、市场和社会共同治理的理念。这种模式的基本要素包括①：（1）主体多元化。市场的力量和公众的潜能得到了前所未有的重视，企业、民间环保组织、大众传媒、个体公民乃至国际组织都成为环境治理的主体。（2）结构网络化。原有的政府主导的“金字塔型”结构被打破，逐渐形成一种平等协商、合作互利的伙伴关系。（3）过程互动化。权力运行或传导不是传统的政府主导的、单一的自上而下的模式，而是包含来自民间社团或私人部门自下而上的上下互动的管理过程。（4）方式协调化。环境公共治理不是控制与被控制的过程，而是相互协调的过程。

当前我国的环境公共治理模式尚不健全，存在诸多的问题：政府与企业的二元治理结构，政府越位、公众参与不足、治理效率低下、环境冲突频发。环境公共治理的最终目标，是要形成由政府、企业和公民社会三个行动者参与的平衡治理结构，这需要我们融合“自上而下”和“自下而上”的治理运行机制，利用法律、行政以及经济激励等综合手段和政策工具，保障治理机制的有效运转。政府、社会组织和公众力量共同努力，促进环境与经济、与发展、与社会互相促进，协调发展。

---

① 曾正滋．环境公共治理模式下的参与——回应型行政体制．福建行政学院学报，2009，(5).

第二章　环境保护与社会发展的国际国内经验

一、世界工业革命以来西方主要国家环境保护与社会发展

从18世纪下半叶起，经过整个19世纪到20世纪初，首先是英国，而后是欧洲其他国家、美国和日本相继经历和实现了工业革命，最终建立以煤炭、冶金、化工等为基础的工业生产体系。到1900年时，世界先进国家英、美、德、法、日五国煤炭产量总和已达6.641亿吨。煤的大规模开采并燃用，在提供动力以推动工厂的开办和蒸汽机的运转并方便人们的日常生活时，也必然会释放大量的烟尘、二氧化硫、二氧化碳、一氧化碳和其他有害的污染物质。同时，释放许多重金属，如铅、锌、镉、铜、砷等，污染了大气、土壤和水域。尤其是这一时期化学工业的迅速发展，水泥工业的粉尘与造纸工业的废液也构成了重要环境污染源。伴随这些国家煤炭、冶金、化学等重工业的建立、发展以及城镇化的推进，出现了烟雾腾腾的城镇，发生了烟雾中毒事件，河流等水体也严重受害。自20世纪20年代以来，随着以石油和天然气为主要原料的有机化学工业的发展，西方国家不仅合成了橡胶、塑料和纤维三大高分子合成材料，还生产了多种多样的有机化学制品，如合成洗涤剂、合成油脂、有机农药、食品与饲料添加剂等。就在有机化学工业为人类带来琳琅满目和方便耐用的产品时，它对环境的破坏也渐渐地发生，久而久之便构成对环境的有机毒害和污染。20世纪50年代起，世界经济由战后恢复转入发展时期。西方大国竟相发展经济，工业化和城镇化进程加快，经济高速持续增长。在这种增长的背后，却隐藏着破坏和污染环境的巨大危机。因为工业化与城镇化的推进，一方面带来了资源和原料的大量需求和消耗，另一方面使得工业生产和城市生活的大量废弃物排向土壤、河流和大气之中，最终造成环境污染的大爆发，使世界环境污染危机进一步加重。

专栏2-1，用简要的办法，回顾和阐述了工业革命之后环境、社会和经济发展过程中的重要历史事件。其中最突出的变化是人口由农村大量流向城市，大规模地改变了人类的生产和消费方式；同时，工业化的发展增强了人类利用和改造自然环境的能力，各种工业活动大规模地改变了环境，导致人类所依赖的生态环

境开始恶化。人们在尽情享受工业化生产和城镇化生活带来的发展成果时，各种环境问题也从天而降，给人们的日常生活甚至生存带来了巨大威胁，引发了各种社会问题。环境与社会问题的关系越来越密切，已成为现代社会的重要特征①②。

专栏 2-1　工业革命以来影响环境与社会的主要事件和行动时间表

时　　间	主要事件和行动
18 世纪 60 年代	• 第一次科技革命，蒸汽机作为动力机并被广泛使用，社会财富随着生产工具的机器化暴增，社会发展进入大工业经济时代。 • 人口大量流向城市和工业城镇，导致生活环境变差并造成局部污染。 • 自然资源，特别是发展中国家的自然资源，被大规模地开发和利用。
19 世纪 20 年代	• 全球人口、人类活动和环境压力急剧增长。 • 欧洲人口快速增长，大量人口移民到美洲、大洋洲和非洲，全球人口分布发生变化③。
1870—1900	• 第二次科技革命，以电力的广泛应用、内燃机和新交通工具的使用、新通讯手段的发明和化学工业的建立为标志，世界进入“电气时代”，石油成为主要能源。 • 西方国家采取了大量措施，颁布了一系列污染防治法规，如英国《碱业法》和《河流防污法》；日本大阪府《工厂管理条例》。美国、法国等国也陆续颁布了防治污染的法规。建立国家公园等对生态环境进行保护的措施越来越多。 • 城市规划进入了新时期，重点关注水供应和公共卫生。
1900—1920	• 地方性的社会组织积极参与自然和景观保护，努力获得自然和文化遗产的长期所有权。 • 在城市规划中，通过改善室内外生活环境，将公益住房的推广与环境、社会发展目标联系起来。

① 20 世纪 60 年代早期到 2012 年间更详细的关键环境事件和行动可向*可持续发展国际研究所*（IISD）咨询 http：//www. iisd. org/pdf/2012/sd_ timeline_ 2012. pdf.

② 世界范围内地区差异评估（考虑环境和社会问题之间的时间轴和关系）见 www. unep. org/GEO <http：//www. unep. org/GEO> 和 Kok，M，等人：发展环境-从全球环境评估中的政策教训。荷兰环境评估局。荷兰：Bilthoven；2009 - www. pbl. nl <http：//www. pbl. nl>.

③ Vries，Bert JM de：可持续性科学。剑桥大学出版社，2013 年。编入全球环境历史数据库中的数据 http：//themasites. pbl. nl/tridion/en/themasites/hyde/.

续表

时　间	主要事件和行动
1920—1950	• 工业化国家煤炭、冶金、化学等重工业以及战时工业大规模地建立和发展，由于缺少针对性的污染防治措施以及对生态效益的忽视，大气、土壤和水域遭受到持续不断的污染；战后城镇化也带动了郊区得到发展。 • 人类历史上第一次环境问题爆发高潮，包括日本足尾铜矿区废水污染农田事件；比利时马斯河谷工业区大气污染事件；美国洛杉矶光化学烟雾事件和多诺拉烟雾事件等。
1950—1970	• 第三次科技革命爆发，原子能、电子计算机、宇航工程、生物工程等领域出现了重大发明和突破。 • 二战后西方大国竞相发展，工业化、城镇化进程加快。工业活动和私人汽车快速增长，健康问题随之急剧增加。 • 各种大气、土壤和食品污染公害事件加剧，包括1953—1965年的日本水俣病事件；1955—1972年日本富山县的骨痛病事件以及1968年在日本九州爱知县等23个县府发生的米糠油事件等。 • 1969年美国制定国家环境政策法令之后，西方国家相继成立环境保护专门机构，并颁布和制定了一些环境保护的法规和标准，以加强法治。 • 国际层面对环境变化（跨境空气污染和水污染；区域和全球问题）的意识得到提升，比如加拿大和美国共同建立的国际联合委员会。
1970—1990	• 非法采伐和开荒、过度捕捞、平流层臭氧耗竭、化学污染、气候变化等一系列问题逐渐被公众所了解并要求采取相应的行动。 • 工业化国家重要流域逐渐得到修复，城市空气污染也得到控制。 • 环保部门以及非政府部门开始用专业化、集成化和系统化的手段解决公共卫生、自然景观和资源等领域的问题。 • 重大工业事故频发，导致了法律法规的强化以及社会自发志愿行动的增加，比如针对化工企业的企业责任等。1970年，上千个组织参与了第一个地球环境日活动。 • 1972年斯德哥尔摩召开联合国人类环境会议。全球环境变化的人文因素计划得以建立。世界自然保护联盟提出“可持续发展”，1987年联合国环境与发展世界委员会对此再次进行重申，可持续发展逐渐被世界各国人民所接受。

续表

时　　间	主要事件和行动
1970—1990	• 联合国环境规划署建立，通过联合国及其机构发挥协调和促进的作用。 • 20 世纪 70 年代早期，人们发现并开始关注全球性的环境变化问题。国际非政府组织（如罗马俱乐部）联合国际的、国家的和当地的公民社会组织参与环境行动。 • 形成了一些区域性多边环境保护协议，如欧洲《远距离越境空气污染公约》，以及一些针对荒漠化和化学物质的协议。
1990—2010	• 人类进入全球化时代，计算机网络技术、信息技术、生物技术、基因工程技术、微电子集成技术等学科高度融合并产业化。 • 1991 年，环境与发展部长级会议在中国召开，通过并发表了《北京宣言》。次年，中国环境与发展国际合作委员会在北京成立。 • 1992 年在巴西里约热内卢召开联合国环境与发展大会。会议通过了《里约热内卢宣言》和《21 世纪议程》两个纲领性文件，标志可持续发展被全球持不同发展理念的国家普遍认同。 • 2002 年，联合国可持续发展世界首脑会议在约翰内斯堡举行，将注意力转到消除贫穷和环境可持续发展上，形成千年发展目标。 • 联合国开发计划署、联合国环境规划署、世界卫生组织和经济合作与发展组织等对国家、地区、全球生态进行系统、综合的评估和展望。 • 通过互联网等各类社会媒体，环境和社会相关问题的传播范围和关注度明显扩大。 • 形成一些国际环保公约，如《维也纳公约》、《蒙特利尔议定书》，达成了一些环境全球治理的共识和原则，如“共同但有区别的责任”等。 • 2009 年在哥本哈根召开世界气候大会，商讨至 2020 年的全球减排协议。
2010 年至今	• 2012 年联合国可持续发展大会——里约峰会（Rio+20），集中讨论两个主题：(1) 绿色经济在可持续发展和消除贫困方面作用；(2) 可持续发展体制框架（包括其确立可持续发展目标）。

从以上历史回顾看出，工业社会以来环境与社会问题的关系变得越来越复杂，两者相互依赖相互作用。环境政策及措施工业化导致的环境污染公害事件层出不穷，按其发生缘由，可分为几类：

（1）因工业生产将大量化学物质排入水体而造成的水体污染事件，最典型的是 1953—1965 年日本水俣病事件。1953 年，水俣湾附近渔村流行一种原因不明的中枢神经系统疾病，称为“水俣病”。1965 年，日本新潟县阿贺野川流域也发生水俣病。日本政府于 1968 年 9 月确认，水俣病是人们长期食用受富含甲基汞的工业废水毒害的水产品造成的。

（2）因煤和石油燃烧排放的污染物而造成的大气污染事件，如 1952 年 12 月 5—8 日的伦敦烟雾事件，即著名的“烟雾杀手”，导致 4000 多人死亡。1952 年的洛杉矶光化学烟雾事件也造成近 400 名老人死亡。此外，1961 年日本东海岸的四日市也发生了严重的气污染事件。

（3）因工业废水、废渣排人土壤而造成的土壤污染事件，如 1955—1972 年日本富山县神通川流域的痛痛病事件。1972 年，名古屋高等法院做出判决，确认痛痛病的病源是神冈矿山的含镉废水。原来，这里的锌、铅冶炼工厂等排放的含镉废水污染了神通川水体，两岸居民利用河水灌溉农田，使镉附集于稻米上。人食用含镉稻米以及饮用含镉水后，逐渐引起镉中毒，患上“痛痛病”。

（4）因有毒化学物质和致病生物等进入食品而造成的食品污染公害事件，如 1968 年日本的米糠油事件。日本北九州的一家食用油加工厂用有毒的多氯联苯作脱臭工艺中的热载体，因管理不善，毒物渗入米糠油中。这年 3 月，成千上万只鸡因吃了米糠油中的黑油而突然死亡。不久，人也因食用米糠油而受害。至 7—8 月份，患病者超过 5000 人，共有 16 人死亡。一时间，恐慌混乱笼罩着日本西部。其次，在沿岸海域发生的海洋污染和海洋生态被破坏，成为海洋环境面临的最重大问题。靠近工业发达地区的海域，尤其是波罗的海、地中海北部、美国东北部沿岸海域和日本的濑户内海等受污染最为严重。海洋污染源复杂，有通过远洋运输和海底石油开采等途径进入海洋的石油和石油产品及其废弃物；有沿海和内陆地区的城市和工矿企业排放的、直接流入或通过河流间接进入海洋的污染物；有通过气流运行到海洋上空随雨水降入海洋的大气污染物；还有因人类活动产生而进入海洋的放射性物质。海洋污染引起浅海或半封闭海域中氮、磷等营养物聚集，促使浮游生物过量繁殖，以至发生赤潮。如日本濑户内海，赤潮频繁，在 1955 年以前的几十年间发生过 5 次，1965 年一年中就发生 44 次，1970 年发生 79 次，而 1976 年一年中竟发生 326 次。赤潮的频繁发生，是海洋污染加重、海洋环境质量退化的一个突出标志。再次，两种新污染源——放射性污染和有机氯化物污染的出现，不仅加重了已有的环境污染危机的程度，而且使环境污染危机向着更加复杂而多样化的方向转化。

（一）1815—1860 年美国工业革命时期

在 19 世纪末期和 20 世纪初期，美国的工业中心城市，如芝加哥、匹兹堡、圣·路易斯和辛辛那提等，煤烟污染也相当严重。1948 年 10 月 27 日晨，在美国宾夕法尼亚州西部山区工业小镇多诺拉（Donora）的上空，烟雾凝聚，犹如肮脏的被单。其实，多诺拉的居民对大气污染并不陌生，因为这里的钢铁厂、硫酸厂和炼锌厂等大厂一个挨着一个，日夜不停地排放二氧化硫等有害气体。但是，像这一次的情景他们却从未见过。因逆温层的封锁，污染物久久无法扩散，整个城镇被烟雾所笼罩。直到第 6 天，一场降雨才将烟雾驱散。这次事件造成 20 人死亡，6000 人患病，患病者差不多占全镇居民（14000 人）的 43%。该事件还影响了当年哈里·杜鲁门和托马斯·杜威之间的总统竞选激战。此时，内燃机经过不断的改进，在工业生产中广泛替代了蒸汽机。30 年代前后，以内燃机为动力机的汽车、拖拉机和机车等在世界先进国家普遍地发展起来。1929 年，美国汽车的年产量为 500 万辆，英、法、德等国的年产量也都接近 20 万—30 万辆。由于内燃机的燃料已由煤气过渡到石油制成品——汽油和柴油，石油便在人类所用能源构成中的比重大幅度上升。开采和加工石油不仅刺激了石油炼制工业的发展，而且导致石油化工的兴起。然而，石油的应用却给环境带来了新的污染。这一阶段，“建立在汽车轮子上的”美国后来居上，成为头号资本主义工业强国，其原油产量在世界上遥遥领先，1930 年时就多达 12311 万吨；汽车拥有量在 1938 年时达到 2944.3 万辆。汽车排放的尾气中含有大量的一氧化碳、碳氢化合物、氮氧化物以及铅尘、烟尘等颗粒物和二氧化硫、醛类、一苯并芘等有毒气体；一定数量的碳氢化合物、氮氧化物在静风、逆温等特定条件下，经强烈的阳光照射会产生二次污染物——光化学氧化剂，形成具有很强氧化能力的浅蓝色光化学烟雾，对人、畜、植物和某些人造材料都有危害；遇有二氧化硫时，还将生成硫酸雾，腐蚀物体，危害更大。这是一种新型的大气污染现象，因最早发生在洛杉矶，又称洛杉矶型烟雾。1943 年，洛杉矶首次发生光化学烟雾事件，造成人眼痛、头疼、呼吸困难甚至死亡，家畜犯病，植物枯萎坏死，橡胶制品老化龟裂以及建筑物被腐蚀损坏等。这一事件第一次显示了汽车内燃机所排放气体造成的污染与危害的严重性。20 世纪六七十年代，核电工程迅速成长。核能在为人类提供巨大的动力和能量时，也产生了核废料以及由这种放射性物质带来的环境污染。更为严重的是，核电厂在运转中发生事故所造成的放射物质泄漏和放射性污染，会对人类造成严重而持久的威胁，美国的“三英里岛（Three-Mile Island）事件”就是典型例证。1979 年 3 月 28 日，美国宾州哈里斯堡东南 16 公里处三英里岛核电厂 2 号反应堆发生放射性物质外泄事故，导致电厂周围 80 公里范围内生态环境受到污染。这是人类发展核电以来第一次引起世人瞩目的核电厂事故，对社会生活、舆论和世界核能利用的发展都曾带来重大影响。

（二）1850—1870年德国工业革命时期

19、20世纪之交，德国工业中心的上空长期为灰黄色的烟幕所笼罩，时人抱怨说，严重的煤烟造成植物枯死，晾晒的衣服变黑，即使白昼也需要人工照明。并且，就在空气中弥漫着有害烟雾的时候，德国工业区的河流也变成了污水沟。如德累斯顿附近穆格利兹（Muglitz）河，因玻璃制造厂所排放污水的污染而变成了“红河”；哈茨（Harz）地区的另一条河流则因铅氧化物的污染毒死了所有的鱼类，饮用该河水的陆上动物亦中毒死亡。到20世纪初，那些对污水特别敏感的鱼类在一些河流中几乎绝迹。

（三）20世纪80年代以来西方发达国家进入后工业化时期

西方国家在环境污染发生初期，采取过一些限制性措施，颁布了一些环境保护法规。如英国1863年颁布的《碱业法》、1876年的颁布的《河流防污法》；日本大阪府1877年颁布的《工厂管理条例》等。此后美国、法国等国也陆续颁布了防治大气、水、放射性物质、食品、农药等污染的法规。但是，由于人们尚未搞清污染以及公害的原因和机理，仅采取一些限制性措施或颁布某些保护性法规未能阻止环境污染蔓延的势头。到20世纪50—70年代初环境污染问题日益加重时，西方国家相继成立环境保护专门机构，以图解决这一问题。因当时的环境问题还只是被看作工业污染问题，所以工作的重点主要是治理污染源、减少排污量；所采取的措施，主要是给工厂企业补助资金，帮助它们建立净化设施，并通过征收排污费或实行“谁污染、谁治理”的原则，解决环境污染的治理费用问题。此外，又颁布和制定了一些环境保护的法规和标准，以加强法治。但这类被人们归结为“尾部治理”的措施，从根本上说是被动的，因而收效不甚显著。这时，西方国家频繁发生的污染公害事件，不仅影响了经济的发展，而且污染了人群的居住环境，损害了人们的身体健康，造成了许多死亡、残疾、患病的惨剧，终于使公众从公害的痛苦中普遍觉醒。

德国作为发达国家，经历了单纯发展工业，自然环境遭受严重破坏的阶段。但通过提出“变黑色工业为绿色工业”的目标，采取一系列措施收到显著效果，自然环境得到恢复。德国的主要经验：第一，加强保护自然生态系统，维护生态平衡的意识。为了保护自然环境和生态系统，德国专门颁布法令规定，凡是被破坏的土地必须还原再造，以恢复原来的自然景观，并根据需要重新全面规划。第二，建立一套涵盖范围广、制定完善的法律体系。上世纪五六十年代莱茵河水污染严重，鱼类濒临绝迹。1972 年德国制定了第一部环境保护法，对生态的保护、废物的处理等严格立法，今天莱茵河水恢复清洁，已达到饮用水标准。德国还先后颁布了《循环经济和废物清除法》、《环境监测法》等，法律规范深入到生产和生活的各个方面。第三，加大高科技投入。目前全世界所开发的生态技术中，德国占到 18%，足可见其对生态技术的重视程度。

20 世纪 50 年代末，当美国环境问题开始突出时，美国海洋生物学家卡逊（Rachel Carson）花费了 4 年时间，阅遍美国官方和民间关于使用杀虫剂造成危害情况的报告，在此基础上，写成《寂静的春天》一书，将滥用滴滴涕等长效有机杀虫剂造成环境污染、生态破坏的大量触目惊心的事实揭示于美国公众面前。该书在 1962 年的出版，引起美国朝野的震动，并推动全世界公众对环境污染问题的深切关注。到 1968 年，来自 10 个国家的 30 位专家在罗马成立“罗马俱乐部”，研究人类的环境问题。1970 年 3 月 9 日—12 日，国际社会科学评议会在日本东京召开“公害问题国际座谈会”，发表《东京宣言》，提出“环境权”要求。同年 4 月 22 日，由美国一些环境保护工作者和社会名流发起的一场声势空前的“地球日”运动，更是令人瞩目。这是历史上第一次规模宏大的群众性环保运动。在学者们和广大公众的强烈要求下，在各国舆论的压力下，1972 年 6 月联合国在瑞典的斯德哥尔摩召开了“人类环境会议”，试图通过国际合作为从事保护和改善人类环境的政府和国际组织提供帮助，消除环境污染造成的损害。会议发布的《人类环境宣言》指出：“保护和改善人类环境是关系到全世界各国人民的幸福和经济发展的重要问题，也是全世界各国人民的迫切希望和各国政府的责任。”《宣言》第一次呼吁全人类要对自身的生存环境进行保护和改善，因为保护自然环境就是保护人类自己。同时，它还要求人们与自然进行有效合作，把保护环境同和平与发展统一起来，作为人类的共同目标去实现。这次会议无疑是世界环境保护工作的一个重要里程碑，它加深了人们对环境问题的认识，扩大了环境问题的范围，冲破了以环境论环境的狭隘观点，把环境与人口、资源和发展联系在一起，力图从整体上解决环境问题。具体到环境污染的治理，则开始实行建设项目环境影响评价制度和污染物排放总量控制制度，从单项治理发展到综合防治。会后，西方发达国家开始了对环境的认真治理，工作重点是制定经济增长、合理开发利用资源与环境保护相协调的长期政策。七八十年代，这些国家在治理环境污染上不断增加投资，如美国、日本的环境保护投资约占国民生产总值的 1%—2%。它们十分重视环境规划与管理，制定各种严格的法律条例，采取强有力的措施，控制和预防污染，努力净化、绿化和美化环境。此外，还大力开展环境科学研究，积极开发低污染和无污染的工艺技术。“在环境科学的研究过程中，从理论上和实践上都摸清了主要污染物质的污染规律。例如，伦敦毒雾是低空煤气污染，有毒物质是CO、CO_2、SO_2；洛杉矶是高空光化学污染，有毒物质是NO、NO_2、O_3；同时，还摸清了光化学污染主要发生在北纬42度上下的范围之内，并具有静风环境条件下的城市。”到80年代，西方国家基本上控制了污染，普遍较好地解决了国内的环境问题。其中，英国的情况具有代表性。1981年，英国城市上空烟尘的年平均浓度只有20年前的1/8，1980年，全英河流总长的90.8%已无重大污染；1982年8月人们在离伦敦24公里的一个堰附近，捕捉到20尾绝迹100多年的大马哈鱼，大马哈鱼的洄游是二次世界大战结束后开始的反污染工作的一个里程碑。1992年6月，全世界183个国家的首脑、各界人士和环境工作者聚集里约热内卢，举行联合国环境与发展大会，就世界环境与发展问题共商对策，探求协调今后环境与人类社会发展的方法，以实现“可持续的发展”。里约峰会正式否定了工业革命以来的那种“高生产、高消费、高污染”的传统发展模式，标志着包括西方国家在内的世界环境保护工作又迈上了新的征途——从治理污染扩展到更为广阔的人类发展与社会进步的范围，环境保护和经济发展相协调的主张成为人们的共识，“环境与发展”则成为世界环保工作的主题。

西方国家环境污染与治理的历史表明，工业革命以来人类对自然的认识经历了一个由否定自然（即无视自然）到肯定自然（即重视自然）的过程，这是人类环境价值观由不科学到科学的转变。在生态危机威胁着人类生存与发展的今天，在许多发展中国家依然重蹈发达国家覆辙的情况下，从道德的高度看待人对自然环境的态度，呼吁全人类树立科学的环境价值观，激发人们保护环境的道德责任感，就显得十分的必要和迫切。

二、新中国建立以来环境保护与社会发展回顾

（一）新中国建立至20世纪70年代末

新中国成立后，我国就已有了对生态文明建设的认识，进行了内容包括植树造林、绿化祖国、美化环境、保持水土、调控资源等方面的生态文明建设。五六十年代，中国的环境污染和生态恶化还没有成为真正意义上的“环境问题”。就

自然环境而言，70 年代初，中国大部分海域环境质量较好。五六十年代进行的大面积农业垦荒并没有对自然生态系统造成不良的影响。就社会生产而言，我国的生产能力总体水平较低，对自然资源的影响也相对较小。比如，五六十年代，森林工业发展缓慢，各种木材和竹材的产量分别为 80 年代初期的 13%和 7%，对当时的森林资源还不会构成很大的影响。用于农业生产的每亩耕地施用化肥量，1957 年为 0. 2 公斤，1965 年为 1. 2 公斤，仅为 1978 年 5. 9 公斤的 3%和 20%，对环境的影响也相对有限。尽管五六十年代，特别是 70 年代的生态文明建设带有萌芽性质，但毕竟迈出了生态文明建设的第一步。1973 年，第一次全国环境保护会议在北京召开。会上，中国政府首次在公开场合承认中国也存在环境污染。这次会议揭开了中国当代生态建设和环境保护的序幕。1974 年，国务院环境保护领导小组成立，各省、市、自治区和国务院的主要部门也相继建立了环境保护机构，全国环境保护机构的建立使中国的生态建设和环境保护事业开始走入正轨。

（二）20 世纪 80 年代以来

改革开放后前二十年中，依靠高投入高消耗的资源战略，中国经济建设获得较快发展。经济建设和生态环境之间的矛盾开始突出。这一时期，我国的生态环境形势很严峻。在环境污染上，工业污染排放总绝对排放量还在增加，环境污染的恶化由点向面、由轻到重。人口的过快增长造成对自然资源的过度开发和消耗，加剧生态破坏。在生态自然资源方面，水土流失日益严重，流失面积约占国土面积的 38%；耕地面积减少，土地质量退化；森林、草原和海洋水产品资源退化减少的现象继续发展；生物多样性受到严重破坏，有 15%至 20%的动植物受到威胁，高于世界 10%的平均水平。社会经济的迅速发展和依旧严峻的生态环境形势促使我国发展战略转变为以经济、人口和资源协调发展为核心的可持续发展战略。这一时期，尽管没有明确提出生态文明，但是可持续发展战略的实施使从国家发展战略的高度来深化认识和全面实施生态文明建设已成为必然。1978 年通过的《宪法》第一次对环境保护作出了规定：“国家保护环境和自然资源，防治污染和其他公害。”1979 年，国家颁布了新中国成立以来第一部综合性的环保基本法——《中华人民共和国环境保护法（试行）》，标志着我国生态环境保护事业逐步走上法制轨道。这一时期，国家陆续颁布了许多重要的生态保护法规。如《海洋环境保护法》、《森林法》、《草原法》、《水土保持法》等。这些法律法规初步形成了中国环保法律体系。1981 年，国务院作出在国民经济调整时期加强环境保护工作的决定，强调人口和经济的发展，不仅要注意经济规律，同时也要注意自然规律，否则就会受到客观规律的惩罚。自然观的变化反映了国家发展观抛弃“大跃进”的理性回归，成为实施生态文明建设的最初哲学基础。1984 年，国务院环境保护委员会成立。1988 年，国家环境保护局改为直属国务院领导。1998 年，政府将原国家环境保护局升格为国家环境保护总局（正部级）。环保部门的设立加强了生态文明建设的组织基础。1991 年，环境与发展部长级会议在中国召开，通过并发表了《北京宣言》。同年，中国环境与发展国际合作委员会在北京成立。中国还加入了修订后的《维也纳公约》、《蒙特利尔议定书》、《里约宣言》和《21 世纪议程》等环保公约。国际交流与合作进一步推进了我国生态文明建设的发展。1992 年 8 月，中国政府提出“转变发展战略，走持续发展道路是加速我国经济发展、解决生态环境问题的正确选择。”1994 年，政府提出建立基于市场机制与政府宏观调控相结合的自然资源管理体系，逐步建立资源更新的经济补偿机制。1998 年，国家制定《全国生态环境建设规划》。在实施生态环境建设的过程中，政府要求将生态环境建设和环境污染治理的重点工程项目纳入国家基本建设计划。要求中央和地方政府都要把防治污染和生态环境建设的资金纳入预算。1999 年，中国国家环境保护总局等 6 个部门启动了以开辟绿色通道、培育绿色市场、提倡绿色消费为主要内容的“三绿工程”。2001 年，中国消费者协会把当年定为“绿色消费主题年”，推动绿色消费进入更多人的生活。2004 年，中国财政部会同国家发展改革委出台了节能产品政府采购政策。2002 年，“可持续发展能力不断增强，生态环境得到改善，资源利用效率显著提高，促进人与自然的和谐”成为全面建设小康社会的四大目标之一。2005 年，中国政府明确提出“建设资源节约型、环境友好型社会”，并首次把建设资源节约型和环境友好型社会确定为国民经济与社会发展中长期规划的一项战略任务。同时提出，在消费环节大力倡导环境友好的消费方式，实行环境标识、环境认证和政府绿色采购制度。中国环境标志认证已经有 65 个认证产品种类、1500 多家企业的 30000 多个型号的产品通过环境标志认证。制定《国家突发环境事件应急预案》建立了生态环境建设的预防和处理体制；2006 年，制定《环境影响评价公众参与暂行办法》建立了健全公民参与生态文明建设的保障机制；国家颁布了 800 余项国家环境保护标准，建立了国家和地方环境保护标准体系。中国财政部会同环保总局制定了《环境标志产品政府采购实施意见》，公布了环境标志产品政府采购清单，要求政府采购优先购买节能产品和环境标志产品。2007 年中共十七大告再次强调：“坚持节约资源和保护环境的基本国策，关系人民群众切身利益和中华民族生存发展。必须把建设资源节约型、环境友好型社会放在工业化、现代化发展战略的突出位置，落实到每个单位、每个家庭。中国环境标志目前已与德国、北欧、日本、韩国、澳大利亚、新西兰、泰国等国家签订了合作互助协议。截止目前，财政部和环境保护部已经发布了两批政府绿色采购清单，共有 14 个种类的 444 家企业进入政府绿色采购清单。2008 年 1 月，中国国务院办公厅向各省、自治区、直辖市人民政府，国务院各部委、各直属机构下发《关于限制生产销售使用塑料购物袋的通知》。通知指出，鉴于购物袋已成为“白色污染”的主要来源，今后各地人民政府、部委等应禁止生产、销售、使用超薄塑料购物袋、并将实行塑料购物袋有偿使用制度。自 2008 年 6 月 1 日起，在所有超市、商场、集贸市场等商品零售场所实行塑料购物袋有偿使用制度，一律不得提供塑料购物袋。2008 年 6 月，以“环境标志发展与绿色采购”为主题的可持续消费国际研讨会让中国环境保护政策、政府采购法规建设和政府绿色采购工作再次成为各方关注的焦点。中国作为人口众多的发展中大国，在现代化进程中占领环境友好、资源节约的绿色技术制高点，是发挥后发优势的关键。华能集团北京热电厂以创建“绿色电厂”为目标，采用液态排渣、低氮燃烧、飞灰复燃等先进燃烧技术，煤耗水平达到 284 克/千瓦时，每年可节约 42 万吨标准煤，污染物排放达到世界先进水平，收到了良好的环境效益、经济效益和社会效益。2010 年通过的中国国民经济和社会发展“十二五”规划强调，2010—2015 年，要加大政府生态建设投入力度，全面强化生态环境治理，强化生态保护立法与执法。推进林业重点生态工程建设，提高造林种草、病虫害防治和森林火灾扑救工作等资金补助或投入标准，尽快将区域性防沙治沙工程拓展为全国性工程；尽快上马沿海防护林、石漠化治理工程，在更大范围内实施封山封沙、禁牧休牧等休养生息措施，确保修复区自然生态治愈，提升自然生产力。推广清洁环保生产方式，治理农业面源污染。保护海岛、海岸带和海洋生态环境。制定国家生态经济标准和评价生态经济效益指标体系。

三、中国改革开放以来环境与社会互动的经验与教训

（1）在科学发展观的指导和要求下，对自然的认识发生了质的变化。胡锦涛指出：“对自然界不能只讲索取不讲投入、只讲利用不讲建设。”建设自然要“坚持保护优先、开发有序，以控制不合理的资源开发活动为重点，强化自然资源保护。”“我们所要建设的社会主义和谐社会，应该是人与自然和谐相处的社

会。”“人与自然整体和谐的社会，要贯穿于建设中国特色社会主义的整个历史过程。”党的十七大还将“人与自然和谐”、“建设资源节约型、环境友好型社会”写入新修改的党章中。这种对自然前所未有的高度重视，标志着我国的自然观从传统的“向自然宣战”、“征服自然”向“建设自然”、“人与自然和谐相处”的实质性转变，标志着生态文明的哲学基础已完全融入国家的发展理念中。

（2）生态文明被正式提出，并且有了明确的内涵和任务。党的十七大首次将生态文明作为一项战略任务和全面建设小康社会的目标正式提出来，提出要建设生态文明，使生态文明观念在全社会牢固树立。胡锦涛指出，建设生态文明，实质上就是要建设以资源环境承载力为基础、以自然规律为准则、以可持续发展为目标的资源节约型、环境友好型社会；从当前和今后我国的发展趋势看，我国建设生态文明必须着力抓好的战略任务是加强能源资源节约和生态环境保护。生态文明建设被赋予明确的内涵和任务，将更有力地推动我国生态建设和环境保护工作进入一个新的历史阶段。

（3）将生态文明作为“四个文明”的现代化建设之一置于“五位一体”的中国特色社会主义总体布局之中。继党的十七大后，2007 年 12 月的中央经济工作会议再次强调，必须把推进现代化与建设生态文明有机统一起来。这标志着在着力建设物质文明、精神文明、政治文明同时，生态文明被确定为社会主义现代化建设的“第四文明”，与其他三个文明建设同等重要。2008 年 12 月，胡锦涛在讲话中将生态文明建设确定为中国特色社会主义事业总体布局之一，要求“全面推进社会主义经济建设、政治建设、文化建设、社会建设以及生态文明建设，更好推进改革开放和社会主义现代化建设”。2012 年中共十八大，再次将生态文明建设与“社会主义经济建设、政治建设、文化建设、社会建设”并列提出。这标志着“生态文明建设”是继经济建设、政治建设、文化建设、社会建设之后的第五大建设，进而使“四位一体”建设演进到“五位一体”建设的总体布局。把生态文明建设放在这样的高度，在我国的历史上还是第一次，反映了建设有中国特色社会主义现代化理论的创新成果。

（4）生态文明建设的手段发生显著变化。经济手段由微观扩展到宏观。比如，按照“探索将发展过程中的资源消耗、环境损失和环境效益纳入经济发展水平的评价体系”的思路，国民经济和社会发展计划中专门提出，要下大力气发展循环经济和清洁生产，用经济的宏观手段调节能源消费结构，促进循环经济产业的发展。2008 年，政府在生态文明建设中进一步发挥宏观经济手段，实施了包括绿色信贷、绿色保险、绿色贸易、绿色税收等在内的一系列宏观环境经济政策。这标志着这一时期建设生态文明的宏观经济手段已经完全贯彻于社会生产的整个过程。在法律手段上也有新的变化。过去的法律手段只规范企业等生产单位，而对政府的宏观规划却没有制约。2003 年 9 月实施的《环境影响评价法》将环境评价范围从建设项目扩大到政府规划，为政府规划要先进行环境评价提出了法律要求。这明显扩大了生态环保法律所约束的范围。

新中国成立以来，生态文明建设取得了巨大的成就，有很多成功的经验值得继承和发展，但同时也走过许多弯路，也有很多教训需要进一步总结。

首先，正确处理长远发展和短期发展、全面发展和单一增长的关系，确立科学的发展观是生态文明建设取得成效的前提。一般地，无论是较早迈入现代化的西方工业发达国家，还是正拼命地追赶西方工业国家的欠发达的发展中国家，在发展问题上都存在着误区，都把发展视为短期内的物质财富的单一增加，选择高投入、高消耗和高污染来换取经济增长，忽视代际间的长远发展。新中国成立以后，我国作为赶超型的发展中国家，也选择了这种短期单一的发展模式。现在出现的自然资源、环境生态问题在很大程度上就是这种模式所导致的发展代价。加上我国的自然资本先天禀赋不足，人口基数过大，这种发展代价将更加惨重。这方面的深刻教训告诉我们，必须正确处理长远发展和短期发展、全面发展和单一增长的关系，确立科学的发展观，大力推进生态文明建设，努力建设资源节约型社会、环境友好型社会，实现科学发展、和谐发展理念的升华。其次，积极探索合理的人口政策，努力降低生态文明建设的压力。新中国成立以来，我国人口政策和计划生育技术政策的确定，曾经历了曲折的过程。特别是在新中国成立初期和“大跃进”时期，鼓励生育，限制节育的政策导致人口剧增。人口剧增加重了对自然环境资源的破坏和掠夺，增大了生态文明建设的压力，其消极影响至今仍然存在。80 年代以来，我国同时将计划生育和环境保护作为基本国策，将人口问题引入生态文明建设，开创了符合我国国情的生态文明建设新途径，为其他发展中国家解决生态问题提供了借鉴。这是我们值得认真总结的经验。但是，我国的人口压力仍然巨大，对生态环境的影响仍然存在。特别是在老龄化日益严重的今天，“未富先老”的国情再次加重了人口对生态文明建设的压力。因此，将人口工作和生态文明建设有机结合，实现二者的良性互动，仍然任重道远。再次，生态文明建设还要求有政治的强有力推动。近些年来，在我国的生态文明建设中，经济和法律手段充分发展，政治推动却相对不足。各级政府在面临经济发展与生态环境保护的二难选择时，常常牺牲生态环境，谋求单一经济指标（GDP）的增长。实际上，生态文明建设事业是政府公共服务中的重要组成部分，涉及全社会方方面面的利益。因此，生态文明建设更需要内化在执政党具体的执政实践中，形成执政党的生态政治。生态文明建设需要政府和权力机关出台必要的政策和强制手段来推进。作为公共政策的制订者与实施者，各级政府应转变执政方式和政绩理念，抓紧调整生态建设的政策导向，加大对生态环境的政策性倾斜；大力推进干部考核改革，不以单纯经济指标（GDP）作为干部考核标准，加强对领导干部任期内生态环境目标责任制执行情况的督察和考核，从而将生态文明建设从指导思想和发展价值目标逐一落实在具体实际中，避免将生态文明建设流于宣传口号之中。最后，生态文明建设还要有充分的社会参与，需要社会建设的有力支持。在生态文明建设中，部队、共青团、妇联、工会等组织充分发挥作用，积极参加生态文明的各项工程建设，广泛组织适龄公民直接参加诸如义务植树活动、林木绿地养护、古树名木领养等生态建设活动。但是，和西方发达国家比起来，公民建设生态文明的社会参与度还是不够，民间资源的调动还很不充分。一些社区组织、环保非政府组织特别是环保专业精英组织没有得到应有的扶持和发展。这说明了生态文明建设还需要与社会建设相结合，需要社会建设的有力支持。“五位一体”的建设模式只是在最高理念上实现了二者的结合，但是在实际的建设过程中，社会建设的各个方面与生态文明建设的对接还有待提高。比如，改变人们的生活方式，实现生态文明生活化，要依靠社会建设；生态文明观念的牢固树立也要依靠社会建设。因此，在实践中如何借助社会建设的发展来推动生态文明建设取得实效，使之固化和扎根于社会生活中，还是一个需要我们总结经验教训，不断探索的新课题。

第三章　环境保护与社会发展关系理论研究

一、环境保护与社会发展关系的理论模型

长期以来，随着人们对环境与社会关系的认识不断深入，针对环境保护与社会发展关系的理论研究形成了生态学、系统论、政治经济学、建构主义、社会转型、整合型等多种理论研究范式。这些研究范式认识到了环境要素与社会发展变化的相互作用，将环境、社会和个体看作统一的系统，认为各子系统之间相对独立又相互联系，可以通过研究这些联系来处理环境与社会问题。

然而，上述范式也存在一些局限性，例如，虽然生态学范式首次在社会学研究中关注环境变量，认识到了环境因素与社会发展的相互作用联系，但其分析框架仍然属于传统社会学范畴；政治经济学范式将环境破坏的原因归咎于人类建构的经济系统和政治系统的不平衡性，认为环境破坏是资本主义制度特有的问题，然而中国所出现的严重环境问题证明了这一理论范式的局限性；建构主义范式重视环境问题的社会属性和社会建构过程对环境产生的影响，但却忽视了环境要素变迁本身的客观性；社会转型范式是针对正处于转型期的中国社会提出的理论模型，这一范式认为需要通过组织创新，优化社会结构，促进社会的民主化以促进中国的环境保护，但并没有形成具体化、制度化、可操作的政策框架。

人类社会的发展史既是一部文明进步史，也是一部人类社会与自然环境的关系史。在漫长的人类历史长河中，人类社会经历了三个阶段。第一阶段是原始社会。这一时期，人类社会对自然环境的破坏极小，人类与自然环境维持着朴素的、原始的共生关系，但生产力极为低下；第二阶段是农业社会。这一时期，人类社会的发展对自然环境有了一定程度的破坏，但生态系统总体上可以自我调节和修复，人与自然整体上维持着相对平衡的融洽关系，局部地区也出现了人口增长超过资源承载能力的状况，但生态系统总体上保持稳定；第三阶段是工业社会。工业革命开启了人类的现代化生活，人类利用自然、改造自然的能力空前提升，创造了前所未有的巨大物质财富。这一时期人类社会的快速发展对自然造成了严重破坏，人与自然的关系变的全面紧张。

回顾人类社会的发展进程可以发现，正是人类社会与环境的不断相互影响和作用产生了环境问题。就形式上而言，自然环境与人类社会的关系也是相互的、双向的关系。随着人类对环境问题的认识不断深入，人类社会与自然环境关系中的变量和要素也在不断的复杂和深化。结合人类社会的发展进程、人类对环境问题的认识过程和相关学者对环境与社会关系范式的研究成果，本研究将环境与社会的理论关系总结为以下四个发展阶段和类型：

首先是环境与社会的直接作用阶段，这种作用关系主要体现在原始社会时期，这一时期人类尚未开始改造和利用自然，人类社会对自然的影响较小，自然环境对人类社会的直接作用也十分明显。

其次是随着人类改造和利用自然能力的不断增强，人类主动的环境行为开始介入，成为环境与社会关系的重要变量，不合理、不科学、不可持续的环境行为造成了全球性的生态环境破坏，人类社会与自然环境的关系日趋紧张，这种关系主要体现在农业社会时期，以及工业社会开始阶段到20世纪中期。

第三是20世纪中期之后，随着人类环境意识的觉醒，世界各国开始采取措施对环境问题进行治理，环境行为和环境治理开始成为影响环境与社会关系的双重变量，这一时期人类社会与自然环境的紧张关系在局部得到缓解，但整体上仍然严峻。

第四是进入20世纪80年代至今，世界各国逐渐形成了生态环境保护的环境价值观念，环境保护和可持续发展也逐步融入了各国政府的发展理念，人类开始从根本上转变经济社会发展模式，探索经济、社会和环境相协调的发展道路。这一时期，环境行为、环境治理和环境价值观念三个变量相互作用，共同构成了环境与社会的最终理论关系框架，这一理论关系框架也是本课题建立逻辑关系、开展理论分析和实证研究的基础。

综上，本章将通过构建“环境与社会直接作用”理论模型、“环境行为变量”理论模型、“环境行为与环境治理变量”理论模型，以及“环境行为、环境治理和环境保护价值观念变量”理论模型对上述不同阶段的理论关系框架进行具体阐述和分析，并结合全球和中国环境与社会发展的实践进程对这些理论关系进行佐证和梳理，最后对中国环境保护与社会发展的未来定位进行分析。

（一）“环境与社会直接作用”模型

环境与社会的关系反映的是人类文明与自然演化的相互作用。人类的生存发展依赖于自然，同时也影响着自然的结构、功能与演化过程。环境与社会的关系体现在两个方面，一方面是人类社会对自然环境的影响与作用，包括人类生存发展从自然界索取资源与空间，享受生态系统提供的服务功能，向环境排放废弃物；另一方面是自然环境对人类的影响与反作用，包括资源环境对人类生存发展的制约，自然灾害、环境污染与生态退化对人类的负面影响。

原始社会时期，早期的人类生存发展很大程度上依赖于周边的自然环境，很少能根据自己的意愿改造环境。在这一时期内，人类改造自然和从事生产活动的手段极其简陋和低下，主要靠采集大自然中的野生食物和渔猎为生。随着对自然界的认识发展和原始经验的积累，人类逐渐以栽培农作物和驯养动物来取代采集和渔猎，开始出现了原始农业和畜牧业。所有的这些活动对整个自然环境而言都可以形成良性循环：被消化的水果的种子继续繁殖，能够逃脱猎手追捕的动物继续生存并繁衍后代，人类活动所留下的废弃物绝大部分都化为其他动植物的生存资源。人类社会本身就构成生态环境系统的一个要素，人类活动受制于生态环境，并无能力去改变环境原有结构，人类只能依附于自然，在这一时期的人类行为并不能对生态环境系统稳定以及环境与社会的关系产生影响。

因此，最基础的环境与社会理论关系模型应当是社会与环境两者的直接相互作用，即人类社会直接依附、从属并受制于自然环境，自然环境以自然灾害等形式反作用于人类社会。如图 3-1 所示。

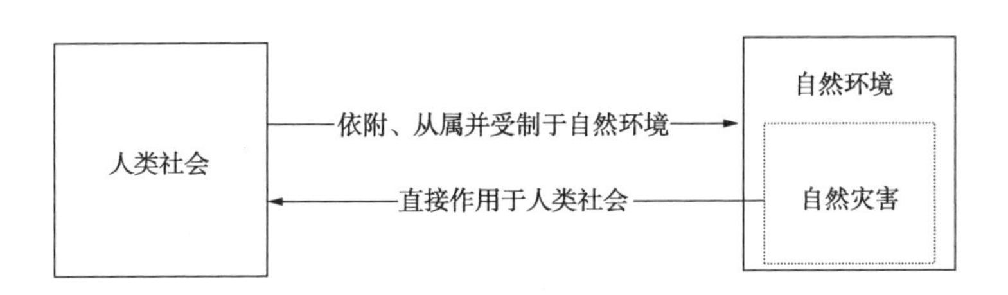

图 3-1 环境与社会直接作用理论模型

（二）“环境行为变量”模型

进入农业社会后，随着人类的思维意识在劳动中不断成熟，人类逐渐改变了原来的劳动方式，那种被动从属于自然界的关系也逐渐改变，人类开始与自然对抗。铁器工具的出现、推广和应用极大地提高了人类改造自然的能力和生产力水平，随着垦荒为田、开渠引水的灌溉农业的发展，局部范围内的自然结构和原有布局受到破坏，开始出现了早期的生态环境问题。然而从总体上来看，由于这一时期人类社会生产力发展水平较低，人类的环境行为对自然环境的作用仍然有限，并没有出现全面性的生态环境危机，人与自然的关系主要表现为局部性和阶段性的不和谐，但整体上是协调的。

18世纪工业革命以来，随着人类社会生产力迅猛发展，利用和改造自然的能力空前增强，工业革命在给人类带来极大的物质财富和科学技术发展的同时，也带来了巨大的生态环境破坏，人与自然环境的关系全面紧张。这些环境问题主要有三种表现。一是由于工业化和城镇化快速发展、人口增加和消费需求持续增长所导致的资源短缺，如水、土地、能源、矿产等资源的短缺；二是大量排放污染物引发的环境污染，如大气、水、土壤污染等；三是资源开发利用不当导致的生态退化，如森林和草原破坏、水土流失、荒漠化、生物多样性丧失、气候变化、臭氧层破坏等。这些问题逐渐成为人类生存和发展的重大威胁。

在中国，人类活动尤其是20世纪70年代改革开放以来的生产生活导致了极大的生态环境破坏，已经成为阻碍经济社会发展的首要问题。这些问题主要表现在：一是主要污染物排放总量大，减排任务艰巨。“两高一资”行业投资增幅虽有下降，但现有产能释放带来的污染物增量压力很大。机动车每年增长1500万辆左右，机动车氮氧化物排放量仍呈增长趋势；二是环境污染仍然十分严重。70%左右的城市不能达到新的环境空气质量标准。2013年年初以来部分地区长时间、大范围、反复出现的雾霾天气，影响17个省（区、市），约占国土面积1/4，受影响人口达6亿，严重威胁到人们正常生产生活，影响了社会的和谐稳定。20%左右的国控断面水质依然为劣Ⅴ类，基本丧失水体功能。一半城市市区地下水污染严重，57%的地下水监测点位水质较差甚至极差；三是环境风险继续增加，损害群众健康的环境问题比较突出。中国正处于社会转型和环境敏感、环境风险高发与环境意识升级共存叠加的时期，长期积累的环境矛盾正集中显现，$PM_{2.5}$、饮用水安全、重金属污染事件和化学品污染问题，引起群众广泛关注。环境污染和生态退化已经成为引致当前中国一些社会问题的重要因素和导火索。

工业革命以来出现在发达国家和中国的生态环境问题，归根结底是由于人类对自然界不合理、不可持续的利用和改造导致的，也是由不受约束的环境行为变量的介入引发的。与此同时，生态环境的恶化又会反作用到社会系统，引起环境健康、环境公平、环境贫困、环境事件等社会问题，形成了环境与社会发展的相互掣肘。环境行为变量介入环境与社会关系的理论模型如图3-2所示。

（三）“环境行为与环境治理变量”模型

环境治理和环境政策对环境行为的调节是影响环境与社会发展关系的又一变

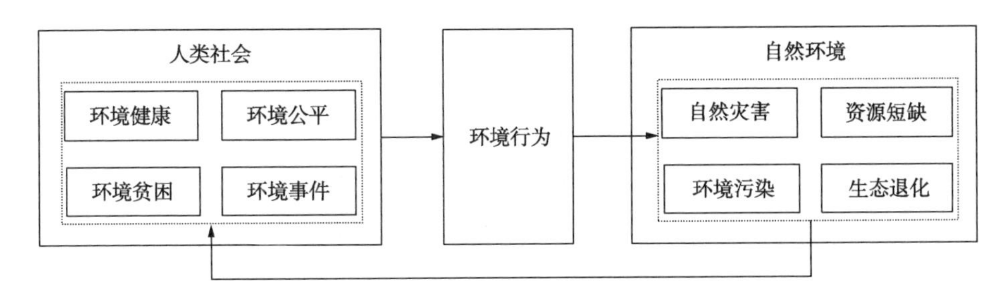

图 3-2 "环境行为变量"理论模型

量和要素。一方面，针对不同的环境行为可以采取有针对性的治理模式和制度政策进行调节；另一方面，通过先进的技术手段可以对出现的环境问题进行治理，遏制环境恶化的趋势，改善环境质量。如果人们采取了合理的生产、生活行为，便能促进自然环境质量改善，进而促进环境问题的解决和社会进步，推动环境与社会协同发展。这一情形是较为理想的环境与社会发展状态，要实现这种状态，必须要从政府、企业和个人三个层面对不文明、不合理的环境行为进行调节和调整。

一直以来，许多国家在协调社会发展与生态环境保护关系、治理生态环境问题方面结合自身实际进行了大量有益的探索。例如，鲁尔工业区曾经是德国乃至欧洲最大的工业区，在 100 多年的采矿炼铁制钢过程中，环境污染严重。从 20 世纪 60 年代开始，德国政府通过调整产业结构，推动绿色转型，发展新兴产业，实施环境污染治理与生态修复工程，使得鲁尔区的环境问题得到根本治理，重现绿色生机，目前生活条件和水平在欧洲工业区中名列前茅，实现了人类经济活动、社会行为和自然环境改善的良性互动。

在理论上，环境治理模式可以划分为"先污染、后治理"的模式、新型环境公共治理模式等，在具体的制度、政策上又存在其他的多种表现形式。"先污染、后治理"模式是当人们的社会活动已经对自然环境造成严重破坏之后，世界各国普遍采取的治理模式。例如，20 世纪 50 年代，英国近二百年工业化带来的生态破坏和环境污染到了无以复加的地步，导致巨大的经济、环境和健康损失。伦敦烟雾事件造成了 12000 多人死亡，泰晤士河也污染严重，鱼虾绝迹，成为"死河"。英国政府痛定思痛，通过了《清洁空气法案》、《河流法》等十几项法律法规，采取严厉的污染控制措施，积极调整经济结构，历经几十年不懈努力，生态环境才得以逐步恢复。这种典型的"先污染、后治理"发展方式，付出了高昂的生态环境和经济社会代价。

新型的环境公共治理模式是由政府、企业和公民社会三个行动者参与的平衡治理结构，三者各自发挥其功能和作用，形成和谐的合作伙伴关系。这种治理模式可以融合“自上而下”和“自下而上”的治理运行机制，行动者之间充分沟通和交流，利用综合手段和法律、行政、经济激励等政策工具保障治理机制的有效运转。在这种模式下，环境保护可以充分参与到经济社会综合决策中，不同部门在环境与发展问题上形成共识和一致的行动基础，环境与发展相互促进和支持，环境保护即能优化经济增长、促进社会发展，同时也能在经济社会发展中解决环境问题。

“环境行为与环境治理变量”理论模型如图 3-3 所示。

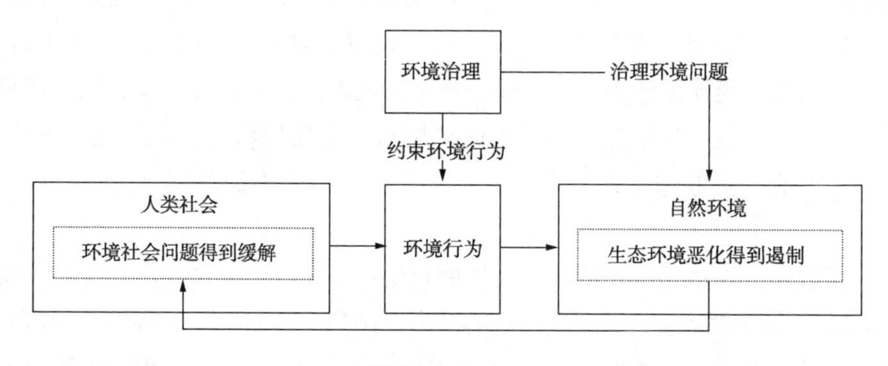

图 3-3　“环境治理与政策调节”模型

(四)“环境行为、环境治理和环境保护价值观念变量”模型

不同的环境价值观念会引导不同的环境行为，并且会采取不同的环境治理模式。环境价值观念可以归纳为以下三类。

首先是唯 GDP 或称极端人类中心主义的价值观念，这种价值观念高估了人类理性的力量，低估了自然界的有限性和自然规律的复杂性，从而把自然界看成是人类可以自由“改造”的对象，强调以人为中心，在人与自然的关系上表现为过分强调对自然的统治和索取，而忽视了对自然的依赖和培育，导致了自然资源和生态承载力迅速衰竭。与此同时，这种价值观念在人类个体和群体之间的关系上表现为过分强调对其他个体和群体的支配，而忽视了社会中个人与个人、个人与国家、国家与国家之间的和谐，进而导致了诸多的环境不公平和社会问题。这类价值观念长期占据统治地位，人们往往采取先污染后治理的模式来解决由此引发的环境问题。

其次是极端生态中心主义的观念，这种观念强调“生态中心主义的平等”，即在生物圈中所有的有机体和存在物，作为不可分割的整体的一部分，在内在价值上是平等的。每一种生命形式在生态系统中都有发挥其正常功能的权利，都有“生存和繁荣的平等权利。”这种价值观念主张停止改造自然的活动，是一种理想化的认识倾向。

第三是20世纪中期以来逐渐形成的可持续发展理念，这种理念在中国主要体现为科学发展观和生态文明的思想。可持续发展理念强调以人为本，明确提出实现人与自然的和谐、协调发展。这种价值观念同时反对极端人类中心主义与极端生态中心主义。极端人类中心主义制造了严重的人类生存危机，而极端生态中心主义却过分强调人类社会必须停止改造自然的活动。生态文明则认为人是价值的中心，但不是自然的主宰，人的全面发展必须促进人与自然和谐。这种理念的本质是为了推动经济、社会和环境的可持续发展与公平公正。在这种理念下，通过创新环境公共治理模式，可以从根本上改善环境质量、促进社会和谐稳定，实现经济、社会和环境的可持续发展。

可持续发展的环境保护价值观念、环境行为和环境治理的相互作用构成了环境与社会发展关系的核心模型，如图3-4所示。

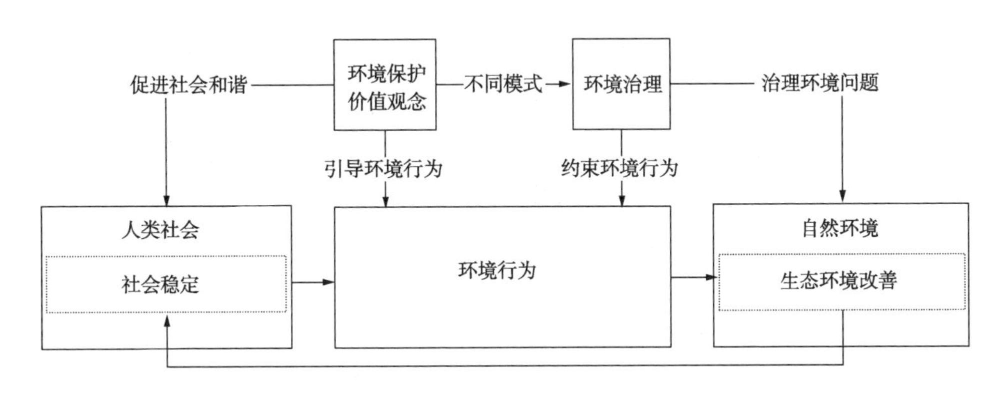

图3-4 “环境行为、环境治理和环境保护价值观念变量”理论模型

从长期演化过程来看，生态环境问题归根结底是由不合理的资源利用方式、经济增长方式和社会行为方式造成的，是人与自然矛盾冲突的外在体现和结果，其本质是发展道路、空间布局、产业结构、生产方式和生活方式问题，其背后蕴含着深刻的政治、经济、社会、文化和制度原因。综合上述理论模型的分析可以看出，要解决社会发展中的环境问题，必须从转变价值观念、创新环境治理模式、调节环境行为三个维度切入展开系统的研究和政策设计。

二、全球环境与社会发展关系认识和实践的演进

全球对环境与社会发展关系的认识和实践主要经历了对环境问题的认识和觉醒，到全面认识环境与经济的关系，再到全面认识可持续发展的环境、经济与社会三个支柱关系的几个阶段。

（一）对于环境问题的认识和觉醒

从20世纪30年代开始，由于长期的生态环境问题积累，导致欧、美、日等发达国家相继发生一系列严重的环境公害事件，付出了极为沉痛的代价。例如1943年的5月至10月美国洛杉矶发生烟雾事件，大量汽车尾气产生的光化学烟雾，在5个月内造成65岁以上的老人死亡400多人。20世纪50年代后期，日本熊本县水俣镇氮肥企业排放含汞废水，造成当地居民甲基汞中毒，1997年官方确认的受害者高达12615人，其中1246人死亡，至今赔偿仍未完结。

严重的生态环境问题，唤起了人类环境意识的觉醒。在这一历史进程中，三部具有代表性的著作产生了巨大影响。第一部是1962年出版的《寂静的春天》，主要讲述农药对人类和环境的危害。该书首次以科普形式介绍了环境污染问题，在世界范围内引起了强烈反响，促使各国政府开始重视环境问题。其代表性观点是“人类一方面创造出了高度的文明，另一方面又在毁灭自己的文明”、“不解决环境问题，人类将生活在幸福的坟墓之中”。第二部是1972年罗马俱乐部提出的《增长的极限》。该书科学分析了地球资源和环境承载的有限性对经济增长的制约作用，给人类社会的传统发展模式敲响了警钟，掀起了世界性的环境保护高潮。其代表性观点是“没有环境保护的繁荣是推迟执行的灾难”。第三部是向1972年人类环境会议提交的报告——《只有一个地球》。该书从社会、经济和政治等角度，评述了经济发展和环境污染对不同国家产生的影响，呼吁各国人民重视维护人类赖以生存的地球。其代表性观点是“不进行环境保护，人们将从摇篮直接到坟墓”。随着这些影响卓著的著作和报告问世，国际环境保护运动逐渐兴起，各国政府开始重视并采取措施解决环境问题。

这一阶段，世界各国的主要是采取积极措施和技术手段集中治理工业化带来的环境污染问题，并在二三十年间逐步遏制住了环境恶化的趋势。

（二）对于环境与经济关系的全面认识

发展经济和保护环境，关系到人类的前途和命运，影响着世界上的每一个国家、民族和个人。20 世纪中期以来，人口急增、能源短缺、全球性气候变暖、臭氧层的破坏、土地退化和水土的大量流失、森林减少、空气、水等的严重污染、物种的灭绝等一系列严重问题随着经济全球化的扩张，成为威胁人类生存和发展的全球性重大问题，而这些问题最先出现在经济发达国家。随着全球环境认识的觉醒，世界各国也逐渐认识到环境问题的本质是经济发展问题，人与自然要相互依存，经济与环境必须协同发展。

1972 年联合国人类环境会议通过了《人类环境宣言》，强调人类的发展必须重视经济发展引发的生态环境灾难，世界各国开始共同研究解决资源环境问题。1987 年第 42 届联大通过世界环境与发展委员会的报告《我们共同的未来》，首次提出可持续发展的概念，并给出了可持续发展的定义，极大推进了世界范围内的环境保护运动。1992 年在巴西里约热内卢举行的联合国环境与发展会议通过了《21 世纪议程》，第一次把经济发展与环境保护有机结合起来，提出了可持续发展战略，并于会后成立了联合国可持续发展委员会。经济的可持续发展开始成为可持续发展体系中的核心问题，即在鼓励经济增长以体现国家实力和社会财富的同时，不仅重视增长数量，更追求改善质量、提高效益、节约能源、减少废物，改变传统的以高投入、高消耗、高污染为特征的生产模式和消费模式，实施清洁生产，倡导绿色消费。

1992 年之后，环境与经济发展的关系逐步被全面认识，世界各国开始围绕环境与经济的关系制定一系列的法规、标准和政策，尤其是开始运用经济手段推动环境保护，在解决环境问题的同时推动经济的绿色化发展。

对于环境与经济关系的处理，德国和日本是典型案例。德国并非一直是生态环境保护的模范，在 1972 年联合国召开首次人类环境大会之前，生态环境保护并不是德国社会的重要议题。二战后，与欧洲其他国家一样，东、西德的主要目标都是推动经济发展。20 世纪 50 年代到 60 年代末，德国经济迅速发展，年均 GDP 增速达近 8%，1968 年成为世界第三经济强国。在经济迅速腾飞的同时，德国的环境污染压力速增，环境问题于 20 世纪 70 年代初集中爆发。1972 年至 1992 年间，世界范围内重大污染事件中有两件发生在德国，即 1983 年西德森林枯死病事件和 1986 年莱茵河污染事件。面临严峻的环境污染问题，德国政府从 20 世纪 70 年代中期开始努力寻找经济发展与环境保护相互促进的增长点，将产业重点由能源消耗型的重化工业逐渐转向汽车、电机等技术密集型产业。同时，德国不断加大环境治理研究和开发的投资，1975—1985 年的十年间环保投资从 6580 万美元增长到 2.4 亿美元。生态环境保护的相关法律法规也不断完善，如 1972 年出台了《废弃物处理法》（此法是德国第一部环境保护法）、1974—1976 年又陆续出台了《大气污染防治法》、《环境影响评价法》、《自然保护法》和《水污染防治法》等。1980 年，德国绿党的出现逐步转变了德国的政治文化，开始引导德国向国际生态环境保护的领导地位迈进。过去的 40 年间，德国的主要政党都已将生态环境保护和可持续发展融入了政治理念。

20 世纪 90 年代初，德国的环境管理走向综合化、可持续化。环境保护写入了《德国基本法》，指出“国家应本着对后代负责的精神来保护自然的生存基础条件”。修订了环境保护法律法规并拓展到土壤保护等新领域，逐步形成了世界上最完备、最详细的环境保护法律体系，联邦及各州的环境法律、法规达到 8000 多部。德国还采用了一整套“德国环境指标”，将环境政策目标转化成具体的可量化指标，规定了一定时期内应当实现的污染物减排和环境质量目标。1990—1998 年期间，德国全国 SOx、NOx、VOCs 和 CO 的排放量分别削减了 76%、34%、47%和 52%。1985—1995 年期间，大气中有毒有害重金属的排放量下降了六倍之多。1998 年联邦政府换届后，德国社民党和 90 联盟/绿党联合政府提出了“生态社会市场经济”的经济发展新模式，第一次把生态作为市场经济的一个要素，提倡“绿色消费”和“绿色生产”。2002 年，德国政府又制定了 21 世纪国家可持续发展的总体框架。通过实施一系列环境法律、标准和政策，推进结构调整和技术进步，德国用了约 30 年的时间走完了“先污染、后治理”的阶段，告别了唯经济发展的时代，严重的环境污染得到治理。

从 20 世纪 50 年代中期开始，日本经济进入了 20 多年的高速增长期，为日本经济腾飞奠定了坚实的经济、物质和技术基础，但同时也导致了严重的环境污染，爆发了水俣病、四日市废气等环境公害事件。面对严峻的环境公害和污染问题，日本于 1968 年出台了《公害对策基本法》，加强产业污染防治和公害防治处理，着重控制和处理工业生产的末端废弃物。20 世纪 80 年代末期日本基本控制住了工业污染和公害问题。进入 20 世纪 90 年代，日本的工业污染物产生量大幅度减少，产业污染及部分自然生态环境问题已基本得到解决，生活型污染和汽车尾气污染等城镇化环境问题，以及全球性问题成为日本生态环境保护的重点领域。日本于 1993 年和 1994 年分别颁布了具有历史意义的《环境基本法》和《第一次环境基本计划》，开始从社会经济的深层次和整体着手，采取综合性对策，解决与公众生活密切相关的环境质量问题。到 20 世纪末，日本已经初步形成了基于循环经济的生产、生活方式，生态环境质量得到了明显提升。大气中二氧化氮、二氧化硫、一氧化碳和颗粒悬浮物等污染物含量都明显降低，其中，一氧化碳达标率在 1995 年就达到了 100%，二氧化氮、二氧化硫和颗粒悬浮物的达标率在 90 年代末期也达到了 90%以上。日本的生态环境问题从 20 世纪 50 年代出现到 70 年代集中爆发，经历了约 20 年。到 90 年代基本解决这些问题也用了约 20 年。这种发展方式大大缩短了“先污染、后治理”的过程，用较短的时间扭转了不利局面，实现了经济与生态环境的协调发展。

（三）对于环境、经济和社会三个支柱的全面认识

环境保护与人类生存和发展有着密切关系，是经济、社会发展及稳定的基础，又是重要的制约因素。2002 年可持续发展世界首脑会议在南非召开，会议通过了《可持续发展世界首脑会议实施计划》，文件突出强调了可持续发展的 3 个支柱即经济增长、社会发展和环境保护相互促进和相互协调的重要性。2012 年在里约热内卢召开的联合国可持续发展大会对可持续发展机制框架做出了新安排和调整，决议实施普通会员制、加强融资，并将可持续发展委员会升级为可持续发展理事会（Sustainable Development Committee），加强公民社会组织（Civil Society Organization）的参与等。会议继续明确了可持续发展的 3 个支柱的平衡性和环境支柱的重要性。

近年来，世界各国开始认识到，环境污染和生态破坏已成为阻碍经济社会可持续发展、威胁人民群众身体健康、制约社会主义和谐社会建设、影响人民真正享受发展成果的重要因素和关键问题。社会的发展和进步、国民经济的可持续发展，要求人民群众的生活质量得到全面改善和提高，不仅要实现经济上的富裕，更要实现在良好的环境中生产和生活；不仅要保障人们的衣食住行、医疗、教育，更要保障人们能够健康生活、长远发展；不仅要维护公众在个人物质财富上的合法权利，更要维护他们依法享有生态环境等公共资源的权益。由此，可持续发展的环境、经济与社会发展三个维度逐步得到世界各国的认可，世界各国尤其是发达国家开始全面考虑推动环境、经济和社会的均衡发展。

新加坡是推动环境、经济和社会协调发展的典型国家。在工业化初期生态环境问题仍处于萌芽状态时，新加坡便前瞻性地谋划和选择了适合其特点的国家环境与发展战略，并立即采取一系列严格的生态环境保护措施。1971 年，新加坡政府提出城市“环型发展计划”，即环绕主岛进行建设。城市实行功能分区，将

工业区与居住区分离，重工业区远离居住区，重污染大型企业建在岛屿上，避免市区环境污染，保护市民健康。从60年代中期到80年代后期，环境保护的工作以政府为主导，加强环境保护的机构建设，建立环境保护的法律框架和管理体系，在土地利用规划指导下开展环境基础设施建设。同时，新加坡还特别重视引导社会公众参与环境保护。经过20多年的努力，新加坡已经建立了比较完善的城市环境基础设施，实现了环境、经济和社会的协调发展，成为举世闻名的花园式城市国家。

（四）中国环境保护与社会发展关系认识和实践的现实阐述

从20世纪70年代中国开始意识到环境问题带来的严重危害，到环境保护列入基本国策，再到提出实施可持续发展战略、科学发展观和建设生态文明，中国对环境保护与社会发展关系的认识也经历了不断探索和深入的过程，大致也经历了三个阶段。

1. 对于环境问题的认识起步阶段

1972年，在许多人认为环境污染只是资本主义的产物、社会主义没有环境污染的时代背景下，中国派团参加了联合国人类会议，并于1973年召开了第一次全国环境保护会议，提出了“全面规划，合理布局，综合利用，化害为利，依靠群众，大家动手，保护环境，造福人民”32字环保方针。把保护环境确定为基本国策，强调要在资源开发利用中重视生态环境保护。1978年，第五届全国人大一次会议通过的《中华人民共和国宪法》规定，“国家保护环境和自然资源，防治污染和其他公害”，资源环境保护成为国家意志。1983年，国务院召开第二次全国环境保护会议，明确提出环境保护是一项基本国策，强调经济建设和环境保护必须同步发展。1984年，中国成立国务院环境保护委员会。1989年，《中华人民共和国环境保护法》正式公布施行，环境保护法律法规体系不断完善。同年，国务院召开第三次全国环境保护会议，进一步明确了环保目标责任制、环境影响评价、“三同时”、排污收费等8项环境管理制度。1990年，国务院颁布《关于进一步加强环境保护工作的决定》，强调要在资源开发利用中重视生态环境保护。

这一阶段是中国改革开放的初期，经济发展与环境保护、环境保护与社会发展之间的矛盾并不突出，人们并未意识到环境问题和社会问题的关系，环境保护工作也主要是围绕工业污染防治开展。

2. 经济发展和环境保护同步推进的阶段

1992 年，中国发布了《中国环境与发展十大对策》，第一次提出要将环境与发展统筹考虑。同年，中国环境与发展国际合作委员会成立。发表了《中国 21 世纪议程——中国 21 世纪人口、环境与发展白皮书》，将可持续发展正式作为国家发展战略。组织实施了“三河、三湖”污染防治、退耕还林还草、天然林资源保护等环境保护和生态建设重大工程。2000 年，启动全国生态省、市、县创建工作。2002 年，党的十六大提出统筹人与自然和谐发展，推动整个社会走上生产发展、生活富裕、生态良好的文明发展道路。

2003 年，十六届三中全会提出科学发展观的重要理念。2005 年，国务院发布《关于落实科学发展观加强环境保护的决定》，环境保护开始实施三个历史性转变，即从重经济增长轻环境保护转变为保护环境与经济增长并重，从环境保护滞后于经济发展转变为环境保护和经济发展同步推进，从主要用行政办法保护环境转变为综合运用法律、经济、技术和必要的行政办法解决环境问题。节能减排作为约束性指标纳入“十一五”规划，提出建设“资源节约型、环境友好型”社会。2007 年，中国共产党的第十七次代表大会首次提出“建设生态文明”理念，并将其作为全面建设小康社会的目标。这一时期还启动了生态文明建设试点示范工作。

这一阶段是中国开始实施可持续发展战略、全面贯彻落实科学发展观和推动环境保护历史性转变、推进环境和经济同步发展的阶段，环境与经济发展的关系在中国开始得到了充分的重视。

3. 开始全面认识环境、经济与社会发展关系的阶段

2012 年，中国共产党的第十八次代表大会明确要求把生态文明建设纳入中国特色社会主义“五位一体”的总体布局，作为执政理念上升为党的意志，明确了“优化国土空间开发格局，全面促进资源节约，加大自然生态系统和环境保护力度，加强生态文明制度建设”四大任务，要求将生态文明充分融入经济建设、政治建设、社会建设和文化建设的各个方面和全过程。坚持在发展中保护、在保护中发展，积极探索环境保护新道路成为推进生态文明建设的突破口和着力点。2013 年 5 月，环境保护部发布《国家生态文明建设试点示范区指标（试行）》，全面开展生态文明建设试点示范。

生态文明建设是中国对生态环境保护理论的扬弃和升华，它明确了解决生态环境问题必须紧紧抓住环境与发展这条主线，并结合中国国情确立了“在发展中保护，在保护中发展”的战略思想，明确了探索“代价小、效益好、排放低、可持续”环保道路的任务。生态文明建设是对世界可持续发展理论的传承和深化，它在继承2002年约翰内斯堡世界可持续发展首脑会议精神的基础上，将可持续发展的支柱从经济发展、社会进步和环境保护，进一步拓展为经济、政治、文化、社会和生态文明五大领域的建设。这其中，将生态文明融入社会建设，标志着中国已经开始认识到环境保护与社会发展关系的重要性，可持续发展环境、经济和社会三个支柱的理念在中国得到了传承和深化。

综上所述，中国对环境保护与社会发展关系的认识和实践取得了很大的进展，然而，从总体上来看，中国目前仍然处于从重经济增长轻环境保护转变为保护环境与经济增长并重、从环境保护滞后于经济发展转变为环境保护和经济发展同步推进的历史阶段，对于环境与社会关系的认识虽然在生态文明建设“五位一体”的布局中得到了体现，但仍然处于起步阶段，并未在具体制度、政策上得到实施。中国经济、社会和环境的政策目标仍然存在差异，环境保护仍然没有成为中国独立的政策领域，距离融入经济社会发展的全过程仍然存在很大差距。

4. 中国环境保护与社会发展关系的未来定位

环境保护与社会发展是可持续发展的核心维度和支柱。结合以上对环境与社会关系的理论分析和对国内外相关实践的梳理，可以从四个不同的层次理解环境保护与社会发展的关系定位。

（1）环境保护与经济、社会发展具有目标的同一性。自1992年全球环境与发展大会确定了可持续发展概念以来，环境、经济和社会作为可持续发展的三个有机组成部分，其共同目标是促进国家和全球的可持续发展，这在国际社会已达成普遍共识。经济发展、社会发展和环境保护是可持续发展相互依存、相互加强的核心支柱，而不是相互矛盾和对立的要素。一个国家要实现可持续发展，应当实现经济、社会与环境政策目标的一体化。

（2）环境保护是独立于经济与社会的政策领域。发展在中国含义包括两个方面，即经济发展与社会发展，环境保护政策被置于社会发展政策之下。但是环境政策与经济政策、社会政策之间既有区别也有联系。应当在顶层设计中确立环境保护的地位，建立独立于经济政策与社会政策的环境规划与政策体系，以利于政府、企业和社会各界更专注于环境问题，从而更好地促进可持续发展。

（3）环境保护应当融入经济和社会发展的全过程。环境保护应当融入经济和社会发展的全过程，在这个过程中需要处理好几个基本关系。一是环境与社会的关系，包括环境与人口、消费和社会服务、环境与消除贫困、环境与卫生和健康、人居环境改善、防灾减灾等；二是环境与经济的关系，包括环境如何优化经济增长、环境与农业与农村可持续发展、工业增长与环境污染防治、交通和通讯业可持续发展、可持续能源生产与消费等；三是资源利用与环境保护的关系，包括自然资源保护与持续利用、生物多样性保护、荒漠化防治、保护大气层、固体废物无害化管理等。

环境保护融入经济和社会发展的全过程，就需要将环境保护主流化到经济和社会发展的综合决策中，在经济和社会发展的重大决策和过程中考虑和体现环境保护的作用和因素。

（4）应当从树立环境保护价值观念、转变环境行为、推动环境公共治理三个维度解决环境与社会发展存在的问题。解决环境与社会发展的矛盾应当从树立环境价值观念、转变社会环境行为、推动环境公共治理三个维度来入手。在我国当前的社会结构下，社会行为包括政府、企业和个体行为三个层面。因此，应当深入研究能够促进政府、企业和个体树立环境价值观念，转变环境行为的新型环境公共治理模式。接下来的章节将围绕这三个维度的理论框架，开展具体的政策和案例研究。

第四章　环境保护与社会发展相协调的框架研究

一、促进企业履行保护环境的社会责任

在我国经济社会转型的关键时期，在经济全球化和气候变化等全球性问题日益增多的背景下，企业履行环保社会责任正在成为全社会关注的焦点，企业竞争正由传统价格、质量、技术竞争演变为更多包含社会责任、商业道德、信誉建设在内的系统性竞争，公民和社会组织对企业履行环保社会责任投以更为严苛的眼光，政府对企业履行社会责任的监管也面临各方面巨大压力，三者之间博弈与合作进入新阶段。企业履行环保社会责任不仅是企业行为，也是社会行为，更涉及发展与环境的关系，一定程度上还是国家发展模式之争。在新形势下，如何促进企业履行环保社会责任、形成各方良性互动，成为必须处理好的重大课题。

（一）我国企业履行保护环境社会责任的现状

1. 近些年来我国企业履行保护环境的社会责任取得了很大进步

一是随着经济发展和环保标准提升，企业对履行环保责任的重视程度加大，自觉性积极性提高，环保投入力度普遍加大。如重点污染防控企业安装在线监测设备，发电企业全部安装脱硫脱硝装置。“十二五”期间我国环保投入将超过5万亿元，占GDP比重将超过2%，其中绝大部分要由企业投入。

二是企业履行环保社会责任与企业发展能够形成良性互动。许多企业都认识到，越是履行社会责任较好的企业，其企业形象和品牌价值就较高，企业的市场地位、竞争能力和社会地位、民众支持就更为巩固。许多企业制订了绿色发展战略，用新的理念、新的方式从事生产经营活动，把履行环保社会责任与企业发展壮大结合起来。

三是越来越多的企业把环保等社会责任融入企业管理之中，力争成为“环保公民”。许多大企业和优秀企业把社会责任要求纳入企业战略决策、日常管理和企业文化建设，并在激励机制、过程管理、节能减排等重点工作中体现出来。

四是企业环境信息披露和履行社会责任公开取得积极进展。很多上市公司开

始披露年度环保投入等信息。加紧建立包括环保在内的社会责任报告制度，内容包括节能指标、环保指标、扶贫捐助、技术进步等。如2010年就有40家中央企业发布了社会责任报告，占中央企业总数的30%以上，主要分布在石化、钢铁、电力、通信、建筑等领域。

五是一些企业尤其是大企业建立了以“社会责任”命名的部门和机构，有的成立了社会责任领导小组，专门负责企业环保等社会责任的落实。

六是企业积极参加社会责任国际交流与合作。许多企业加入联合国可持续发展等相关组织之中，有的积极参与国际标准化组织（ISO）社会责任标准的制订工作。

七是政府大力推进企业履行环保社会责任。在确定单位生产总值能耗和污染物排放总目标的前提下，各级政府综合采取经济、法律、行政等手段，强化环保标准，加大淘汰落后产能、结构调整、上大压小等工作力度，使用电价、污水处理费、生态补偿基金等差别化价格手段促进企业加强环保，利用财税金融等政策、设立引导基金支持发展循环经济、环保生态产品的生产，对企业环境违法行为的处罚力度不断加大。

八是社会组织和个人在监督和推动企业履行环保社会责任中的作用越来越大。

2. 当前企业行为与环境保护中存在的矛盾和问题

一是企业履行环保等社会责任的意识和行动普遍滞后。根据2011年中国社会科学院发布的《企业社会责任蓝皮书》对国有重点企业100强、民营企业100强、外资企业100强共300家企业的调查，中国企业社会责任发展指数普遍较低，平均得分仅19.7分，除少数领先企业外，多数企业处于旁观阶段，国有企业好于民营和外资企业，责任实践领先于责任管理，市场责任好于环境责任和社会责任。我们重点选取在履行环保社会责任方面有代表性的三类企业来看：第一类是上市公司，环保意识普遍不强。据年报披露，近年来上市公司环境污染事件持续增加，2012年一季度发生的环境污染事件超过了2009年、2010年全年的数目，2000多家上市公司有77家发生了113起环境污染事故。第二类是在华国际品牌企业，履行环保责任状况也不容乐观。据统计，国际时装品牌在华生产企业中，只有1/3采取了比较严格的污染防控措施。第三类是污染排放相对较重的重化工企业。国际上跨国公司新建重化工企业时，环保及相关投入占建设资金的比重一般达到15%—35%，而我国只有少数企业能够达到10%。

二是由企业行为引发的环境社会事件处于快速上升期。我国已经进入环境风险突显、环境污染事故多发高发期，每年环境事件以约30%的速度增加，其中主要是企业环境事件。环境事件与违法征地拆迁、劳资纠纷一起，成为引发社会矛盾的“三驾马车”。企业环境事件的发生呈现出几个特点：

第一个特点是集中爆发。目前企业普遍环保标准较低，相互关联度高，又往往集中于工业园区，因而处于环境事件易发高发期。同时，一些慢性、隐性问题引起的环境事件可能经会在今后一个时期集中爆发，企业排污造成的后果有：儿童出生缺陷、癌症村的出现、疑难病症、生物多样受损、物种濒危等等。比如陕西凤翔县儿童血铅中毒，就是在企业污染若干年后被发现，并发酵成为群体性事件。

第二个特点是环境事件演变即时化、多样化、严重化。当今时代，由于信息传播手段多样化即时化，尤其是微博、手机等新兴媒体的兴起，使得企业环境事件信息能够迅速传播，加上外部因素复杂，以及谣言等推波助澜，使企业和政府控制力度减弱，形势迅速复杂化。2013 年伊始，在中东部地区发生的大范围雾霾天气，其背后也是的主要成因之一是企业污染排放普遍性、长期性超标。同时，由环境事件引发的社会事件等次生事件层出不穷，加之对社会不满、失业、流动人口增多等社会矛盾因素较多，参与者既有利益相关者，也有利益无关者。这些因素使得企业和政府对环境事件应对起来往往措手不及，难以下手。

第三个特点是“西增东不减”。从环境事件与经济周期的关系看，我国的每一轮投资热，都会伴随着环境事件的增多。就中西部来看，大多尚处于工业化过程中，大规模资源开发、大规模投资上项目、规模大规模污染排放仍然难以完全避免，随着东部产业加速向中西部转移，“东污西移”也是必然趋势，环境污染的负效应仍在不断显现，因此企业环境事件驱动发生的增多是由环境污染本身增多导致的。就东部来看，经济处于转型发展中，很多地方已经完成大规模工业化，进入中上等收入和服务业大发展阶段，就环境事件趋势来讲应当是减少的，但由于这些地方群众环境意识增强，对环境事件的容忍度下降，因而环境污染的相对减少并不能表现为企业环境事件的减少，甚至有增多的趋势。如浙江每年环境信访量达五六万件之多，充分说明群众对环境的关注度在上升。

三是政府在企业发展与环境保护之间处于两难，甚至由于环境事件处置不当引发政府信任危机。在针对最近一些环境事件的舆情调查中，对地方政府公信力的质疑始终排在前三位。有的质疑政府处置环境事件的能力，有的认为，环境发生的根源于政府监管不力。还有的认为，地方政府与部门在环境事件涉及利益较多，因而姑息迁就，养痈为患。有的涉环境影响项目因群众反对或发生群体性事件不得不下马，典型的如江苏启东“王子排海工程项目”、四川什邡“钼铜项目”，以及最近发生的浙江镇海炼化 PX 项目等；有的企业环境污染长期得不到治理；有些涉企业环境危害谣言得不到有关权威部门的澄清。这些使得人们往往将污染事件驱动归结到政府身上，由此造成了政府公信力降低。比如湖南紫金矿业污染事件中，该企业贡献的税收占当地县财政税收的60%以上，处置起来难免投鼠忌器。环境事件引发了政府的决策方式、发展理念等的深刻改变。

3. 企业环境问题产生的原因和发生机理分析

我国企业环境事件多发的背后有其深刻的发展阶段背景，凸显了环境与发展、经济与社会、人与自然之间的矛盾。我国仍是一个发展中国家，从宏观看，工业化尚未完成，城镇化正在加速推进，总体经济结构、消费结构、产业结构都还处于中低端。对环保的投入也处于低水平，仅约占 GDP 的 1%左右。从微观看，我国现有企业是从国有企业、集体企业、乡镇企业、私营企业甚至个体户等多种类型企业发展而来，企业素质、管理水平参差不齐，原有生产方式大多是粗放型，发生环境问题有其必然性。企业环境问题产生是一个长期积累过程，对环境责任的认识也是一个长期过程，上升到自觉履行环境保护责任也必然是有一个较长过程，同时人们对环境问题的关注和诉求、经济社会发展对于环境问题的要求，都有一个过程。

国际经验证明，当一国处于经济起飞阶段，企业环境事件的发生处于上升期，随着经济发展达到工业化中后期，企业环境事件才会到达顶点并走向下降。1995 年美国普林斯顿大学的格罗斯曼和克鲁格提出了“环境库兹涅茨曲线”。根据这一假说，经济增长初期大规模经济活动既需要大量资源投入，也会产生大量污染排放，环境质量恶化，从而引发大量环境事件，随着新技术应用、经济结构优化、清洁能源使用等，环境污染逐步减少，环境质量得到改善。这一过程形成了环境质量和经济增长之间的“倒 U 形”曲线关系，环境呈现先恶化后改善的趋势。它反映了从清洁的农业经济到污染的工业经济再到较为清洁的服务经济的结构演变过程。现实中，美欧日等国的经济发展和工业化过程也验证了这一点。规律的客观性也使我们必须要经历这一阶段。

应当看到，由于后发优势、环境技术进步、设备工艺流程改革等因素，加上我国较早重视发展中的环保问题和企业环保责任，我国单位产出的能源资源消耗强度和污染物排放强度，都远低于发达国家的同样发展阶段。但也要看到，伴随改革开放 30 多年来我国经济持续快速增长，工业化和城镇化推进之快也是前所未有的，发达国家在一两百年工业化里程中不同发展阶段出现的环境问题，在我国几十年的时间里集中出现，历史遗留的环境问题尚未解决，新的环境问题又接踵而至，加之我国基础设施不足，环境软硬功件建设滞后于经济社会发展，政府和企业对环境问题的管理也缺乏经验，监管体系不完备。在全球化时代，公众对环境事件的要求也是全球同步的，加之我国现阶段社会矛盾问题多，因而企业环境事件的增多有其必然性，环境风险的集中爆发，也会使环境问题由局部问题演变成全局问题。

我国经济发展正处于向工业化中后期发展的过程中，潜在经济增长率下降，劳动力供求拐点正在来临，累积的财力增长和技术进步，尤其是发展观念的变化，使得环境在工业化、城镇化中得到改善成为可能。我国作为后发展国家，有条件也有能力找到这样一条有别于发达国家曾经走过的发展道路，为人类文明进步探索新路。新世纪以来，我国提出了“科学发展观”的理念，要求加快转变经济发展方式，推进经济结构优化升级和产业转型，加大环境治理。“十二五”规划提出了单位 GDP 能耗下降 16%、单位 GDP 二氧化碳排放下降 18%、主要污染物排放总量化学需氧量和二氧化硫下降 8%、氨氮和氮氧化物下降 10%的目标，并作为约束性指标。我们借鉴国际上的成功经验，汲取一些失败的教训，发挥后发优势，可以避免重复“先污染、后治理”的老路，也完全有可能通过我们的努力部分地打破发达国家规律，提前到达“倒 U 形”环境曲线的顶点并开始下降。从目前东部情况看，随着产业升级和环境污染有效控制，空气污染、水体污染、植被破坏等初步得到控制，随着环境保护工作力度加大，总体看环境问题恶化的趋势在减弱，环境事件有可能得到较好控制。

从宏观看，企业环境事件发生增速的迅速上升，除客观因素外，还有四个方面的原因：第一，群众环境意识和环境诉求的增强。主要表现为公民意识、维权意识、健康意识大大增强，对生存环境的要求提高，因为环境问题既是民生问题，也是民权问题，基本的环境是政府应当保障和提供的基本公共产品。随着群众维护环境权益和追求环境公正意识的增强，环境维权迅速增加。有媒体甚至预言：“中国即将出现一波环境保护运动潮”。由企业引发的环境事件普遍演变成为环境社会事件，有的甚至成为打砸抢等暴力事件。但总的看，群众环境维权趋于理性化、专业化、组织化，参与者往往是既是一般社会人员，也有律师、医生、记者、工程师等较高层次的人员，组织程度较高。比如大连 PX 项目抗议现场，数千人静坐后志愿者现场清理，活动结束后没有留下一片垃圾，这体现了环境事件中民主意识、公民意识、法律意识、文明意识等的增强。

第二，我国环保约束惩戒制度还不完善。主要表现为是相关法律和强制性标准缺失以及法律执行的不力。由于发展阶段等原因，我国企业总体处于中低端，生产方式大多是粗放型的，其环保制度和标准总体上处于低水平，政府监管也处于粗放型，但随着时代发展，环保制度、标准和监管不适应群众新期待的问题逐渐凸现出来。从环保法律约束和惩戒机制看，相关法律缺位，原则多而细则少，可操作性差。尤其是“守法成本高、违法成本低”的问题十分突出。如2011年6月哈药集团被披露多种环境违法问题，其中恶臭气体硫化氢超标近千倍，而罚款仅123万元，为企业年收入的万分之一。违法成本低，守法成本就高，而守法企业由于环境成本高，反而可能遭遇“逆向淘汰”。同时环境民事赔偿法律不健全，环境民事案件面临立案难、举证难、审判难、执行难等“四难”。

第三，发展理念滞后，发展方式仍然粗放。主要表现为政治意愿、政治决心与发展路径依赖的矛盾。许多地方政府在“发展第一”还是“环保优先”之间难以取舍，“在发展中保护、在保护中发展”的理念难以落实。在对发展与环保关系的认识上，有的政府与群众出现了脱节。表现之一是“政府算加法、群众算减法”。政府过于看重重大项目带来的经济效益，认为有利于增加财政收入、促进民生改善，热衷于招商引资，甚至饥不择食，对环保责任则相对忽视，而群众更加担心环境污染给当地带来的损害。表现之二是“与民赛跑”，有的地方政府希望趁群众环境意识觉醒、环保运动兴起前，抓紧时间上几个大项目。表现之三是“虎头蛇尾”，一些地方在环境事件发生后，首先想到的不是做过细的工作，而是不能屈服于民意，因而采取强力手段，体现出权力的傲慢，而后在强大社会压力下，又只会选择轻易放弃项目。表现之四是“花钱摆平”。有的地方认为针对企业环境的个体行为较多，只要做好相关补偿，就能解决环境矛盾，即使是发生涉环境事件，也能通过花钱买平安。这些表现既不能在企业环境事件处置中形成良好机制，从根本上防止环境问题再次发生，又会造成各方利益受损、互不信任加剧的“多输”局面。

第四，环境问题成为国际社会关注和博弈的重点领域。主要表现为环境问题渗透进入国际谈判、竞争优势甚至发展模式之争、内政主权之争中。气候变化、环境影响等问题越来越多地成为国际多双边讨论的重点，有的发达国家无视其自身历史上排放较多的事实，自己不愿承担应有责任的同时，反而要求发展中国家承担更多的责任，有的企图借此约束发展中国家崛起，同时又在道义上占据制高点。因而环境问题经常成为在联合国气候大会等场合国家斗争的焦点。同时，也要看到，我国一些环境事件发生的背后，有着境外组织和势力的插手，使环境事件复杂化、国际化。如2012年美国驻华大使馆、驻上海总领事馆相继公布北京、上海等地的$PM_{2.5}$数据，引起社会广泛议论和针对相关部门的不满情绪，虽然有其客观性，但环境数据涉及社会公共利益，属于政府公权力范围，外国机构这样做的用意何在，值得深思。绿色和平组织等国外 NGO 也在一些地方环境事件中起到了一定作用。

第五，企业环境事件频发的背后，最根本的还是深刻的利益交错格局。主要表现为企业发展、政府履行职责与群众环境权益维护三者之间的矛盾、既得利益集团与国家长远根本利益之间的矛盾日益突出。很多涉及环境污染的企业，同时又是当地利税大户，为政府贡献了大量的税收和就业，政府在短期利益和环保利益之间难以割舍。更有甚者，很多地方官员在热衷“亲商”的过程中，与企业发生了利益关系，甚至结成“利益共同体”，政府官员成为污染企业最大的“保护伞”，对企业环境污染问题“睁一只眼闭一只眼”。有的地方官员在企业环境问题爆发之前得过且过，抱有侥幸心理，不愿轻易在自己任期引爆这个火药桶，而往往把小问题拖成大问题，苗头性问题发展成趋势性问题，最终难以收拾。在重大项目环境影响评价、环保监督机制建立和环保成本付出的过程中，更涉及政府部门、企业、当地群众、环境评价机构、社会组织以及媒体等多方的利益，有的甚至关系企业生死存亡、群众生命健康、政府媒体公信力等，加上外部势力干扰，就使涉环保问题的利益博弈更为激烈。因此，建立一种企业、政府、社会组织、公民等共同遵守的履行环保责任的有效机制，就成为经济社会发展必需而又紧迫的制度保障。

以上是从外部和宏观角度看制约企业履行环保责任的因素。

从企业自身来说，制约其履行环保社会责任的主要因素以下方面：其一，最主要的因素是成本因素。长期以来我国很多企业能够在国内市场和国际市场取得成功，很大程度靠成本低廉，包括劳动力、土地，当然更包括环境成本低甚至不支出环境成本，很多地方政府也以此吸引外国投资企业。增加环境投入，势必加大成本。企业浪费资源、排放达不到环保标准，其实质就是希望将内部成本外部化，将相应承担的责任交给政府和社会。今后随着环境约束增大和企业转型升级条件的改善，企业环境成本的投入持续增加将是必然趋势。同时，环保标准和环保责任与企业运营的矛盾也将增大，很多项目将因此不能上马，甚至有些项目和企业将因达不到环保标准而被淘汰。

其二，最突出的问题是外部约束乏力，也就是“守法成本高、违法成本低”，因而企业不愿意履行环保责任，也存有侥幸心理。包括法律法规不完善、相关环境标准缺失，最终是执行不力。有些企业存有侥幸心理，在环保监控下故意偷排、漏排，而一旦被发现也只是交少量罚款了事。

其三，最直接的原因是企业忽视社会责任。我国很多企业尤其是中小企业过度以营利为导向，缺乏做“百年老店”的战略眼光，企业化思想和企业文化十分松散，同时行业自律和监督也十分薄弱。很多企业认为，环保是政府部门的事，不愿和抵制履行环保社会责任的行为普遍严重，不能自觉转化为企业行为。如对安装的环境在线监控设施，很多企业想方设法逃避运行，只是在检查应付了事。有的企业重视商业信用而忽视环保信用等社会信用，宁愿在广告、展会上花大钱、投巨资，也不愿在环保和社会责任上加大投入。

其四，最为社会所关注的是企业环保与社会信息沟通不畅，甚至信任缺失。从涉环保事项决策机制看，政府和企业关门决策，环境影响评价等相关信息向社会公开透明不够，对群众意见重视不够，也没有完善的民意征求和社会沟通交流机制，企业、政府和公民、社会组织相互信任缺失，往往引发民意对立，小事件酿成大冲突。当然，也有些企业对污染造成的社会危害故意视而不见，热衷于和当地政府、少数居民达成妥协，而不是采取积极有效措施。

其五，对企业具有最直接影响最可怕的就是对环境事件应对不力，造成环境问题扩大化甚至发生次生事件。从环境事件发生后的应对机制来看，早期不重视、准备不充分、应对无章法、后续不彻底，恶化了处置环境，谣言满天飞，在环境事件处置和舆情应对上处于被动应付，增大了社会风险，使得环境事件及其次生事件一发再发，呈持续高发增长态势。

（二）建立促进企业履行环保社会责任的长效机制

促进企业履行环保社会责任，从宏观和根本上来说，要加快转变经济发展方式，建立有利于科学发展的体制机制，使企业在保护环境的同时获得可持续发展动力。就当前来看，企业环境问题实际是外部性问题，其造成的社会成本如政府治污花费、对人类健康的损害等往往由外部承担，从作为市场主体的角度看，企业并不愿意承担环保责任，因而必须建立相应的外部约束的体制机制，并使其内化为企业采取符合环保要求的行为。同时企业环保社会责任也是内部问题，是企业道德和社会责任的重要组成部分，是企业长远生存发展的精神动力和思想保证，也应当使其内化成为企业文化建设的主要内容。因此，促进企业履行环保社会责任，必须从“软”和“硬”两方面建立长效机制，既要强化制度，又要强化文化，既要强健其“体魄”，又要强健其“精神”。

1. 建立健全促进企业履行环保社会责任的制度体系

要综合采取法律、经济、行政、文化等多方面手段，依靠法律法规和相关制度设计，加快构建起便于企业遵循、对企业行为形成软硬约束、促使企业发展与环境保护相统一的环保社会责任制度体系。也就是说，促进企业履行环保责任，既要靠政府行政和政策调节，又发挥市场的积极作用，强化道德和文化约束，但最根本的是要靠制度的力量。这既需要加快环保领域相关改革，更需要加快制度创新，还要加大执行和落实力度，尽快推动企业履行环保责任有大的进步，促进环保事业有大的发展。

首先，要在发展理念上进行变革调整。党的十八大首次把科学发展观列为党的指导思想，并把生态文明建设与经济建设、政治建设、文化建设、社会建设列为社会主义建设总体布局的重要组成部分。我们将把生态文明建设融入整个现代化建设之中，加快转变经济发展方式，探索“在发展中保护、在保护中发展”的道路，同时也提出了绿色发展、低碳发展、可持续发展等理念，实现改善民生、发展经济、保护生态的共赢。这为每一个企业和个人自觉履行环保责任、建设生态文明提供了理论的指南。

其次，在法制层面，要重点加强环境领域立法，弥补相关法律空白。近年来，我国环境立法步伐加快，但总的看，立法的步伐跟不上加强环境保护的需求，使一些领域出现的环境污染处理无法可依，为一些企业不履行环保社会责任留下了空间。要抓紧修订《环境保护法》，使之成为环境治理的“宪法”；修订《环境影响评价法》等，以增强企业履行环境责任和进行环境影响评价的权威性，加大执法和处罚力度，对造成公共环境污染事件的企业，除依法追究相关人员刑事责任外，也要提高罚款额度，同时对企业法定代表人也给予经济处罚，并增加没收、拆除、限令停产、关闭等强制性措施。

同时，根据现实需要，加快出台相关的专项法律，如针对反复出现且趋势日益加重、群众和社会普遍关注的大范围雾霾天气，以及形成污染的复合型、结构性和压缩型特征，抓紧修订《大气污染防治法》并出台具体实施细则；针对农村土地面源污染日益加重，抓紧出台《土地污染防治法》；出台针对化工企业的《重金属污染防治法》、《化学污染防治法》，等等，使这些专项法律成为约束企业环保行为的重要根据，也是支持企业履行环保责任的重要导向。

在制度机制层面，要着力建立市场机制与宏观调控相结合的环境保护政策体系和体制，建立有效的环保激励约束机制。当前需要采取的措施：

第一，支持各级政府加大环保规划和投入力度。进一步加强环保规划的严肃性、约束性、可行性。鉴于当前环保问题的长期性、严重性，建议在相关专项规划和区域规划中明确一定时期内环保投入占 GDP 的比重，目标可考虑设定为 GDP 的 2%—4%左右。明确这一目标后就能够促进企业、政府以及社会组织为达到这个目标而努力。

第二，加快完善激励企业履行环保责任的相关政策措施。要制订鼓励企业在环保和节能减排技术研发等方面加大投入的财税、金融、产业等支持政策；完善环境和资源价格机制，理顺环境成本和价格；抓紧出台环境保护税费改革办法，建立规范的环境保护税，取代排污费等各种环保收费；如针对一个地区和企业实行环保“以奖代补”政策，完善财政奖惩机制等。

第三，建立市场机制和增量调控机制，便于企业和社会投入环保。当前很多重点排放型企业也希望履行环保责任，但往往难以找到合适途径，一旦出现环境事故又难以单独承担。这是一个现实存在的问题。如：（1）建立排放权市场交易机制；（2）建立区域性的生态建设和环保补偿机制；（3）针对企业难以单独应对和处理环境事件难的现状，建立环保责任保险机制，即由企业参加、商业保险机构承担的环境保护责任险，对可能形成的环境损害建立共同经济责任，可考虑采取强制险；（4）同时建立由政府或相关组织管理的环境保护基金，对潜在的环境和生态风险进行预防和治理；（5）大力发展节能环保产业，鼓励企业使用清洁技术、环保设备和清洁能源，优化生产流程。

第四，强化政府对环保准入和监管条件设置。如建立严格的企业排放污染物申报和收费制度；如强化证券市场、债券市场、银行信贷等对企业履行环保责任的审查，作为上市、融资融券等的必须准入条件，实行一票否决等。

在行业和区域层面，应当建立更加严格的环境排放标准和监管体系。环境标准落后和监管执行力度弱，是我国企业环保责任落实不力的重要原因。一是要加强环境质量标准体系建设。我国环境质量标准有些标准已经落后，比如近期细颗粒物 PM_{10}已经由 $PM_{2.5}$取代，比如“三鹿奶粉”事件中取消的所谓“国家免检”制度，比如近期对“雾霾”预警标准的修订，都是根据形势而变化，得到了社会和群众认可。要根据群众对环境的需求和经济社会发展条件，参考国际标准限值，加紧修订和增加具体标准，使环境标准与群众感受相一致。要科学确定符合国情和经济社会发展条件、体现区域特点和顺应民意的环境基准，完善环境质量评价体系，同时加强行业的环境自律建设，增强行业协会等的环境自我约束能力。二是全面建立并完善环境信息定期向全社会的发布制度，建立涉环保重大项目向全社会公示制度。我们已经公布新《环境空气质量标准》。长期以来，我国同一地区、同一流域公布的环境质量数据不同，环境质量评价不一，在社会产生了负面影响，要统一监测点设置和监测数据发布。三是建立跨行政区域、跨流域的环境保护联防联控制度，完善区域和流域环境保护规划，建立健全区域环境污染补偿机制。

在企业层面，要完善企业污染防治责任制度，坚持“谁污染谁治理”，实行生产者责任延伸制度，重点解决“违法成本低、守法成本高”的问题。明确企业作为环境事件处置第一责任人的责任，不仅要对排放污染物负责，而且要对污染物排放造成的公共环境影响负责，承担对受污染损害的单位和个人的赔偿责任。完善企业污染限期治理制度，堵塞漏洞。在企业内部建立环保责任制度及向职工代表大会报告环保并接受监督的制度，明确涉污企业应当开展自我监测和接受有关部门监测，有责任安装并保证运行环境监测设备设施，并依法公开监测数据。对弄虚作假或擅自拆除或闲置的，应予重罚。企业应当依法加强环境污染风险防控，定期开展环境污染事件应急方案制订和演练。在突发环境事件处置中企业应积极配合和主动消除环境事件的公共影响，在应急处置工作结束后应当及时评估事件造成的环境影响和损失，提出长效解决办法。

在社会层面，建立企业环境行为社会评价机制。加强企业环境信息的采集、整理，推进国家环境信息监测与统计能力建设，建立企业环境信息档案库，建立企业环境信息向全社会公开发布制度。建立企业污染物排放状况向全社会发布制度。发挥行业协会针对企业环境行为的监督与约束作用。建立企业环境信用评价制度，针对企业环境行为进行评级，然后进行分类管理、监督管理，建立对环保失信企业的法人、实际控制人、主要股东及关联人的惩戒制度。同时，要将企业环境信用评价作为审核企业发债、上市、银行贷款等资格的重要依据。

2. 完善重大项目环境影响评价机制

我国《环评法》实施以来，环境影响评价制度已经形成较为完善和充实的体系，特别是实行“三同时”（环境保护设施与主体工程同时设计、同时施工、同时投产使用），在推动环境影响评价制度化方面取得了重大进展。但由于我国环保欠账多，在环评制度实施过程中也暴露出一些问题：民意公信力降低，一些企业项目在环评通过后依然遭到居民强烈反对，居民对环评结果普遍不信任；环评走过场，环评机构难以做到公正公平，群众难以真正参与环评；企业环评通过后依然发生重大环境污染，反过来降低了环评权威性。环评制度已经站在一个关系生死存亡的关键点上，能否真正完善环评制度、重塑公信力，事关企业、政府、环评机构三者的信誉和发展。

环评问题的产生有着环评机制设计不合理、有关各方不够重视等多方面的原因，但最根本的是存在“利益链条”，从而损害了公众环境权益，是典型的“私权损害公权”行为。环评报告由业主单位聘请具有资质的机构完成，这自然会造成“谁出钱替谁做”的做法；环评机构与环评项目的多少有直接利益关系，主要靠环评费维持运营，做得严了下次就不会有更多企业来找他做环评，做得松可能存在问题，处于两难，“两头受气”，同时“环评违法成本低、环评通过收益高”的想法，使其不惜代价造假；企业为了尽快或低成本上项目，自然不希望环评过严，甚至会买通环评机构和专家；有的地方政府为了项目尽快上马，放松了环评监管，甚至给环评机构和环保部门压力，帮助企业过环评。这就使得环评成了“软约束”、“走过场”。重塑环评机制，就必须打破利益链条、完善约束机制，真正实现环评的独立性、公正性、权威性。

首先要着力切断业主单位、环评机构和政府部门之间的利益链条。环评机构要真正从政府部门中独立出来，摆脱错误政绩观干扰，使环评机构成为公益性事业单位；其经费既来源于企业缴纳的环评费用，又不与企业挂钩，而是与其环评的公信力挂钩，可考虑建立环评基金，由涉及环评的企业普遍缴纳。同时鼓励发展民间环保公益组织，赋予其环评资质。对环评机构及其从业人员，要进行分类考核、资格评定与分类管理。其次，要进一步完善环评机制。切实增强环评的权威性，坚决实行一票否决，同时完善环评否决机制。比如“三同时”和允许企业补环评的规定，一定程度上纵容企业相对忽视环评，应明确环评在先，项目上马在后；比如环评在发改委、规划、国土等部门审批之后，造成了项目既成事实再环评的结果，环评比较被动，建议环评应同时进行；比如建立环评结果的追溯机制，加强环评结果对环评机构的“硬约束”，建立健全环评责任制，环评单位和个人要对环评报告负责，建立长期“环评信用制度”。建设单位、审批单位以及有关人员也要对环评负责；再比如，加大环评违法的处罚力度。第三，要加大环评的公众参与力度。制订公众参与项目环评的具体实施细则，进一步明确环评公众参与的比例、程序、对象选择、权威性等，使其具有真正的代表性、公正性和影响力，加大环评结果向全社会公示的力度，增强环评专家论证的可信度，对环评公众意见的答复和整改措施要向社会公布。

3. 构建重大项目环保社会风险评估机制

我国自 20 世纪 80 年代开始实施环境风险评价，近 30 年来逐步形成了比较成熟的环境风险评价制度。但从实施看，这一机制总体与环境影响评价制度有较多重合，且偏重于技术风险、环境承载力、环保可靠性等。随着工业项目集中上马、群众环境维权意识觉醒和环保群体性事件频发，企业项目环保问题成为关注度最高、涉及面最广、敏感性最强的社会问题，环境风险评估亟需建立与环境影响评价机制同步进行的社会风险评估机制。许多论证成熟、建成后多方受益的项目，往往由于社会风险评估不充分、不准确，导致重大群体性事件发生，直接造成项目下马，政府、企业、居民多方受损，近期的镇江炼化 PX 项目、什邡钼铜项目、启东制纸项目均是如此。近年来，一些地方积极探索建立企业涉环保重大项目社会风险评估机制，取得不少值得重视的经验。如贵州省的社会稳定风险评估“六步工作法”，把充分调研、科学论证与民主决策结合起来；江苏省专门出台环保项目公众“强制听证”制度，提高项目的公众知情权、参与权、表达权和监督权。各地还普遍制订了“分级审批、分组预警”机制。这些制度起到了有效化解矛盾、降低风险的作用。但是总的看，企业涉环保项目社会风险评估机制还不够完善，覆盖面较小，而且屡屡发生预判不准、“评估失灵”、风险爆发、工程夭折的企业环境事件和群体性事件。随着“环境维稳”形势日趋严峻，亟待构建有效的环保社会风险评估机制，使其成为经济发展与社会和谐的“减压阀”、“防火墙”。

一是把环保社会风险评估放在突出重要位置。环保社会风险评估的核心是维护人民群众的根本利益，因此应当放在重要位置。对可能引发社会风险的企业涉环保重大项目，实行社会风险评估“一票否决”制度；对涉环保重大项目实行社会风险评估“前置审批”，在提交环境影响评价报告的同时，同步提交和审核“社会风险评估报告”；凡是涉及群众环境权益的重大决策、重大政策、重大项目、重大改革，均纳入环保社会风险评估等。

二是完善环保社会风险评估体系。重点健全环保社会风险评估对象和内容。评估对象：主要是重大项目实施、重大涉及环境的决策、重大改革事项等。评估体系的内容包括：（1）程序合法性评估。是否按照环保相关要求履行必要的程序，否则视为违法上马。（2）政策合理性评估。环保项目和政策是否充分考虑相关制约因素，时机是否成熟等。（3）方案可行性评估。项目是否取得群众的理解和支持，引发较大社会稳定事件的可能性等。（4）诉求可控性评估。群众对项目实施的反映和要求，其风险是否在可控范围内等，其前提是充分的知情。

三是完善环保社会风险预警与控制体系。建立分级预警机制，存在严重社会稳定风险、可能引发重特大维稳事件的项目为一级预警，不准实施；存在较大稳定风险、短期难以解决的为二级预警，暂缓实施；存在一定风险但在可控范围的为三级预警，分步实施；稳定风险较小的为四级预警，准予实施。完善环保社会风险控制体系。重视打“苗头”，把矛盾解决在萌芽状态；随着事态升级，要建立相应工作预案，提出环保社会风险解决措施，对项目可能涉及社会风险和可能出现问题的要有预先处置方案等。

四是健全环保社会风险民意沟通机制。首先政府和企业要转变观念，环境风险由全社会共同承担，环境项目决策也应由全社会参与，社会风险评估的主体不只是政府和企业，也应包括民众，让好的项目成为民众共同意愿，切忌抱有防、瞒甚至欺骗等心态。在重大项目决策前，通过座谈会、听证会、社会公示等多种形式，充分征求和吸纳民意，制定周密评估报告，并邀请人大、政协、行业协会以及社会各界代表对社会风险评估报告进行审评。要将重大项目可能带来的风险和利益真正透明公开，使民众从社会风险评估中了解、信任和支持该项目。

五是强化社会风险评估与化解工作机制。首先要加强领导与协调，建立党委政府领导、政法综治等部门牵头，多部门合作、协调联动工作机制。其次要在社会风险评估时，同步推进矛盾排查与调处，及时化解纠纷。第三要强化监督问责机制，对履行评估不严格、造成“评估失灵”的干部严肃处理，对不重视社会风险评估结果的决策者严格问责。第四可考虑引入第三方评估机制，与政府和企业的评估同步推进。

4. 建立企业与政府、社会组织、个人的环保沟通机制

西方发达国家历史上对于环境事件的处置，往往是民意倒逼的结果。民众环境意识的觉醒、新闻媒体的渲染、社会组织力量的推动，使得政府不得不提出应对措施，实际上他们对环境事件的重视和行动总是滞后的。比如在工业革命几十年后，英国人才向议会提交了第一份水污染报告；20 世纪 50 年代伦敦烟雾事件后，议会颁布了第一部空气污染防治法案——《清洁空气法》。此后，民间环境保护组织也如同雨后春笋一般迅速发展起来。但这也推动政府更加重视环境安全，对工业环境的评价更加严格，更加重视环境与经济发展的关系，等等，同时对环境事件的处置更加尊重民意、尊重舆论，更加重视发挥社会组织的作用。

如何在处理企业环境事件中形成企业与政府、社会组织与个人的良性互动，需要各方都做出努力。政府、公民、社会组织、媒体相向而行，最大限度取得共识。一是企业尤其是企业家要自觉接受政府、社会和公众对企业履行环保社会责任的监督。倍加珍惜自己的环保声誉，宁肯环保上多投入，自觉履行好环保责任。一旦发生环境问题，就要配合有关方面诚恳务实地处置，重视各呼吁和诉求，取得理解和支持。要增强项目环保措施和设施的透明度，就项目环境问题，不仅有政府审批，更要向公众及媒体宣传说明，与环保组织合作，普及相关环保知识，避免误解和对立。要重视发挥好企业内部工会等组织的监督作用。企业家是企业的领军人物，其价值取向对企业社会责任建设至关重要，现代企业家不仅要有市场发现和经营管理能力，还要具备履行社会责任的素质和能力。

二是政府要切实履行好环境监管职责，推进环境决策科学化、民主化。着力建设责任政府、法治政府、受群众信任的政府。推动环境法律法规和标准建设，依法处理环境事件。完善环境事件问责制。党和国家对此一贯高度重视和严肃认真，坚决实施问责制。如松花江污染事件后，启动了问责制，使当时的环保总局局长引咎辞职；三鹿奶粉“三聚氰氨事件”后，国家质检总局局长、河北省副省长等辞职。完善监管机制，解决好政府监管不力的问题。

三是发挥好公民和社会组织在处置环境事件中的积极作用。有些人认为公民和社会组织在环境事件中的作用是负面的，把他们对立起来，实际上他们的作用发挥好了，完全可以成为推动事件解决的积极因素，而害怕躲避、企图绕过去的思想，最终会成为企业环境问题解决的障碍。要建立公民有序表达环境诉求的渠道。建立重大环境决策民意沟通和听证论证机制，完善信息公开、专家咨询等机制。支持民间环保组织发挥监督、督促等作用，积极有序发展环保志愿者队伍，推动社会组织在环境事件处置中有序有效发挥作用。

（三）构建企业突发环境事件的应对机制

之所以关注这一问题，是因为在当前企业环境事件多发频发成为客观存在的情况下，如何在建立和完善应对环境事件的机制中促进企业履行环保社会责任，是一个十分现实而紧迫的问题。同时也要看到，企业环境事件发生后，就不是单纯的企业行为和企业责任，往往迅速由局部性扩展到全局性、由个体性延伸到社会性，演变成为公共危机事件，并很可能发展衍生出次生社会事件。因此，企业环境事件的处理与社会发展稳定密切相关，应对企业环境事件，是企业、政府、社会、公民共同的事情，都要承担起应负的责任，并加强相互沟通协商，共同应对危机。

1. 企业突发环境事件应对的经验教训

经验教训之一：加强信息公开透明度，及时发布权威信息，不要刻意隐瞒，否则就会造成民意对立、处置环境恶化。

案例1——广东大亚湾“核泄漏”事故：2010年6月15日，香港电台披露深圳大亚湾核电站发生“核泄漏”事故。广东核电集团先是否认此次事故，随后发表声明承认发生核泄漏，但是比较轻微，不会对公众构成影响。但媒体质疑为何事故发生近一个月未公布。此后，中广核和有关方面反复解释澄清，采取了相关措施，并提出今后将本着安全、透明的原则，通报后续情况，媒体关注才逐渐平息。这本是企业正常规范下的一起小情况，但引起恐慌事故，主要是由对核电不了解、事故处置不透明不公开引起的。由图4-1可见，信息公开是企业环境事件发生后首要关注的问题，占65%以上。在近些年发生的事故中，信息不透明始终是媒体、网民等质疑的重要因素，政府、企业都应以此为戒，高度重视企业环境事件发生后的宣传引导。

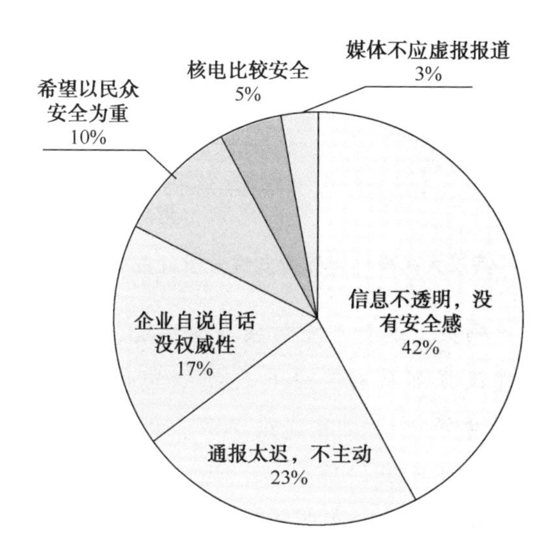

图4-1　广东大亚湾核泄漏事件言论倾向性（抽样调查100份）

案例2——武汉雾霾天气事件：2012年6月11日10时，湖北武汉出现大面积雾霾天气，同时即出现了化工厂氯气泄漏、武钢锅炉爆炸等谣言。武汉中心气象台发布大雾黄色预警，但并未解释具体原因，使质疑和关注度持续升温。当天下午和第二天上午，湖北环保厅等有关部门陆续作出科学解释，同时武汉警方拘留了两名造谣者，事件得以较快平息。15日又出现大雾，民众情绪稳定。处置的主要经验是：遵循“黄金4小时”原则，利用传统媒体和微博等新媒体平台，迅速传递事实真相，多部门共同回应，赢得了市民和网民的信任，促进了事件迅速解决。由图4-2可见，此事件中信息透明是大家的主要关注点。

经验教训之二：针对企业环境问题要正确审慎决策，事件发生后要果断处置，理性引导，迟钝、拖延、松软只会导致事件恶化，最终多方利益受损。

案例3——江苏启东项目环境群体事件：2012年7月25日，江苏启东有网民发布消息称，当地拟新上“王子制纸排海工程项目”，建设内容包括2条年产40万吨的高档铜版纸生产线、1条年产70万吨的木

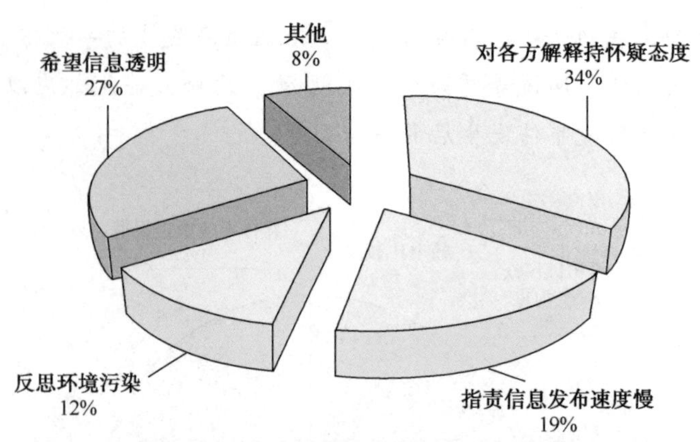

图 4-2 武汉大雾事件媒体倾向性分析（抽样调查 200 份）

浆生产线以及 2 座码头，这会严重污染当地环境、危害人体健康，启东市百姓对此项目建设强烈反对，并自发组织向政府请愿。26 日，启东市常务副市长宣布暂停排海管道建设工程，但并未获得民意支持。28 日，启东市民开始在市政府门前抗议，场面一度失控，政府机关被冲击，有官员被扒掉衣服羞辱，有车辆和电脑被砸。同日，南通市政府宣布永久取消启东市有关王子制纸排海工程的项目，平缓了民众情绪。但 30 日再爆出“记者在启东采访被打”、“打死一名 17 岁高中生”等传闻，媒体关注再度升温，由环境事件升级为打记者事件。警方拘留了散布谣言者，并由有关部门发布权威信息，事件逐步趋于平息。这次环境事件中，政府决策未充分听取民意，调研不充分，与群众沟通不力，对企业环境问题失之于宽、失之于软；事件处置中也存在不足，出现了群体性暴力行动，最终使符合环保要求的项目不能上马，应对媒体过程中也出现反复（图 4-3，图 4-4）。但处理过程总体上温和、理性，未过度使用警力，最终结果满足了群众的核心利益需求，促进了事件的较好解决。

经验教训之三：对企业环境事件越早处理越主动，把事故消灭在萌芽状态，防止小事拖大、大事拖爆，防止善后安置和后续处理不当，问责不力，以致发生环境事件的次生事件。

案例 4——四川什邡项目环境群体事件：2012 年 6 月 29 日，四川宏达集团钼铜多金属资源加工项目在什邡市开工，引起人们对其造成环境污染问题的质疑和反对。6 月 30 日和 7 月 1 日，分别有上百名市民和

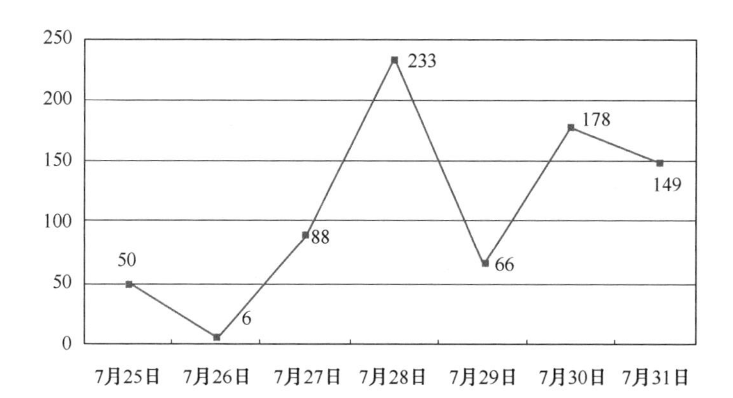

图 4-3　江苏启东事件媒体关注度走势

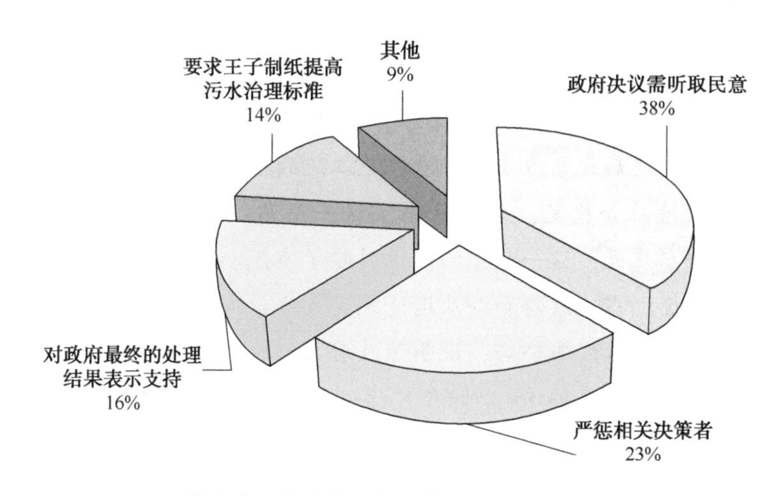

图 4-4　江苏启东群体事件网友观点倾向性分析（抽样 187 条）

学生上访示威，要求停建。遗憾的是，这一苗头性问题并未引起重视，政府和企业没有采取有效应对措施，也没有在第一时间到达现场去和群众沟通协商。7 月 2 日示威群众迅速增多至数千人，特警动用催泪瓦斯和震爆弹对过激人群予以驱散。晚间，什邡市政府宣布停止这个已经通过环评、投资上百亿的钼铜项目建设。这起事件中，政府和企业起初处于被动，后来宣布项目停止后，积极救护学生和市民，对发生死亡等不实信息及时辟谣，对非法集会进行通告，释放部分强制带离的人员，逐渐取得了主动权（图 4-5）。

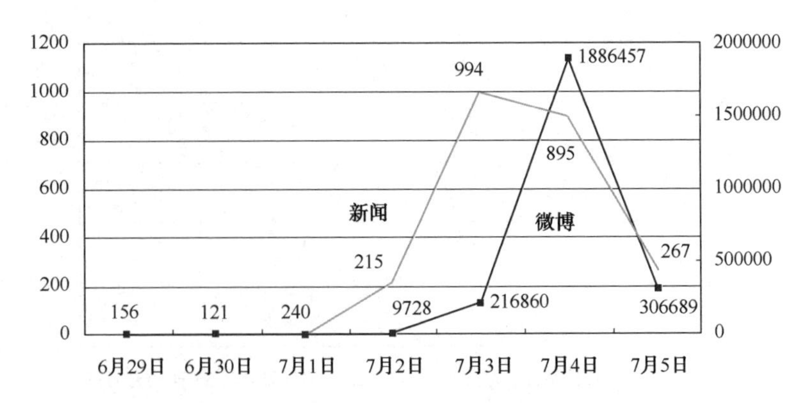

图 4-5　四川什邡群体事件舆论关注度走势

经验教训之四：加强政府和企业环境事件危机公关能力，高度重视做好舆论应对工作。

案例 5——吉林石化污染危机：2004 年 11 月 13 日吉林石化发生爆炸，预计一周之后化学污染物将沿松花江流到哈尔滨。最初，地方政府和企业为保持社会稳定，合作隐瞒了这一消息，称由于大规模管道维修，全市停水三天。这一明摆的假说引起了市民和媒体猜疑，很快引起了大范围恐慌。当地政府和企业迅速改变策略，决定向全体市民说明情况，在当地电视台发布公告，说明了政府善意隐瞒事实的原因，请大家保持信心，表示将每隔 15 分钟在广播电台、电视台、政府门户网站向大家直播，公布采取的措施，并将所有的大学生派到机场、火车站等，劝阻市民不要出去。一系列周密公关安排迅速稳定了局势。当年年底，《时代周刊》将哈尔滨政府集体评选为“全球十大风云人物”。

案例 6——丰田广告危机：2004 年丰田汽车为其新推的丰田霸道打了两则不恰当广告，分别描述丰田车拖着一辆破碎的东风卡车和一座石桥上的石狮子向其敬礼，由此引起各大媒体的口诛笔伐，使丰田陷入民意危机。丰田迅速召开新闻发布会，一方面说明公司高层召开紧急会议追查两条不恰当广告发布的情况，并诚意致歉，表示下不为例，另一方面强调丰田将一如既往地在中国种树、整治沙漠、捐助小学、捐助母亲等，从而缓解了这次危机，很好地以价值取向引导了事件的解决。

经验教训之五：切断政府与企业利益链，政府在企业环境事件中要站在公众的立场上处置危机，摒弃错误的政绩观。

案例7——紫金矿业污染事件：2010年7月12日，福建上杭县境内中国最大的黄金生产企业——紫金矿业集团公司突然发布停牌公告，宣布旗下公司涉嫌重大突发环境事件，造成了汀江流域严重水污染，企业负责人以环境污染事故罪被刑事拘留。经调查，县政府是该企业的最大股东，占上杭县财政收入70%以上，不仅如此，该县还有数十名官员在企业入股、挂职或任职。事实上紫金矿业污染由来已久，政府和企业故意纵容企业污染，发生事故后又隐瞒事实。政府有关人员和企业结成利益共同体，对企业的污染睁一只眼闭一只眼，同此导致了重大环境污染事故的发生，也导致了政府公信力的丧失。事实上，近年来发生的环境事件中，对政府失职渎职的问责声音不断升高，监管不到位的背后是隐形的利益链。

案例8——湖南浏阳镉污染事件：2009年7月30日，湖南浏阳市镇头镇发生村民因镉污染问题围堵镇政府和派出所事件。经查，长沙湘和化工厂为污染责任人。各类媒体纷纷质疑，该化工厂已经非法排污生产6年，期间群众反映不断，部分村民体内镉超标，周边树木枯死，河里鱼大批死亡，为什么直至发生严重污染才处理？这不仅是企业环境保护责任的缺失，无良老板利令智昏，更是政府不作为、监管失灵的直接结果，也是“环境换经济发展”的错误政绩观的体现，况且在处置中有避重就轻之嫌，要求严厉问责相关人员，呼吁政府和企业关注民生，重视民权。

在这些企业环境事件的处置中，有一系列重要经验启示。一是要坚持以人为本、民生为重。救人始终是环境事件后第一位的任务；政府处置要始终站在人民的立场上，而不是某个利益相关方。尊重群众的生存权、发展权、环境权、知情权。二是要发挥密切联系群众的优势，全民动员应对危机。要把信息和政策真正交给群众，迅速建立与群众、媒体对话的信息平台，真正“看懂群众的脸色”。如湖南浏阳市在总结处置镉污染事件时认为，如果早一点“走进群众，交根交底，走进心灵，交亲交友”，就不会发生上千名群众游行、围堵地方权力机关的结果。同时，树立“共渡难关”意识。同是PX项目，在厦门遭遇抵制下马，而迁建漳州经受住了民意检验，3年后平稳完成建设并进入试投产，关键在于取得了群众的理解支持。三是政府、企业、社会组织、个人和媒体形成合力，“专辅结合”，综合采取经济、法律、行政、技术等手段。比如新安江污染事件，发动全体市民和各界力量，发起了“杭州保卫战”，就达到了很好的效果。四是要力争掌握环境事件处置的主动权。在事故苗头阶段就要采取有效措施，一旦事故形成，就不能强硬以对，要充分思考和回应民意，并适时引导舆论导向，慢慢掌握主动权。如上海磁悬浮风波中市领导的“冷处理、徐图之”、东莞垃圾焚烧厂的“公众环评”，都是很讲究策略的处理方法。五是要坚持依法、理性处置环境事件，不可采取强硬措施、过度使用警力甚至采取非法手段等激化矛盾，政府和企业在公众面前放低姿态才有利于事件平稳解决。

2. 完善应对突发企业环境事件的程序性机制

突发性的企业环境事件往往迅速升级为公共危机事件。根据学者斯蒂文·芬克的理论，应对公共危机往往划分为四个阶段：预警期、手术期、住院期、疗养期。尽管我们无法精确掌握危机的发生发展，但在这四个阶段，需要根据事态发展，及时果断而又有序有效地作出应对。

——按照应急预案，在第一时间内启动环境事件应急响应机制。环境等有关部门和人员立即进入应急状态，到达现场，启动相关处置工作；按照事件轻重缓急作出应急响应级别判断；政府负责人召集有关部门紧急会议，听取汇报，准确作出判断，对相关工作进行动员和部署。同时向上级报告事件情况。

——迅速展开以控制事态以核心的各项工作。一是制定并实施应对方案，明确工作方向和重点，拿出相关措施。二是抽调和组织人力赶赴现场，动员相关力量执行方案，并做好相应物资、设备保障。三是建立信息畅通渠道，进行动态监测，及时掌握事态发展。四是动用公安、武警等力量保护和管控现场，防止失控、事态扩大，并造成进一步的损失。五是如果是群体性冲突事件，则应充分听取群众诉求，作出满意的答复。六是充分考虑可能的事态发展、可能引发的次生事件和采取措施后的后果，作出针对下一步可能情况的预案，防患未然，确保万无一失。

——建立应急指挥体制机制。以政府有关负责人为主成立应对企业环境事件领导小组，必要时成立一线指挥部，所有工作应在领导小组和指挥部下进行。领导小组应设事件处置组、专家组、物资财力保障组、信息保障组、安全管理组等。

——将全力保障群众生命财产安全放在第一位。（1）组织消防、公安等力量队伍，全力以赴救援环境事件中的死伤人员，无论什么人都应先行救治。（2）组织医疗人员赶赴现场展开救治。（3）做好相关受害群众的安置工作和抚慰工作。（4）对可能的危险，应及时动员群众转移避险。

——针对环境事件提出科学合理的解决方案。

——高度重视环境事件舆情应对工作。及时、准确、实事求是地发布事件进展动向，做好正面引导，掌握主动权，把大家的注意力引导到齐心协力共同应对事件上来；充分展示诚意，邀请各方媒体共同关注，做好媒体接待并为其采访提供方便。关注网络微博等新媒体动向，利用好新媒体平台，发布信息，引导舆论。及时公布权威真相，不含糊、不隐瞒、不拖沓。在事情未完全弄清楚的时候，不急于下结论；对不实猜测报道和谣言及时澄清，打击故意散播虚假信息等。

——做好社会面上的工作。重视做好市场供应稳定工作，保证交通通讯秩序不乱，加强社会治安，防止趁火打劫，防止别有用心的趁机作乱。时刻关注社会面上的情况，坚持稳定至上。在全力应对危机的同时，要千方百计做好稳定社会、稳定人心、稳定市场等各项工作，不能顾此失彼，防止危机引发新的事件，同时不利于环境突发事件的解决。

——依法公正做好善后处理工作。一是妥善安置受害群众，对利益受损的作出补偿、赔偿等。二是对相关责任人、责任单位作出处置，给社会作出一个交代。三是深刻反思事故发生的原因，实事求是地给出调查报告和结论。四是完善相关制度，从根本上杜绝事故再次发生。

在处置环境事件中应当坚持以下原则：第一，首先的是保障人民群众的生命健康安全。无论环境事件责任在谁，原因是什么，都要把救人放在第一位。坚持以人为本。危机发生时，首要的就是做到“对人民负责，让人民知情”。这是各级政府执政为民宗旨的体现，也是应对事件的第一要义。重点保障学校、医院、养老福利院和商场等人群集中的地方，保障老弱病残、低收入群体等。第二，紧紧抓住“三个第一”，即“第一时间、第一现场、第一负责人”。要在危机爆发的第一时间作出反应，掌握事件处置和危机管理的主动权，延迟一分，就会带来更大损失和威胁。当然这是建立在平时有预警、有准备和对事件作出准确判断的基础上。第三，迅速对环境事件作出科学研判。应对环境事件如同打仗，刘伯承元帅说，“五行不定，输得干干净净”，在环境事件发生后，要对其性质、程度、影响、背景等迅速作出判断。第四，坚持依法处置。尽管政府在特殊时期需要采取特殊措施，但必须建立在法律法规授权允许的范围内，否则容易留下后患。对危机期间的违法行为，更必须依法严厉打击。第五，及时研究分析、认真回应人民群众在环境事件中的利益诉求。

3. 如何应对和处置企业环境事件中的谣言

近年来，与环境相关的谣言呈现越来越多的趋势。大多数环境事件发生后，都会出现不同数量、不同版本的谣传；在应对企业环境事件过程中，也会持续不断地发生相关的不实传言。甚至有许多的所谓环境事件本身就是谣言，比如2009年6月发生的河南杞县所谓“钴60泄漏爆炸事件”，2010年6月的“广东大亚湾核泄漏传闻”等。谣言虽小，看似无影无踪，但却给经济社会乃至政府公信力带来了实实在在的危害。应当高度重视，抓紧建立完善的处置和应对机制。

所谓环境谣言，就是在环境事件发生后，有些人出于各种目的，往往通过各种媒介，包括微博、门户网站、电子邮件、手机短信等新媒体和传统媒体，有意或无意间通过虚构、联想、夸张等手段不正确传播，夸大其词、散布虚假信息，还有的甚至无中生有，以致造成社会混乱，人心惶惶，使政府经常处于被动应付地位，增加了处理环境事件的难度。

应当看到，环境谣言增多的背后，有其深刻的经济社会背景。首先我国环境事件处于上升期，多发易发，这是环境谣言产生的大背景。其次人们对生活质量提升有了新的期待，不仅要求享受较高的物质文明、精神文明、社会文明，更把喝上干净的水、呼吸新鲜的空气、更加安全的食品等，作为更重要的民生问题，对生态文明建设提出了更高的要求。第三，环境因素与每个人利益息息相关，环境信息对公众有天然的贴近性、普及性。当前反复出现的环境问题，使人们对环境事件的敏感度增加，稍有风吹草动就会产生警觉，由于环境知识的普及教育不足，环境法制不健全，人们在应对环境谣言时具有盲目性。第四，一些地方政府对环境事件处理固守传统思维，对现代信息传播规律并不适应，公开机制不透明，加上处理做法欠妥，往往使自己陷入被动应付，公信力逐渐缺失。第五，有些机构和个人怀有各种目的和动机，借机生事。“舆论的阵地如果我们不去占领，敌人就会占领”。应对和处置谣言，对于政府、社会和每个人来说，实际上也一场十分严肃的斗争。

由于谣言具有隐蔽性、炒作性、攻击性、诱惑性等特点，处置起来难度很大。互联网等新媒体的出现为谣言的快速大范围扩散提供了比过去更加便利的条件，导致其负面效应和社会影响得到前所未有的放大，进而引起人们情绪波动和态度的转变。真真假假的“网络民意”对政府的处理产生了倾向性影响，甚至带有“绑架”性质。一旦环境事件在舆情的发酵下引发社会事件，便会使事件性质发生根本变化，处置起来就不是单纯的环境事件了，就可能影响社会安定和民意基础。

现阶段政府对网络等谣言的处置难度在逐步增大。一方面政府取证并解释清楚需要一定时间，这一“时间差”带来了谣言传播的空间。同时互联网、微博等使谣言传播速度大大加快甚至即时化、现场化、全面化。另一方面在政府公信力降低的情况下，许多分辨能力较弱的群众往往对谣言抱着“宁可信其有，不可信其无”的态度，使正确的信息难以取得应有的效果。美国学者卡斯·R. 桑斯坦认为：“谣言通过社会流瀑和群体极化进行传播，我们倾向于相信别人，尤其是我们对事件一无所知时。”对于谣言，既不能采取“鸵鸟政策”，置之不理，也不能一味采取“堵”的办法。这两种处理方法都会使事件进一步恶化。

处置环境事件谣言，从根本上讲，我们要对谣言涉及的环境等问题进行整治。一是政府要创造良好的社会舆情环境，广泛听取民意，积极解决群众反映的环境问题，构建起邪不压正的应对谣言长效机制。要加强公信力建设，树立权威信息源，经常发布经济社会发展重大信息，形成良性互动，赢得人民群众的信任和期待。二是建立各类常态的舆情信息收集、预警、分析工作机制，运用各种有效手段及时监管网络信息传播渠道，有效压缩谣言生存空间，铲除谣言产生的土壤。三是加强对人民群众相关知识的普及，提高媒介认知力和信息鉴别力，使“谣言止于智者”。平时工作做到家了，一旦有事时，即使有人煽动也难以兴起大的风浪，政府讲话就不怕没人信。四是加强环境法制建设，完善环境立法；对恶意引起环境谣言的，必须依法严惩，以儆后人；加强法制宣传，提高群众对谣言的“免疫力”，自觉地做到“不信谣、不传谣”，同时发动群众，动员群众批判谣言，发现恶意传播谣言者及时举报，使其失去生存的土壤。

从环境事件发生后的应对机制来讲。环境谣言事件发生后，第一，根据事件性质、严重程度，启动相应的响应机制。经验表明，同事故救援一样，谣言发生后的舆情应对也有一个“黄金 4 小时”。如果反应麻木，就会使谣言蔓延，失去解决问题的时机，即使加以澄清，负面影响的后果也已经造成。同时，面对谣言也不要荒乱，要在细致论证、缜密思考后作出正确的决策。第二，“谣言止于公开”。政府相关部门要在第一时间站出来，主动出击，向社会发布真相，增强信息透明度；针对谣言内容加以澄清，作出正确、客观、权威的解释。及时建立信息发布机制，充分利用包括网络在内的各种媒体全面、及时地发布信息，公布事件的真相，正确引导舆论的方向。第三，对谣言涉及的环境事件，要坚持以人为本，做出详细解释，同时提出客观合理的解决方案，使人民群众感到可信、可期待。第四，要保持主要传播渠道的畅通，及时跟踪网络舆情形势的发展变化，及时跟踪并公布环境事件的发展动向，以及政府采取的对策，通过新闻发布会等公开渠道，解疑释惑，更好地了解民众的需求，解决群众的心理危机。第五，善于打应对谣言的“人民战争”。要善于动员各类媒体、新媒体积极释放正面信号，积极引导网民等正确认识谣言、瓦解谣言，自觉采取正确行动，加强对网上“意见领袖”的引导，使正面舆论占领阵地。

二、环境保护与完善基本公共服务体系

生存和发展的基本载体是环境，环境状况与人的生活质量息息相关。基本的环境质量是一种公共产品，是一条底线，是政府应当提供的基本公共服务。

（一）环境保护纳入基本公共服务意义重大

1. 这具有充分的理论依据

界定基本公共服务的概念是开展本研究的基础。按照公共产品理论，公共产品是私人产品的对称。公共产品是指每个人消费这种物品或劳务不会导致别人对该种产品或劳务的减少的产品。公共物品具有3个典型特征①：（1）效用的不可分割性。指公共服务是面向全体公众的，其效用为整个社会成员所共享，不能限定为某个个人或群体单独享用。（2）消费的非竞争性。指某群体对公共服务的使用不影响、不排斥另一群体对同样公共服务的使用，即公共服务的边际生产成本和边际拥挤成本为零。（3）受益的非排他性。即在技术上无法将拒绝为公共服务付款的个人或群体排除在公共服务的收益范围之外。

在界定公共服务的基础上，我们来进一步界定基本公共服务。本研究认为，基本公共服务，指建立在一定社会共识基础上，由政府主导提供的，旨在保障全体公民生存和发展基本需求的公共服务。环境保护的产品是环境质量。基本的环境质量是在环境容量允许的情况下，能满足人类的基本需要的水清、天蓝、地干净的宜居环境。基本的环境质量，符合公共产品非竞争性、非排他性的基本特征，符合基本公共服务的基础性要求。而提供这一产品的环境保护活动或服务，就是环境基本公共服务。

2. 这符合发展的层次性规律要求

公共服务的范围，是随时间和发展阶段的变化而不断改变的。从国际经验看，普遍是根据综合实力和人们基本民生需求来确定适合本国实际的基本公共服务项目。加拿大把教育、医疗卫生和社会服务作为联邦政府基本公共服务均等化的主要项目；印度尼西亚把初等教育和公路设施列为政府财政均等化的内容。另外，即使同一国家不同历史阶段的基本公共服务范围也是不一样的。如美国在

① 孙学玉等．公共行政学．社会科学文献出版社，2007.221.

“进步时代”以前，食品、药品监管并没有纳入公共服务的范围，“进步时代”以后，这方面内容成为公共服务的重要组成部分。我国基本公共服务的范围，经历了从科、教、文、卫逐步递进到住房保障、环境保护等领域的历程。“十二五”规划，明确提出“十二五”时期基本公共服务的范围为：公共教育、就业服务、社会保障、医疗卫生、人口计生、住房保障、公共文化、基础设施、环境保护 9 个方面。

3. 这是强化环保工作的紧迫需要

目前，我国严峻的环境形势，与基本公共服务均等化的要求不相适应，与人民群众的需求不相适应。从供给情况看，环保领域投入长期不足，历史欠账多。农村仍有 8000 多万人饮水不安全，30 %的县城没有建设污水处理设施等。“十一五”时期，我国城市污水处理率从 2005 年的 52%提高到 2010 年的 77%。“十二五”规划要求，到 2015 年城市污水处理率达到 85%。大城市灰霾天数接近全年的 30%—50%。

从城乡情况看，农村环保公共服务严重滞后，可及性差。如全国城市污水处理率 80%，而同期县城污水处理率 45%；县级政府所在城镇和设市城市的饮用水水源地达标比例为 80%，而农村饮用水水源地水质达标比例仅为 59%。从区域情况看，不同地区环保公共服务差距比经济差距还要大。东部经济发达地区环境基本公共服务水平相对较高，如山东、江苏、浙江、北京等城镇生活污水处理率在 85%以上，而中、西欠发达地区水平较低，如广西、贵州、湖南、海南、青海 5 省（区）不足 45%。

总之，环境保护纳入基本公共服务意义重大。有利于完善公共服务体系、改善民生；有利于缩小城乡区域差距、推进均衡发展；有利于落实环保责任、强化国家环境意志。

（二）环保基本公共服务不足的成因分析

1. 对环保的民生属性认识不够

总的来看，环境问题已成为全面建成小康社会的重大制约因素。从观念上，更多地从发展角度理解环保，相比于教育、卫生、住房，对环保民生属性认识不够。人们生存和发展的基本载体就是环境，环境状况与人的健康状况息息相关。让百姓喝上干净的水、呼吸清洁的空气、吃上放心的食物是民生的基本需求。目前，我国影响和损害群众健康的环境问题还不少，一些地区污染排放严重超过环境容量，突发环境事件高发。2012 年，全国七大水系水质总体为轻度污染，1—

3 类水质断面比例占 63.9%，劣 V 类水质占 12.4%。受工业污染、固体废弃物、过度和不合理使用化肥等影响，土壤环境也呈现总体质量下降。近年来，不少地方出现重金属污染事件，每年因重金属污染的粮食高达 1200 万吨。我国人均国民收入已超过 6000 美元，进入中上等收入国家行列，人民群众对提高生活水平和质量有了更多期盼和要求，优良的环境越来越成为城乡居民的普遍追求。必须转变观念，认真回应人民群众的迫切愿望，切实抓好环境保护。

2. 缺乏对环保公共服务的范围界定

目前缺乏概念界定，导致职责不清，范围也有必要进一步拓展。例如，国家“十二五”规划纲要，只是以专栏的方式提出，环境公共服务主要包括污水处理、垃圾处理、环境监测评估、饮用水源地保护等 4 个方面。2012 年，中国首次制定了国家基本公共服务体系“十二五”规划，这个规划再次重复了国家“十二五”规划的表述，并没有进一步阐述。

国家“十二五”规划纲要：“十二五”时期我国环境保护公共服务的范围和重点一是具备污水、垃圾无害化处理能力和环境监测评估能力；二是保障城乡饮用水源地安全。

国家基本公共服务体系“十二五”规划：本规划的范围确定为公共教育、劳动就业服务、社会保障、基本社会服务、医疗卫生、人口计生、住房保障、公共文化等领域的基本公共服务。“十二五”规划纲要还明确了基础设施、环境保护两个领域的基本公共服务重点任务，这些内容分别纳入综合交通运输、能源、邮政、环境保护等相关“十二五”专项规划中，不在本规划中予以阐述。

国家环境保护“十二五”规划又增加了“农村环保”的内容。但这三个规划，只是笼统的规定，没有界定边界。概念界定的模糊性，直接带来政府责任的模糊性，导致政府在环保公共服务提供上的缺位。有的基层干部认为，环保是企业、个人的事，政府有余钱才干环保，没钱就可以缓一缓。

3. 财税体制增大环境公共服务供给难度

我国长期实行偏重与发展的财税体制，对“民生财政”、“绿色财政”重视不够，对教育、卫生、环保等领域投入不足。

从现实情况看，我们国家和发达国家环保投入的差距甚大。发达国家环保投入占 GDP 的比重，一般在 2%到 3%左右，有的超过 3%。比如，20 世纪 70 年代美国大约是 2%，日本在 20 世纪 80 年代末期超过 3%，是 3.4%，德国 2.1%，英国2.4%[1]。我们国家这几年投入力度有所加大，但投入比重还比较低。目前，我国环保方面投入占GDP比重不到2%，大概是1.6%—1.8%左右。我国环保投入比重一下提高到发达国家的程度在操作上有难度，但要朝着这个方向去努力。未来10年内，环保投入占GDP比重要逐步提高到2%，甚至达到3%。如果按2%测算，未来10年GDP是7%的增长率，大概需要投资10万亿。

近些年来，环境公共服务投入的增长速度落后于财政支出增长速度和整体经济增长速度。同时，“财权在上、事权在下”的体制，造成地方环境公共服务供给动力不足。1994年分税制改革以后，增值税实行中央75%、地方25%分成；所得税实行中央60%、地方40%分成，存在财权重心上移的情况。上级政府将数额大、易征收的税收上收或共享，将较分散、数额小、难征收的税种留给地方，基本上地方政府大多缺乏稳定、大额的主体税种。与之相反，政府层级之间的事权确出现了下移的趋势，一些上级政府的法定事权也通过责任考核和一票否决等分解给了下级政府，基础教育、医疗卫生这类公共服务的责任过多地由地方政府承担。据专家综合测算，地方政府用不到全国50%的财权，却要干全国80%的事权[2]。

4. 环保公共服务责任落实不到位

缺乏明确的环保监测、评估和考核体系。环保领域“守法成本高、违反成本低”的问题十分突出。根据2008年修订的《水污染防治法》，对污染河流等一般水体的，最高罚款50万元，污染饮用水水源地的最高罚款才100万元。根据2000年修订的《大气污染防治法》，对污染大气的各类违法行为，一般罚款1至20万元，最高罚款50万元。对地方政府领导，偏重于GDP考核，环保方面的指标未纳入考核体系。受地方保护主义的影响，一些地方环保机构不同程度存在不敢碰硬、不敢执法、不愿执法的现象。不少基层环境执法队伍对开发区不敢查，对重点保护企业不敢查，领导不点头的不敢查，有的甚至还替企业说情。

（三）完善我国环保基本公共服务的建议

完善环保基本公共服务，要借鉴国外经验，结合本国实情，按照保基本、强基层、增投入、建机制的原则，逐步建立环境基本公共服务体系，并推进均等化。具体包括：

---

① 引自《人民日报海外版》，2006年第5版.

② 张萧然. 源头根治土地财政依赖症. 国产经新闻，2011.

1. 合理确定环境基本公共服务的范围和标准

借鉴国际经验，考虑需求与可能，合理确定我国环境基本公共服务的范围和水平。一是环境基础性服务，包括配备污水处理、垃圾处置等设施；二是基本民生性服务，包括保障公众清洁水权、清洁空气权及宁静权等；三是环境安全性服务，包括环境事故应急等；四是环境信息服务，包括保障公众环境知情权和环境监督权。需要指出的是，环境基本公共服务不仅仅是指物化的产品或服务，还包括制度安排、法律、宏观经济政策等。

2. 增强基层环境公共服务硬件和软件建设

县、乡、村，是我国环境公共服务体系的薄弱环节。在继续做好省、市两级环境基本公共服务的同时，把环保工作重心下移。一方面，加强基层环境污染设施、设备和环保人员配备。强化村庄和饮用水水源地环境综合整治，统筹建设城乡生活污水和垃圾处理设施，在有条件的地区推行城乡供水一体化。另一方面，全面推进监测、监察、宣教、信息等软件建设，提高农村污水处理率、垃圾处置率、饮用水达标率。

3. 建立有利于环境公共服务体系均等化的公共财政体系

财政是政府提供公共服务的重要手段，清晰的事权划分是财政间支出体制划分的依据，也是实现财政分配均等化目标的前提。建议将环保与教育、医疗、社会保障等一道纳入公共财政预算，逐步提高环境基本公共服务支出所占比重，规定财政性环保支出要达到 GDP 的比例。可借鉴财政教育投入的经验，规定每年必须达到的投入目标。要确保环保投入合理增长，保证环境保护预算支出增长幅度高于财政经常性支出增长幅度。应在合理划分中央和地方财政支出事权的基础上，完善财政转移支付制度，增强县以下基层政府环境公共服务能力。安排专项资金，对达不到环境基本公共服务水平的地区予以重点支持。

4. 完善环境公共服务提供方式

环境基本公共服务由政府负责，不等于由政府包揽。应加强制度创新，扩大公众参与，最大限度调动社会各方面的积极性。通过实行政府购买、管理合同外包、特许经营、优惠政策等方式，逐步建立政府主导、市场引导、社会充分参与的多元供给机制。比如，环保的监测、评估、宣教、污水处理，都可以调动社会力量来做。放宽准入限制，鼓励和引导社会资本参与环境基本公共服务供给。

5. 构建政府环境基本公共服务绩效评价考核体系

有监管、有考核，各项制度才能有效落实。应研究制定科学的评价体系，加强各级政府环境基本公共服务监测评价，把环境基本公共服务数量和质量指标纳入政府绩效考核体系。建议进一步强化政府在环保方面的责任，除考核 GDP 外，增加城市空气质量优良率、垃圾处理率、污水处理率、森林覆盖率等指标，并建立追责机制。同时，积极引入外部评估机制，建立多元化的绩效评估体系。例如，$PM_{2.5}$等空气质量的监测，不能政府一家说了算，引入独立监测机构。

三、创新社会管理促进环境保护研究

环境问题，既是一个自然问题，也是一个经济问题，还是一个民生问题和社会问题。加强环境保护，不仅关系经济发展全局，也关系民生健康福祉，关系社会和谐稳定。

本研究报告主要把环境问题作为一个社会问题来加以研究，立足于通过创新社会管理来加强环境保护，深入分析我国在环境保护的社会管理方面存在的突出问题，提出创新社会管理加强环境保护的总体思路、主要做法和政策措施，包括积极支持和引导社会公众参与环境保护，充分发挥社会组织在环境保护中的重要作用，创新环境保护的社会管理体制机制，制定有利于环境保护的社会政策等，试图在有关社会管理与环境保护的理论研究上取得新的进展，对中国政府创新社会管理促进环境保护提出决策建议，起到咨询和参谋的作用。

（一）创新社会管理对于加强环境保护具有重要意义

随着中国经济快速发展，中国社会发生了巨大变化。总体上中国正处在快速的工业化、城镇化过程中，处在社会大发展、大变动、大转型时期，整个社会快速地从过去传统的农业社会、农村社会向现代工业社会、城市社会转变。一个突出的标志就是，出现从农村到城市的大规模的社会流动，个体私营等非公有制经济和民间组织发展起来，产生了越来越多的自谋职业者和自由职业者，特别是形成了一个世界上非常罕见的庞大的农民工群体。2012 年，中国农民工总量达到 2. 6 亿人以上，其中流动到外地就业的农民工 1. 6 亿多人。中国城镇化率达到 52. 6%，城镇人口超过农村人口，实现了中国社会结构的一个历史性变化。

中国在社会发展、社会管理和环境保护方面取得了历史性的成就。

——社会发展成就辉煌。主要表现在几个方面：一是社会事业快速发展。这些年，更加重视民生和发展教育、卫生、文化、体育等社会事业。教育方面，在城乡普遍实行免费义务教育，高等教育进入大众化阶段，大学毛入学率达到 25%。职业教育规模不断扩大，培养了一大批高技能实用人才。卫生方面，大规模推进医药卫生体制改革，职工医保、城镇居民医保、新农合参保人数超过13亿人，覆盖率达到95%以上，建立起了覆盖城乡的全民医保网。文化方面，致力于保护文化遗产，加快发展文化事业和文化产业。体育方面，在伦敦奥会取得了突出成绩，全民健身运动蓬勃发展。二是人民生活水平迅速提高。2012年，全国人均GDP达到6000美元以上，进入中等收入国家行列。我国一些地方如北京、上海、天津人均GDP已超过1万美元，达到1.3万美元以上，开始进入高收入国家水平。10年来，城镇居民收入和农民收入年均增长分别达到9.2%和8.4%。随着收入的增加，消费上升到新的阶段。明显的标志是，汽车、住房、旅游三大消费热点快速发展。现在，城市家庭每百户拥有汽车21.5辆。国内旅游人数达到29.6亿人次，出境旅游人数达到8318万人次。中国的网民人数达到5.64亿人，互联网普及率达到42.1%，中国成为世界上手机拥有量最多的国家。三是公共服务水平不断提升。社会就业规模不断扩大，每年城镇新增就业都在1000万人以上。城镇登记失业率在4%左右，登记失业人数在1000万人以下。在建立全民医保的同时，建立了以社会保险为主、包括社会救助、社会福利、慈善事业在内的社会保障体系，初步形成了覆盖全民的社会保障安全网。

——社会管理得到加强。经过30多年的改革开放，我国经济体制改革取得了巨大进展，已经建立起比较完善的社会主义市场经济体制。适应社会主义市场经济体制的建立和完善，使我国社会管理体制改革也取得了重要成就。主要表现在：一是随着农村改革和城市改革的进展，逐步打破城乡分离的“二元社会结构”，朝着建立城乡统一的社会管理体制迈出了重要步伐。二是随着自由择业和社会流动的发展，产生了越来越多的自谋职业者和自由职业者，许多人已经从“单位人”转变成为“社会人”，对于社会流动的管理积累了新的经验。三是随着工业化和城镇化的发展，许多农村人逐渐转变成为城市人，同时也形成了庞大的农民工群体，对农民工的公共服务不断取得进展。四是随着人们工作性质和社会生活的发展变化，社会保障体系逐步建立和不断完善，由以前的单位保障和家庭保障逐步向社会保障转变，人们工作生活的社会化程度不断提高。五是随着个体私营等非公有制经济发展，一些民营经济和民间组织逐步发展起来，与此同时，一些公益性社会组织包括环保组织从无到有，逐步发挥其应有的作用。

——环境保护不断进步。这些年来，中国高度重视生态建设和环境保护，大力推进节能减排，积极应对气候变化，参与国际合作交流。应该说，在环境保护方面取得了重要进展。在经济结构调整中，淘汰了一大批落后产能，包括钢铁、有色、水泥、纺织、造纸等行业，技术改造提高到了一个新水平。在生态建设方面，实施天然林保护、植树造林、退耕还林、退牧还草、防沙治沙等工程，治理水土流失。在污染防治方面，治理江河湖海水污染、城市空气污染、农村面源污染和重金属污染等。最近5年来，中国单位GDP能耗下降17.2%，化学需氧量、二氧化硫排放总量分别下降15.7%和17.5%。全国森林覆盖率提高到20%以上。

中国在社会发展、社会管理和环境保护方面取得积极成就的同时，也面临着突出的问题。

一是社会发展相对落后。我国发展面临的一个突出问题，就是经济社会发展不协调，经济与社会发展"一条腿长，一条腿短"的问题突出。相对于经济快速发展来说，社会发展滞后，就业、社会保障、教育、医疗、文化等方面的发展满足不了人们不断增长的需要，尤其是在广大农村地区和中西部地区更加突出。作为拥有13.5亿人口的发展中大国，吃饭和就业问题始终是我们面临的头等大事。按照联合国的标准，中国还有1.28亿人生活在贫困线以下。每年城镇需要就业的劳动力在2500万人左右，其中高校毕业生有700万人，农村还有富余劳动力1亿多人，社会就业压力巨大。中国社会保障体系还不健全，保障程度低，标准不统一，覆盖范围有限，随着老龄化的快速发展，养老问题也成为一个突出的社会问题。城乡差距、地区差距、收入分配差距还呈扩大之势，一些低收入群众生活还比较困难。中国处在快速的社会转型期，社会利益格局调整，进入社会矛盾凸显期。在快速工业化、城镇化过程中，出现了征地拆迁、住房紧张、食品安全、环境污染等人民群众反映比较多的问题。

二是社会管理比较滞后。社会管理还赶不上经济社会快速发展的需要，面临着突出的矛盾和问题。总体上看，社会管理体制还相对滞后，与社会主义市场经济体制不相适应，还没有建立起适应现代社会流动、充分激发社会自身发展动力和活力、具有中国特色的现代化的社会管理体制。中国社会还处在大变动、大转型的过程之中，还没有形成比较稳定的成熟的社会结构。工业化、城镇化快速发展造成大量的社会流动，突出表现为中国存在世界上非常罕见的庞大的农民工群体，对于一个处在高度流动中的社会管理还缺乏经验。由传统的农业社会和农村社会急剧转变为工业社会和城市社会，还没有建立起一套社会普遍遵守的社会规则体系，表现为比较突出的社会失范和无序状态。在社会管理方面，政府包揽的社会事务太多、干预太多，管了许多不该管、管不了、也管不好的事情，社会组织发育不足，社会自身发展还缺乏动力和活力。

三是环境保护问题突出。总体上看，环境恶化的趋势还没有得到根本改变。中国作为一个有13.5亿人口的大国，人均资源占有量低，环境承载能力弱，这是中国的基本国情。随着经济社会快速发展，人口、资源、环境的矛盾越来越突出，资源环境问题成为制约中国发展的最大瓶颈。我国现在是世界上煤炭、钢材和水泥的最大生产国，能源的第二大消费国。以煤炭为例，中国的能源结构中以煤为主，占到能源消费的70%。现在煤炭的消费量已经达到36.5亿吨，相当于其他国家煤炭消费的总和。我国也已经成为世界上二氧化碳排放量最多的国家。能源资源消耗多，环境污染严重，尤其是空气污染、水污染的问题突出。今年以来，北京等中国北方地区多次出现大面积的严重雾霾天气。我们在发展中越来越深刻地认识到，节约能源资源、加强环境保护，已经成为中国面临的一个重大战略性任务。必须从根本上转变经济发展方式，下大力气加强生态建设和环境保护，建设生态文明，建立一个绿色的社会、环保的社会、节约的社会。

环境保护是一项综合性的重大系统工程，不仅取决于一个国家的经济发展程度，也取决于一个国家的社会发展程度，需要一个国家的政府、市场、社会三个方面共同努力。在一个经济社会高度发达的国家，环境保护做得好，必然是这三个方面共同努力的结果。相反，一个处在经济社会发展中的国家，环境保护搞得不好，肯定是这三个方面其中之一出了问题。

如果从经济、社会发展的角度，从政府、市场、社会三个方面来分析，中国在发展过程中的环境保护走过了曲折的道路，既有失败的教训，也有成功的经验。从刚开始致力于经济发展，不重视环境保护，环境方面出现了问题，到在经济发展的同时重视环境保护，但也只是重视政府的管理作用；随着市场经济的发展，越来越认识到，除了发挥政府的主导和调控作用之外，还要发挥市场配置资源的基础性作用，政府可以制定相应的经济政策来促进环境保护；再后来，随着经济发展起来以后，更加重视社会发展，注重保障和改善民生，注重解决人民群众关心的环境问题，但在环境保护中社会的作用发挥得还不够，还没有很好地通过制定社会政策来促进环境保护。现在，中国的发展已经到了一个新的阶段，就是通过环境保护优化经济社会发展、建设生态文明的阶段。我们在实践中越来越认识到，需要从经济社会发展的综合性的角度，来统筹考虑进一步加强和促进环境保护，这就是充分发挥政府、市场、社会三方面的作用，在不断加强政府、市场作用的同时，更加重视发挥社会的作用，将环境保护建立在更加广泛、更加坚实的基础之上，这是中国加强环境保护的必然选择。

从根本上说，创新社会管理，充分发挥全社会的作用，对于加强环境保护，建设生态文明，具有重大的战略意义。

第一，加强环境保护，建设生态文明，必须充分发挥社会公众参与的积极性、主动性和创造性。环境保护是亿万人民的共同事业，涉及到社会生活中的每一个成员、每一个公民、每一个家庭。它不只是政府和企业的事，不只是通过市场规则和利益机制就可以完全解决的问题。环境保护在很大程度上涉及社会价值观念、生活习惯和行为方式，这些往往比经济利益更难以改变。环境保护，需要每一个社会成员树立环保的理念和生态文明的价值观，需要改变每个人不文明、不环保的生活习惯和行为方式，并且把环境保护变成每个人的自觉行动。这就需要长期的引导、宣传、教育，需要辅之以必要的奖励和惩罚措施，以激励正确行为，限制错误行为，最终内化到人们的生活习惯和行为方式之中。这对于一个国家、一个社会的环境保护和生态文明建设，具有极为重要的意义。

第二，加强环境保护，建设生态文明，必须充分发挥社会组织特别是各类环保组织的重要作用。随着经济社会发展和生活水平的提高，人们的环保意识不断增强，许多有志于从事环保工作的志愿者积极行动起来，加入到环保队伍中来，他们建立起各种类型的环保组织，对环境保护起到了重要的促进作用。参加环保组织的人员，都是社会中的环保积极分子，一般具有强烈的环保意识和环保责任感，同时又具有一定的环保知识和技能，对于引领一个社会的环境保护，带动全社会的环保事业，能够发挥不可替代的影响力。对于政府来说，加强环境保护，必须充分发挥环保类社会组织的重要作用，为他们创造必要的环境条件，以形成"人人关心环保、参与环保"的有效保证。

第三，加强环境保护，建设生态文明，必须最大限度地发挥城乡社区的基础性作用。环境保护要得到有效落实，必须深深地扎根于基层。一个国家、一个社会的大环境是建立在千千万万的一个一个小环境的基础之上的。没有一个一个社区的好的小环境，不可能有整个社会良好的大环境。因此，加强环境保护，必须从最基层做起，发挥好城乡社区在保护环境中的重要作用，尤其是每一个家庭都应该投身到环境保护中来，建立一个一个良好的社区环境。这是整个社会环境保护成功的可靠保证和最有力支撑。

第四，加强环境保护，建设生态文明，必须形成完善的社会体制机制，制定切实有效的社会政策。一个社会要正常运转，规则非常重要。一个社会要搞好环境保护，非有具体可行的规则不可。这就需要制定有利于促进环境保护的具体的社会规则，并且监督这些规则的落实。比如，不能随地吐痰，不能随手乱扔垃圾，要通过规则的制定和落实变成人们的自觉行动，违犯规则就要受到处罚，受到社会舆论的谴责，形成一种公众监督制约的保护环境的社会氛围。在社会政策方面，比如可以鼓励人们少开汽车，减少一个人开一辆车，国外就有这方面的规定，多人车可以走快车道，一人车只能走边上的慢车道。总之，制定相应的社会政策，建立完善的社会体制机制，对加强和促进环境保护，建设生态文明，具有至关重要的意义。

我们应该充分认识创新社会管理，发挥全社会在环境保护中的重要作用，对于加强和促进环境保护的重大意义，更加重视从社会方面来加强环境保护，充分调动和发挥全社会参与环境保护的积极性、主动性和创造性，共同把中国的环境保护推向前进。

（二）创新社会管理加强环境保护的总体思路

中国经过多年的探索发展，在环境保护方面已经形成了一整套的理念、发展战略和政策措施。概括起来，就是“一个指导思想，一个发展战略，一个基本国策，一个生态文明，一个‘两型社会’”。

一个指导思想，就是树立和落实科学发展观。统筹人与自然和谐发展，处理好经济、社会发展与人口、资源、环境的关系。坚持以科学发展为主题，以加快转变经济发展方式为主线，加快调整经济结构，加强能源资源节约和生态环境保护。再也不能走过去那种粗放型增长的老路，不能以大量消耗资源、污染环境为代价来换取经济增长，必须走一条节能环保的发展道路。

一个发展战略，就是实施可持续发展战略。把可持续发展上升到国家战略的高度，使得中国经济社会发展的良好势头能够长期保持下去，不仅能够福泽当代，而且能够惠及子孙，给子孙后代留下天蓝、地绿、水净的美好家园，实现中华民族永续发展。

一个基本国策，就是实行节约资源和保护环境的基本国策。把节约资源和保护环境上升到基本国策的高度，就是要作为一项国家长期稳定的重大政策，始终坚持、毫不动摇地贯彻落实下去，建设美丽中国。

一个生态文明，就是建设社会主义生态文明。把生态文明建设放在更加突出重要的位置，融入经济建设、政治建设、文化建设、社会建设各方面和全过程，做到环境保护与经济社会发展同步并重，促进环境保护与经济社会协调发展。

一个“两型社会”，就是建设资源节约型和环境友好型社会。形成节约资源和保护环境的产业结构、增长方式、消费模式，发展循环经济，促进生产、流通、消费过程的减量化、再利用、资源化，把节约资源和保护环境落实到全社会每个企业、每个单位、每个家庭。

中国共产党第十八次全国代表大会报告对创新社会管理加强环境保护也提出

了明确要求，指出必须加快推进社会体制改革，围绕构建中国特色社会主义社会管理体系，加快形成党委领导、政府负责、社会协同、公众参与、法治保障的社会管理体制，加快形成政社分开、权责明确、依法自治的现代社会组织体制，加快形成源头治理、动态管理、应急处置相结合的社会管理机制。加强和创新社会管理，提高社会管理科学化水平，必须加强社会管理法律、体制机制、能力、人才队伍和信息化建设。加强基层社会管理和服务体系建设，增强城乡社区服务功能，强化企事业单位、人民团体在社会管理和服务中的职责，引导社会组织健康有序发展，充分发挥群众参与社会管理的基础作用。建设生态文明，必须树立尊重自然、顺应自然、保护自然的生态文明理念，全面促进资源节约，推动资源利用方式根本转变。坚持预防为主、综合治理，以解决损害群众健康突出环境问题为重点，强化水、大气、土壤等污染防治。加强生态文明宣传教育，增强全民节约意识、环保意识、生态意识，形成合理消费的社会风尚，营造爱护生态环境的良好风气。

2013 年初召开的第十二届全国人大一次会议，审议通过了《国务院机构改革和职能转变方案》，提出转变政府职能，必须处理好政府与市场、政府与社会的关系，深化行政审批制度改革，减少微观事务管理，该取消的取消、该下放的下放、该整合的整合，以充分发挥市场在资源配置中的基础性作用、更好发挥社会力量在管理社会事务中的作用。改革社会组织管理制度，加快形成政社分开、权责明确、依法自治的现代社会组织体制。重点培育、优先发展行业协会商会类、科技类、公益慈善类、城乡社区服务类社会组织。成立这些社会组织，直接向民政部门依法申请登记，不再需要业务主管单位审查同意。坚持积极引导发展、严格依法管理的原则，促进社会组织健康有序发展。完善相关法律法规，建立健全统一登记、各司其职、协调配合、分级负责、依法监管的社会组织管理体制，健全社会组织管理制度，推动社会组织完善内部治理结构。

根据这些新的规定和要求，创新社会管理促进环境保护的总体思路应该是：处理好政府、市场、社会三者之间的关系，在充分发挥政府的主导作用和市场配置资源的利益导向作用的同时，更加重视发挥社会力量在环境保护中的重要作用。加强和创新社会管理，构建中国特色的环境保护的社会管理体系，充分发挥社会公众参与环境保护的积极性、主动性和创造性，充分发挥社会组织特别是各类环保组织的重要作用，充分发挥城乡社区在环境保护中的基础性作用，形成全社会共同参与环境保护的比较完善的社会体制机制和社会政策体系。

（三）引导支持社会公众积极参与环境保护

保护环境既是政府的责任，也是社会公众的责任。一个社会的环境保护如何，首先与其公民的环保意识和环保行动密切相关。公民参与是现代社会的基本特征，也是环境保护能够做好的根本保证。现代社会的环境保护实践证明，政府、企业与公众共同参与环境保护，构成了“三位一体”的环境保护可持续发展模式。对于环境保护来说，这三者缺一不可，而社会公众在环境保护中发挥着最广泛、最基础的作用。

社会公众参与环境保护的程度，直接体现着一个国家环境意识、生态文明的发育程度，也体现着一个国家重视和保护公民权利的程度。引导支持社会公众积极参与环境保护，是贯彻落实环境保护这一基本国策，增强公众环境保护意识，形成覆盖城乡的环境保护网络的重要举措。社会公众参与的原则，就是依靠群众的原则，依靠群众一直是我国环境保护的基本方针。实践证明，仅仅依靠环保部门推动，环境很难取得彻底改善，广泛推动公众参与才是解决环境问题的根本途径。

1. 国外社会公众参与环境保护的做法和经验

国外发达国家在经济社会发展的基础上，环境保护做得比较好，一个重要的因素就是公众参与环境保护，每个人都自觉地保护生态环境，形成了规范化、制度化的环境保护参与模式。

——建立公众参与环境保护的体制机制。

法国非常重视环境保护，并且建立起环境保护方面的公众参与的体制机制。2008 年 3 月，法国实行环境保护方面的大部门制，将原来的环境、可持续发展与领土整治部更名为环境、能源、可持续发展与领土整治部，其职能范围扩展到资源、领土与居民、能源与气候、可持续发展、风险预防与交通等更广泛的领域。法国的公众参与环境保护的主体主要包括环境保护协会、专家机构和专业性办公室（局）等。法国公众参与环境保护的实施程序包括：（1）公众调查程序。环境保护领域的公众调查的目的是告知公众，收集他们的喜好、建议以及反对意见，然后将其作为影响评价的公众意见。（2）公众辩论程序。法国依法成立有全国公众辩论与听政委员会，该委员会由 20 余人组成，其中 1/3 是中央和地方的政界人士，1/3 是不同行业的资深专家，1/3 是各类社会团体和非政府组织成员代表。围绕环境保护方面的议题进行公开的辩论。（3）地方公民投票。经过公众调查和公众辩论程序以后，接下来就是进行地方公民投票。1/5 注册的选民代表能够向市政委员会对相关决策进行投票，投票结果直接影响到决策。总体上看，法国的公众参与环境保护主要包括几种形式：知情、咨询、商讨、最后是共同决策，形成了一套完整的制度保障公众参与的决策机制。

——开展广泛的环保宣传教育，普及环保知识和技能。

巴西教育部和矿业能源部从 1992 年起联合在中小学实施“学校节能教育计划”。1999 年，巴西总统颁布了“全国环保教育政策”。2002 年颁布的“全国环保教育政策”细则规定，环保教育应成为学生整个教育过程中的一个重要和持续进行的组成部分，全国各级和各类公立、私立学校都要将环保教育列入教学大纲。从 2003 年起，教育部每年培训 3.2 万名从事环保教育的教师，在 1.6 万所中小学成立由教师和学生代表组成的环保委员会。教育部和环保部提供环保教育教材和录像片，并在全国交流推广中小学开展环保教育的经验。

日本学校教育重视培养学生的节约观念。当一批孩子毕业时，学校还会呼吁学生及其家长将能够继续穿用的校服捐赠给学校，供新生使用。一些学校还邀请垃圾处理公司的员工开着垃圾车为学生实际演示清理垃圾的操作，并指导学生进行垃圾分类比赛。

欧洲一些国家先后推出了宣传绿色消费理念的学校教育计划。如法国中小学通过让学生清扫校园、以甜菜汁为原料制作有机燃料、管理学校垃圾等，提高学生的环保意识；奥地利许多中小学开设了环保课，定期聘请环保专家讲授环保和垃圾回收知识，如介绍何种购物方式能够有利于垃圾减量以及如何正确进行垃圾分类等。

——重视保障公众参与环境保护的知情权和监督权。

德国 2005 年出台的《环境信息法》要求，联邦管理部门和相关社会机构有义务公布环境信息。该法规定，对公众有关环境信息的咨询必须在一个月内作出答复；有关机构应免费开放环境信息资料，供公众查阅；联邦管理部门应充分利用互联网，广泛发布环境信息。环境部的负责人表示，只有人们了解环境信息，才能参与公共决策并实行有效监督。

韩国邀请民间人士参与对大江大河的环境检查，在制定和修改环境政策时同相关企业、专家和环境团体协商。为了消除居民对水质的不信任，成立了由民间团体和有关专家共同参与的“自来水市民会议”，对水质进行实时监控。韩国“绿色环保联合会”有几万名会员，形成一个强大的监督网，对有损环保的事例在新闻媒体上曝光。各地方政府还由地区居民和民间团体为主导组成了 231 个“民间环境监视团”。一些污染环境的大型工程，因为当地民众和团体的反对而被叫停。

——社会公众参与环境保护成为一种自觉行动。

日本虽然是一个地域狭小、自然灾害频发和多山的国家，但生态环保却做得非常好。穿行在日本的城市之间，高速公路许多时候是在山洞和桥梁之间通过，所见山峦到处都是森林覆盖，一片郁郁葱葱。日本的森林覆盖率达到64%，是世界上森林覆盖率最高的国家之一。日本非常重视绿化，即使在繁华的东京，高大的树木、整齐的草坪、大片大片的绿色也尽收眼底，似乎有土地的地方就被绿色的植物所覆盖。访问日本真正地感受到什么是整洁干净。无论是城市还是乡村，给人的第一感觉就是干净。据介绍，日本是世界上垃圾分类管理最严格的国家，家家户户自觉对垃圾进行分类，按时定点收集，甚至街道上的垃圾筒也不多，人们出门都自觉地带着塑料袋，把准备扔的垃圾保存起来，放在有收垃圾的地方。在任何地方，都看不到乱扔垃圾和随地吐痰的问题，也没有如美国纽约地铁乱写乱画的现象。

日本更是十分强调节约的国家，全民具有强烈的节约意识。日本的饭菜都是分量较少，刚好够一个人吃即可，即使是大家一起聚餐，所点食物也是够吃就好，不会出现吃不完浪费的现象。日本人认为，他们的资源和食物有限，虽然现在已经非常富裕，但视浪费为犯罪，良心上感到不安。交通节能也是日本节能的一大领域，政府大力发展公共交通，国民自觉不开汽车，现在骑自行车在日本又流行起来，既节能，又减少污染，还可以锻炼身体，受到人们的普遍欢迎。日本应该说是世界上节能减排做得最好的国家，其能源使用效率相当于中国的15倍之多。

2. 中国公众参与环境保护取得的进展和存在的问题

近年来，随着我国公众环境意识的不断增强，公众参与环境保护的积极性日益提高。2008年4月，中国社会科学院社会学研究所和中国环境意识项目组联合公布了《全国公众环境意识调查报告》。报告列举了包括环境污染在内的13项社会问题，结果显示，10.2%的被访者将环境污染列为当前我国面临的首要社会问题，有9.1%的被访者将环境污染问题列为第二重要的问题，有13.2%的被访者将其列为第三重要的问题。调查组经加权计算，环境污染问题在13项社会问题中列第四位，公众对其关注度仅次于医疗、就业、收入差距问题之后，而居于腐败、养老保障、住房价格、教育收费、社会治安等问题之前。《报告》显示，随着中国环保事业进一步发展，中国公众对于环境保护的重要性、必要性、责任感和紧迫感均有显著提升。

社会公众关注环境问题，特别是对一些突出环境问题提出了不少批评意见，引起了政府决策部门的重视，有些问题得到了及时解决，有些意见成为政策决策的重要参考依据。应该说，社会公众的积极参与对加强环境保护，起到了重要的促进作用。

与此同时，社会公众参与环境保护还面临不少问题，总体上公众环保意识和行动还处在较低的水平。中国环境文化促进会组织编制，被誉为中国公众环保意识与行为“晴雨表”的国内首个环保指数《中国公众环保民生指数》中列出了两组数据：一方面，86%的公众认为环境污染对现代人的健康造成了很大影响；另一方面，公众的环保意识总体得分为 57.05 分，环保行为得分为 55.17 分，没过及格线。数据显示，2006 年，环境问题与医疗、教育问题成为公众关注的三大热点问题，关注人数超过了 40%以上；63%的公众认为现阶段我国环境问题非常严重和比较严重，只有 8%的人认为不太严重和不严重；只有 24%的公众对我国总体环境表示满意和比较满意，32%的被访者对总体环境质量表示不太满意或不满意。公众参与环保活动最直接的方式是节电、节水和节约用煤等，主动参与公益环保活动的很少；有 80%的公众对环保只停留在关注层面，只有 6.3%的公众在最近 3 个月参加过环保活动；知道“12369”环境问题免费举报电话的只有 16%。由此看出，我国公众环保关注度很高，但认知度较低，参与能力不强，评价能力较弱。

由长沙环境保护职业技术学院组织开展的“环境保护公众参与调查”，列出了一些具体的问题：（1）你的生活垃圾会分类后再处理吗？（2）你会随处丢垃圾吗？（3）你曾经参加过某个环境保护组织吗？（4）看到别人乱丢垃圾你会主动制止吗？（5）你会主动监督身边发生的污染行为吗？（6）你对长沙环境保护公众参与的现状满意吗？调查结果显示，长沙市民具有较强的环保意识，同时也暴露出公众参与不足的实际情况。如对随地乱扔垃圾和公众参与不足表示关心的比例较高，而在垃圾分类处理、参加环保组织、制止乱扔垃圾以及污染监督方面所做的工作不够。调查分析结果表明，当前市民环保公众参与存在许多不足之处，既有社会公众层面的问题，也有政府作用层面的问题。从社会公众来看，主要问题有：一是环境意识高与主动性不强的矛盾。市民对环境污染的认识程度比较高，绝大多数市民认为当前环境污染对社会、经济、生态构成了巨大威胁，肯定了公众参与在环境保护中的重要性，但在付诸环境保护的实际行动中却做得不够。面对环境保护和污染治理，部分市民认为是政府的事，持观望态度，社会责任感不够强。二是公众参与层次低。当前长沙市环境保护公众参与的行为多为阶段性，甚至突击为主，譬如利用“六五”世界环境日做一些环境宣传活动，缺乏连续持久的深层次公众参与行动。长沙市环境保护公众参与未深入到环境决策层面，公众参与具有一定的被动性和边缘化特点。三是公众参与的组织性不强。现实中大多数市民环境保护的参与行动普遍是自发的与分散的，主要表现为盲目性大，随意性强，人数有限，参与的频度不够，参与不彻底。四是公众参与的经济基础相对薄弱。普通市民的经济收入水平相对偏低，每天为生计忙碌，加之都市生活节奏加快，无多余时间和精力投身环保公益活动，这也是导致公众参与主动性不强的直接原因，同时也反映了经济状况决定公众参与的水平。

从政府作用层面来看，主要问题有：一是环保宣传教育力度不够。大多数30岁以上的市民，没有接受过良好的系统环境教育。长期以来环保宣传教育深度不够力度不足，市民整体环境意识的养成和环境素质的提高任重道远。二是公众参与环境保护方面的信息比较缺乏。公众希望了解更多环境保护方面的信息，但信息的公开透明还显不足，尤其是涉及环境污染负面的信息常常难以得到，这限制了公众的知情权。三是公众参与环境保护的监督权难以保证。公众参与的监督权虽然在法律层面上得到了规定，但在参与的条件、形式、要求等方面缺乏明确细致的法律规定，一旦遇到具体的环境问题，公众无从知道该用何种方式参与，更难以把握采取什么样的参与方式最合理、最合法。四是环境利益协调机制还未建立起来。公众参与环境保护过程中，往往会遇到一些环境利益纠纷，如果缺乏有效的协调机制，很可能导致各方以自身利益最大化为出发点，引发群体性事件。

3. 推动社会公众参与环境保护的政策建议

总结国内外在公众参与环境保护方面的经验教训，我们深刻地感受到，随着我国经济社会发展和人民生活水平的提高，环境问题日益凸现，群众的环保诉求日益强烈。中国需要在社会公众参与环境保护方面进一步采取积极行动，大力培养和增强全社会的环保意识和生态文明理念，广泛开展环保宣传教育，使环保观念深入人心，形成人人理解环保、关心环保、参与环保的良好社会氛围。充分调动社会成员参与环保的积极性、主动性和创造性，广泛深入开展全民环保行动，使环保成为每一个社会成员的自觉行为。

第一，广泛开展对公众参与环境保护相关知识和技能的宣传普及工作。采取各种形式，通过电视、网络、报刊、广播等传媒，组织、举办报告会、展览、征文、知识竞赛等活动，广泛宣传节约资源、保护环境的知识，提高全社会的环保意识。让广大公众认识到，保护好生态环境，既是政府的责任，也是社会的责任，是每一个公民应尽的责任和义务。公众关注环保，既要维护个人权益，更要树立“责任公民”意识。通过每一个单位、企业、学校和整个社会的广泛的宣传教育，使环境保护观念深入人心，不断增强公众的环境保护意识，引导公众树立生态文明的理念，提高环境保护的知识和技能，使社会公众广泛参与到环境保护中来，为建设美丽中国贡献出每一个人的智慧和力量。尤其要在基础教育阶段增加环保知识内容，突出操作性、趣味性，促使青少年从小养成环保意识和习惯。

第二，进一步加大对公众参与环境保护的信息公开程度。环境信息公开是环境知情权的重要内容，而知情权是公众参与环境保护的重要前提。信息不畅通不仅让公众无法实际参与环境保护，更不利于其参与环境保护的积极性。公众只有在了解环境信息的基础上，才能实际有效地参与环境保护工作。环境信息应以公开为原则，不公开为例外。政府要加强环境信息的发布，逐步把城市环境质量、大气污染、城市噪声、城乡饮用水水源水质等环境信息以及污染事故信息，通过电视、广播、报纸等大众新闻媒体及时向公众公开，以维护公众的环境知情权。这样，不仅可以对污染者产生强大的警示和约束作用，而且有利于公众自觉遵守环境保护法规，有利于形成保护环境的公众舆论和公众监督。政府要逐步扩大环境信息的公开性和透明度，并要建立公共环境事件信息披露制度，对谎报、瞒报者要追究责任。特别是在突发环境事件发生后，要尽快把事实告知公众，要公众在第一时间拥有知情权，让公众知道污染源在哪里，污染是怎么产生的，它将给社会公众的生产和生活带来什么影响和危害，应该如何防护和治理。

第三，加大对公众参与环境保护的法律保障力度。近年来，虽然在《环境保护法》、《清洁生产促进法》、《环境影响评价法》等各项法律中，对公众参与环境保护的知情权、参与权、检举权、监督权，都作了一些规定。但是，目前公众参与环境保护的权利在参与的具体条件、方式、程序上还缺少明确细致的法律规定，公众一旦遇到具体的环境问题，不知道应该用何种方式参与。要加快公众参与环境保护法律法规的建设，尤其在公众参与的程序、方式等方面应有明确、具体和细致的规定。还要保护公众的环境权益，对于环境权益受到损害的群体，要依法补偿。对涉及公众环境权益的立法，要充分听取公众的意见。

第四，建立健全社会公众参与环境保护的体制机制。各级政府都要重视来自公众的声音，多层次地搭建政府与公众座谈、对话的平台，对涉及公众环境权益的政策、规划与建设项目，要采取多种形式，充分听取公众的意见，建立公众参与环境保护的相关制度。要使公众通过各种途径和形式，参与环境保护法律法规、规划、政策和标准的制定，使公众能够对政府有关部门实施规划、政策、措施以及对企业行为是否符合环境保护标准进行有效监督。凡是涉及公共利益的项目，均应建立公众参与机制，完善和积极落实投诉处理制度，明确公众的环境权益。建立环境保护问卷调查、群众信访和听证会等制度，使公众拥有合法、有效、便捷的渠道来表达切身利益，参与环境决策与管理。建立公众环境保护问责制度，即对政府一切涉及环境的行为及后果都能够追究的制度，实现广大公众对环境保护有效监督。建立公众意见回应制度，对反馈信息给予及时处理，既能增强公众参与的积极性，也使得公众参与不流于形式，起到真正有效的政府与公众互动的效果。

第五，培养社会公众参与环境保护的良好习惯和行为准则。环境保护不仅要靠法律、制度等外在的规范约束来进行，也要靠生态道德、生态文明意识等内在的自觉自律来开展。通过有效引导，让生态文明意识成为大众文化意识，让绿色消费、适度消费成为全体公民的自觉行动。引导公众养成保护环境的生活方式，养成符合生态文明和可持续发展要求的良好的道德准则和行为习惯。在各行各业大力普及有关资源能源节约的窍门和方法，引导公众从我做起、从身边小事做起。制定并实行鼓励公众参与环境保护的激励措施，调动公众参与环境保护的积极性，从而形成人人关心环境、人人保护环境的良好社会风尚。

（四）充分发挥社会组织在环境保护中的重要作用

社会组织在环境保护中承担着重要的职能，在提升公众环境意识、促进公众环保参与、改善公众环保行为、开展环境维权与法律援助、参与环保政策的制定与实施、监督企业的环境行为、促进环保国际交流与合作等方面，都发挥着越来越重要的作用。在全世界应对气候变化谈判会议中，往往可以见到环保组织的身影，他们积极参与应对气候变化政策的制定，批评一些消极行为的国家所采取的政策措施，形成一种广泛的社会舆论压力，起到了重要监督和促进作用。

1. 国外及香港地区环保组织的发展状况及其作用

国际上的环保组织，一般属于民间组织，称为 NGO（Non-Governmental Organization），定义为“志愿性的以非营利为目的的非政府组织”。环保非政府组织从 20 世纪 70 年代开始起步，已经发展成为全球环境治理中重要的力量。国际上大的环保组织如地球之友、绿色和平、世界自然基金会等，都集结了大批优秀专业人士，积累了丰富经验并拥有专业性、多层次、多功能的社会动员和服务网络。

“绿色和平”成立于1971年，由加拿大人大卫·麦克塔格特等人创办，总部设在荷兰阿姆斯特丹，在全球30多国家开展活动，28个办公室分布于世界各地，共有300多万支持者和相当数量的志愿者。其资金筹集方式和组织形式堪称NGO的经典：只接受300万支持者的捐款，不接受任何政党、财团的捐助。每个具体项目中基金会的赞助只占小部分，60%—70%来自个人捐款，以保持组织及每个项目的独立性。进入21世纪，“绿色和平”的工作方向和工作方式都有所改变。目前的主要工作为：森林保护、气候变暖与再生能源、有毒物质污染、海洋生态保护、核武器和核能、生态安全六个方面，在中国大陆主要项目是与高校合作开展对生态农业和生物安全研究。由于“绿色和平”是联合国认可的大型NGO，在国际法律和公约实施方面参与了大量工作。

随着非政府组织的发展，政府和非政府组织的关系问题成为现代公共管理的核心课题之一。英国是处理政府与非政府组织关系比较成功的国家。在英国，各类非政府组织有近20万家，其中有18万家在英国慈善委员会进行登记注册。慈善委员会是英国统一负责各类非政府组织登记注册和监督管理的国家监管机关，享有执法权和包括制定相关法规等部分立法权。非政府组织在慈善委员会进行登记注册的好处，首先是能够得到政府每年提供的总额达33亿英镑的资金支持，它们占到全英非政府组织年度运作经费的约30%；其次，英国对待开展慈善活动的非政府组织实行减免税和退税制度，每年因减免税和退税优惠的税收总额高达30亿英镑；还有，非政府组织经过登记注册成为政府认可的慈善组织，能够有更高的公信力，可以从社会募集大量的慈善捐款，其年度募款总额，也达到30多亿英镑。为了促进政府和非政府组织的合作，英国政府和各类慈善组织之间签订了“政府与非政府组织合作框架协议”，其中详细规定了政府对待非政府组织的态度、原则、义务，以及双方合作的基本框架和内容。

欧美各国在各级政府部门、专家系统和科研部门之外，有规模很大的非政府组织，政府、专家和公众参与三支力量互为补充。环保运动的开展是自下而上的，对出现的环境问题，由民间组织向法院起诉、向议会呼吁、游说，最终通过立法，实现对污染和生态破坏的治理、补偿、监督和控制。因此，公众参与和NGO成为与完善的法律制度一起在环境保护中发挥着同样有力的作用。欧美各国环保NGO数量众多，宗旨各异。其组成有按地区组织的，也有针对某一具体问题组织的，例如保护本地区的湿地、保护某一种动物或某一片树林。大的组织如世界自然基金会（WWF），在全球几十个国家有1000万会员，在美国有120

万会员，在北京设有分部。环保 NGO 在开展公众教育、参与环境保护和治理、促进立法、协调跨部门、跨地区的环境问题等许多方面发挥了重要作用。目前，美国一些大学开设了 NGO 的管理和法律之类课程，许多青年人从大学毕业后，自愿到 NGO 做几年非功利的工作，回报和服务社会。美国目前有 1 万多个非政府环保组织，其中人数较多的 10 个组织的成员已有 720 万人。环境保护运动的高涨对政府行为产生了巨大影响。

与西方国家的环保 NGO 相比，类似的组织在日本的规模较小，历史较短。"日本野生鸟类协会"据说是日本最大的环保 NGO，有会员 5 万多人。"日本世界范围自然基金"如将其全体会员都计算在内也有 5 万人。"日本自然保护协会"有 2 万会员，日本还有 5000 多个规模较小的环境保护团体。这些团体虽然会员人数少，但在基层环保工作上都发挥着重要作用。日本环保 NGO 也活跃在中国等第三世界国家。

韩国著名的民间环保团体——"绿色之家"有 530 万名成员，主力军是来自全国 213 所中小学的青少年，由老师担任顾问。"绿色之家"经常组织中小学生到溪流河旁、山林中去实地考察，捡拾旅游废弃物。从 1998 年夏季开始，韩国教育部号召全国 830 万中小学生过"环保暑假"，不要在海滩、山林、小溪等避暑地区乱丢废弃物，尽量不用一次性器皿等。

环保组织在一些发展中国家也得到发展。如在非洲肯尼亚，非政府组织成为环保的积极倡导者和实践者。如"绿带运动"主要开展植树造林、环保教育与宣传、环保合作、生态旅游等活动，在肯尼亚和非洲其他地区组织种树超过 3000 万棵。该组织创办者、肯环境部副部长马塔伊因此被授予 2004 年诺贝尔和平奖。"保护环境俱乐部"为首都市民提供环保新闻和电视节目，开办了陈列环保出版物的图书馆。该组织发动了"垃圾就是财富"活动，提倡贫困人口回收垃圾获得收入。"非洲可持续发展基金会"为当地人尤其是青年和妇女提供可持续利用资源的宣传培训，出版一份刊登环保和垃圾处理新闻的双月刊。这些非政府组织都有自己的定点联系或支持企业。

在中国香港，"地球之友"成立于 1983 年，是一家全球性慈善机构在亚洲地区的分会，后独立成为香港最具影响力的 NGO 之一。资金主要靠企业和个人捐助及义工的支持。"地球之友"以提高广大民众环境意识，关注生态环境问题为自己的职责。目前拥有 1000 多名个人、100 多所学校和其他团体的会员，多名职员担任特区政府咨询委员会的职务，义务向政府提供专业意见。通过调查研究、环境教育和社区活动，推动香港地区环境质量的改善，聚集了一批如著名电影明星钟楚红、家庭妇女代表钟庄德芳等“铁杆”环保志愿者。

随着环保方面的国际交流与合作的发展，许多海外环保 NGO 已通过各种方式进入中国。如“绿色和平”致力于把国际社会关心的问题引入中国，协同中国政府一起反对跨国公司在转基因食物方面不负责任的行为，同时把中国国内民众迫切关心的环境问题如“沙尘暴”反映到国外，以引起国际社会的重视，以协调解决全球性环境问题。香港地区“地球之友”与中国环保部合作建立地球奖，为在环境领域有突出贡献的个人和青年团体带来了资金支持和社会的承认，同时它还支持大陆学生环保组织的发展。国际野生生物保护学会投入了很大力量，对东北虎和扬子鳄的保护进行研究。德国“拯救我们的未来基金会”提供的“绿色使者”流动教学车（羚羊车）穿梭于城乡之间，给孩子们送去环境启蒙的课程，并提供环保的理念、资金和工作方式。世界自然基金会（WWF）和国际爱护动物基金会（IFAW）在我国开展保护藏羚羊的国际合作，承担了从会议、资金及装备支持到面向海外宣传等多重任务。“全球绿色资助基金”给中国的民间环保网站、学生环保组织提供了资金支持。

2. 中国环保组织发展及存在的问题

中国环保组织一般定义是以环境保护为主旨，不以营利为目的，不具有行政权力，为社会提供环境公益性服务的民间组织。或者说，是指非政府的、非营利的、自主管理的、非党派性质的，致力于解决环境问题的社会志愿者组织。

中国环保组织的发展大体经历了三个阶段：一是自 1978 年起到 20 世纪 90 年代初，中国环保民间组织处在产生和起步阶段。1978 年 5 月，中国环境科学学会成立，这是最早由政府部门发起成立的环保组织。1994 年“自然之友”在北京成立。二是从 1995 年至 20 世纪初，环保组织把环保工作向社区和基层延伸，进入了发展阶段。1995 年，“自然之友”组织发起了保护滇金丝猴和藏羚羊行动，这是我国环保民间组织发展的第一次高潮。1999 年，“北京地球村”与北京市政府合作，成功进行了绿色社区试点工作，中国环保民间组织开始走进社区，把环保工作向基层延伸，逐步为社会公众所了解和接受。三是近 10 多年来，环保组织的活动领域逐步扩大，进入了相对成熟的阶段。2003 年的“怒江水电之争”和 2005 年的“26 度空调节能行动”，让多家环保民间组织开始联合起来，为实现环境与经济发展目标一致而行动。“绿色江河”先后完成了对青藏铁路沿线藏羚羊种群数量调查、长江源冰川退化监测等十多个研究的项目，其中给铁路部门建议在藏羚羊的主要迁徙路线上为其开设迁徙通道、哪些地方不能设置营地、什么时段不能施工以便为藏羚羊让道等，都得到采纳。世界自然基金会（WWF）评价，“绿色江河”是中国第一个对国家决策产生影响的民间环保组织。中国环保民间组织已由初期的单个组织行动，进入相互合作的时代。环保民间组织活动领域也从早期的环境宣传及特定物种保护等，逐步发展到组织公众参与环保，为国家环保事业建言献策，开展社会监督，维护公众环境权益，推动可持续发展等诸多领域。许多环保组织开展环保宣传教育，倡导环境保护理念，组织环保志愿者行动，进行环保调查研究，提出环保政策建议，做了大量卓有成效的工作。

中华环保联合会在全国范围内首次组织开展了“中国环保民间组织现状调查研究”工作。中国环保民间组织主要集中在北京、天津、上海、重庆及东部沿海地区，其次是湖南、湖北、四川、云南等生态资源丰富省份，其他地区的环保民间组织相对较少。人员组成的特点：一是年轻人多。80%左右为30岁以下的青年人，70%的负责人在40岁以下。二是学历层次高。50%以上拥有大学以上学历，13.7%拥有海外留学经历，90.7%的负责人拥有大学以上学历。三是奉献精神强。经调查，95%以上的从业人员是为环保事业而不是为发财致富，91.7%的环保志愿者不计任何酬劳，有的把自己的工资和储蓄拿出来支持环境公益事业，有的甚至奉献了自己宝贵的生命。

据统计，2011年，民政部登记注册的生态环保类组织共有7900多家，其中社会团体近7000家，民办非企业800多家，基金会64家。中国环保民间组织分四种类型：一是由政府部门发起成立的环保民间组织，如中华环保联合会、中华环保基金会、中国环境科学学会、中国环境文化促进会，各地环境科学学会、环保产业协会、野生动物保护协会等。二是由民间自发组成的环保民间组织，如自然之友、北京地球村，以非营利方式从事环保活动的其他民间机构等。“自然之友”是中国人数最多的环保民间组织，会员总人数已超过10万人。中国环保组织平均全职人员在25人左右。三是学生环保社团及其联合体，包括学校内部的环保社团、多个学校环保社团联合体等。四是国际环保民间组织驻华机构。环保类组织主要以政府部门发起成立和学生环保社团为主，两者占到总数的90%，民间自发的环保组织还相对较少。

我国环保民间组织具有几个基本特征：一是正规性，多数组织均进行了注册登记或拥有其他合法身份。二是民间性，不拥有任何行政权力。三是非营利性，不以谋求利润为目的。四是自治性，在国家法律规定范围内独立开展环保活动。五是志愿性，在组织建立、管理和开展活动中充分体现自愿参与原则。六是公益性，为社会公众提供环保公益服务。环保民间组织通过为社会公众提供环境公益性、互助性服务，反映和兼顾不同社会群体的环境权益，缓和社会矛盾，维护社会稳定，起到了社会“调节器”和“稳定器”的积极作用。

中国民间环保组织与政府之间建立起良好的合作关系。多数环保组织遵循“帮忙不添乱、参与不干预、监督不替代、办事不违法”的原则，寻求得到政府的支持、帮助与合作。经调查，61.9%的环保民间组织与政府之间有直接沟通的渠道，64.6%的与政府之间有密切的合作关系，32.1%的与政府之间保持非合作亦非对抗的关系，3.3%的与地方政府在环境污染方面的保护主义存在矛盾关系。

这些年来，中国环保民间组织从无到有，不断获得发展，在环境保护工作中发挥了积极作用，已经成为推动中国和世界环境事业发展不可缺少、不可替代和不可忽视的重要力量。它起到了政府职能所不易做、不便做的拾遗补缺作用，起到了政府与社会之间的沟通、交流和融合作用，起到了监督政府、保护百姓环境权益的作用，起到了宣传群众、引导群众、组织群众参与各种环境活动以及咨询和服务等方面的作用。

一是宣传与倡导环境保护，提高全社会的环境意识。开展环境宣传教育、倡导公众参与环保，是我国环保民间组织开展最普遍的工作。环保民间组织和环保志愿者走进工厂、社区、学校、乡村，通过发放宣传品、举办讲座、组织培训、出版环保书籍、在各种媒体上开设环境宣传窗口，以及开展环保公益活动等，采用多渠道、多方式向社会和公众宣传、传播环保理念，提高公众保护环境的责任意识，增强保护环境的自觉性。如50%以上的环保民间组织都建立了自己的网站，目的是向社会和公众传播环境知识、宣传环境主张、提高全社会的环境意识；“北京地球村”在中央电视台开设了专栏《环保时刻》；中国环境文化促进会每年组织万人参与环境文化节，宣传人与自然和谐的环境文化。

二是开展环境问题调查研究，为环境保护建言献策。民间环保组织围绕节能减排、保护江河水源、保护野生动植物、开展低碳生活、应对气候变化等广泛的内容，开展调查研究工作，为政府相关部门提出政策建议。怒江梯级水电开发在社会上存在很大争议，“绿色流域”、“绿家园”和“自然之友”等多家环保民间组织围绕怒江水电开发，组织邀请各方面的专家和当地群众进行数次研讨，提出保护自然生态、实行合理有序开发的建议，并通过云南省政协提出了“保护怒江、慎重开发”的提案，环保民间组织代表也联名向中央有关部委呈送了公开信，最终各方达成了怒江开发要充分论证的共识。

三是开展环境问题监督，维护公众环境权益。环保组织的另一重要作用就是监督政府实行环境主张，落实环保举措，促进环境问题的解决。2002年，重庆市决定在主城区建三十万千瓦燃煤发电厂，市民反映强烈。重庆市绿色环保联合会组织市民召开研讨会，建议政府停建以牺牲重庆市主城区空气环境为代价的工程。2003年底，重庆市政府采纳了建议，停建该工程。环保民间组织在维护社会和公众环境权益方面发挥了积极作用，有力地推动了政府把环境知情权、参与权、监督权和享用权真正赋还给公众，把公众对“四权”的真实意见反馈给政府。

四是保护珍稀濒危野生动物，保护生物多样性，建设人与自然和谐相处的绿色家园。我国是世界上生物多样性最丰富的国家之一，由于多种原因，生物多样性遭受破坏的景象惊人，一些珍稀动植物濒临灭绝。中国环保民间组织不遗余力地加以保护，作出了卓越贡献。如“自然之友”发起了对滇金丝猴的保护行动；“绿色江河”在西藏可可西里建立了我国第一个民间自然保护站——“索南达杰”自然保护站。“自然之友”会长梁从诫先生专门致信英国前首相布莱尔，敦促英国政府制止伦敦藏羚羊绒的黑市交易。南京学生环保社团“绿石”组织志愿者守护国家二级保护动物——中华虎凤蝶，直至幼蝶顺利孵化。

中国环保民间组织在推动环境事业发展中发挥了重要作用，作出了积极贡献，但也面临很多问题，制约了环保民间组织的健康有序发展。

一是对环保民间组织的认识不到位，缺乏支持政策措施。一些政府部门对环保民间组织的积极作用缺乏正确认识，存在“怕添乱、惹麻烦”以及重管理轻发展、重限制轻扶持的思想，缺乏积极主动促进环保民间组织发展的热情；一些企业，尤其是污染企业担心环保民间组织的发展壮大，对其监督的力量加大，怕增加成本，影响发展；公众对环保民间组织了解的还不够深，环保志愿服务的意识还很淡薄。因此，导致了对我国环保民间组织作用的认识不到位，政策支持力度不够，影响其生存和发展，给环保组织开展活动、吸引人才、筹集资金、招募志愿者带来了困难和阻力。

二是登记注册难，环保组织设立和运转受到很大限制。根据现行的《社会团体登记管理条例》规定，任何一个民间组织注册必须先找一个政府部门做业务主管单位，然后才能到民政部门登记注册。据统计，全国只有20%的民间组织按照规定在民政部门登记注册。中华环保联合会曾对2768家环保NGO组织身份进行调查，调查结果显示，2768家环保NGO组织中，在各级民政部门注册登记率仅为23.3%。据清华大学专门从事中国NGO研究的邓国胜教授调查，52%的环保NGO提出，他们最大的愿望是希望降低登记注册门槛。北京市已迈出改革的一步，规定社会组织注册无须再找主管单位，可以到民政部门直接登记。

三是环保组织经费缺乏，工作条件困难。虽然有良好的环保愿望和行动，但缺乏必要的经济条件。环保民间组织资金最普遍的来源是会费，其次是组织成员和企业捐赠、政府及主管单位拨款。76.1%的环保民间组织没有固定的经费来源。我国环保民间组织的工作条件总体上比较差，办公场所欠缺。60%以上没有自有的办公室。政府部门发起成立的环保组织60%以上由主管部门提供办公室；学生社团办公场所多设在校团委，没有专门的办公室；民间自发组织的办公场所，主要依靠租赁和借用；国际环保民间组织驻大陆机构，一半拥有自己的办公场所。为了加强与社会的沟通与交流，53.2%和47.5%的环保民间组织拥有自己的网站和内部刊物，80%以上的组织拥有计算机，学生环保社团拥有率只有27.1%。

四是参与政策制定有限，社会监督渠道不畅。一些政府部门和企业对环保民间组织实施环境监督，心存戒备和疑虑，持消极态度，导致环保民间组织不能正常参与环境的一些政策研究、法规建设、污染防治、公众参与等重要活动。再加上环境听证制度、公开制度、公众参与制度不健全，不能实行及时和有效的监督。

五是社会公众参与环保组织困难，国际交流受到很大限制。社会公众对环保组织了解有限，对于参与环保组织的权利、责任和义务还缺乏认知，致使公众参与环保组织的活动还不普遍。近些年来，国际民间环境交流非常活跃，得到了世界各国的广泛关注。我国环保民间组织的数量虽然不断增加，但获得联合国咨商认证资格的极少，再加上专业能力和国际交往水平有待提高，尚不能充分利用国际民间环境交流合作的平台，宣传中国政府的环境主张，维护中国的环境形象，争取更多的环境权益。

3. 进一步支持和促进环保组织发展的政策建议

2010年12月，环境保护部制定了《关于培育引导环保社会组织有序发展的指导意见》，积极培育与扶持环保社会组织健康有序发展，促进各级环保部门与环保社会组织良性互动，发挥环保社会组织在环境保护中的作用。《指导意见》明确了培育引导环保社会组织的基本原则和总体目标，即积极扶持、加快发展，加强沟通、深化合作，依法管理、规范引导，积极培育与扶持环保社会组织健康、有序发展，促进各级环保部门与环保社会组织的良性互动，发挥环保社会组织在环境保护事业中的作用，力争在“十二五”时期，逐步引导在全国范围内形成与“两型”社会建设、生态文明建设以及可持续发展战略相适应的定位准确、功能全面、作用显著的环保社会组织体系，促进环境保护事业与社会经济协调发展。

为培育和引导环保社会组织健康、有序发展，必须加强政策扶持力度，改善环保社会组织发展的外部环境，加强能力建设，引导环保社会组织健康、有序发展。环境保护部要求，各地要做好以下工作：制定培育扶持环保社会组织的发展规划；拓展环保社会组织的活动与发展空间；建立政府与环保社会组织之间的沟通、协调与合作机制；奖励表彰优秀的环保社会组织与个人；加强人才队伍建设，开展多方面、多层次的业务培训；加强规范引导，促进环保社会组织的自律；促进环保社会组织的国际交流与合作等。

中国环保组织发展正面临新的机遇和有利条件，关键是要提高对发展环保组织的思想认识，积极引导环保组织健康有序发展，充分发挥社会公众参与环境保护的重要作用。坚持积极引导发展、严格依法管理的原则，促进环保组织健康有序发展。贯彻“积极引导、大力扶持、加强管理、健康发展”的方针，改革和完善现行民间组织登记注册和管理制度，研究制定有利于公众参与、公益捐助等政策鼓励措施，切实解决我国环保组织发展中面临的困难和问题，为我国环保民间组织的健康发展提供有利的法律和政策环境。

第一，积极引导和规范发展各类公益性环保组织。政府有关部门要实行监督管理与引导服务并举，为各类环保组织发展创造条件。特别是帮助解决他们面临的注册难、经费难、社会参与难的问题。国家政策规定，重点培育、优先发展行业协会商会类、科技类、公益慈善类、城乡社区服务类社会组织。成立这些社会组织，直接向民政部门依法申请登记，不再需要业务主管单位审查同意。要按照这些要求，积极为各类环保组织登记注册、依规设立创造条件。学习借鉴国外环保组织管理的经验，政府有关部门可以为环保组织开展活动提供必要的服务，包括实施专业培训，提供适当补助，推出合作项目，进行宣传鼓励等，充分发挥他们的积极作用。

第二，发挥行业协会在环境保护中的职能作用。我国各行各业有许多行业协会，而且机构设置和运作都比较正规完善，联系着行业内的众多企业，具有熟悉行业情况、了解国家政策、参与行业管理的优势。在促进企业环保方面，行业具有特殊重要而不可替代的作用。如组织企业开展环保培训，推广节能环保技术，进行环保交流合作，对企业环保进行监督管理和服务，推动企业环保健康发展。

第三，积极鼓励和引导城乡社区参与环境保护。社区是社会的基本单位，家庭又是社会的细胞。一个社会环境保护的状况，与社区生活密切相关。要通过城乡社区，包括城市住宅小区、农村村庄，以及每一个家庭，发挥他们在环境保护中的基础作用。如城市社区承担垃圾分类、废品回收、园林绿化等作用，还能够宣传动员每一户家庭开展节电、节水、节气等广泛具体的节约活动。农村社区在环境保护和治理中担负着重要作用。如苏州在建设社会主义新农村活动中，开展“六清六建”（清理垃圾，建立垃圾管理制度；清理粪便，建立人畜粪便管理制度；清理秸秆，建立秸秆综合利用制度；清理河道，建立水面管护制度；清理工业污染源，建立稳定达标制度；清理乱搭乱建，建立村容村貌管理制度），取得了良好的效果，对于发动公众参与环保，建设社会主义新农村发挥了重要作用。

第四，广泛开展环保组织的国际合作交流。环保无国界，环境问题已经成为一个全球性问题。环保需要国际合作，共同加以应对解决。除了政府之间的环保合作交流之外，民间环保组织交流可以发挥重要的国际合作作用。如国际生物多样性保护、海洋生态保护、应对气候变化等，都需要开展广泛的国际合作。民间环保组织在这些方面，都可以发挥不可替代的重要作用。政府要积极鼓励和支持民间环保组织参与国际交流合作，同时给以必要的引导和管理，更好地发挥民间环保组织促进中国环境保护、建设人类共同家园的积极作用。

（五）创新环境保护的社会管理体制机制

加强环境保护，必须充分调动社会公众广泛参与环保的积极性，引导环保组织健康有序发展，更好发挥全社会力量促进环保的重要作用，这些都需要创新社会管理体制机制，创造有利于环境保护的良好的社会环境。

1. 建立和完善社会管理与服务制度

现代社会管理是依法管理，要把对环境保护的管理建立在法治的基础之上。环保组织是依法设立、依法自治，政府是依法监督、依法管理。一是要建立依法监管为核心的政府管理制度。对使用公共资源、提供公共服务的非政府组织，进行必要的行政监管和社会监督是国际惯例。有的国家如英国，实行登记和监管相统一的监管体制；有的国家如德国、日本，实行的是登记和监管相分离的监管体制。鉴于中国现行法规仍然实行双重管理，建议在环保部和民政部充分协商一致的基础上，对现行双重管理体制进行部分改正，将登记职能集中在民政部门，而将监管职能统一到环保部门。各级环保部门建立环保公益组织监管中心，统一行使对环境保护领域各类非政府组织特别是环保公益服务组织的监管职能。要在培育发展的基础上，逐步实行包括备案、登记、认定的三级准入和分类分级管理制度。对于组织化程度低、规模小、人员少的社区环保等组织，应实行备案制度；对于规模大、人员多、活动广、组织化程度高的组织，实行法人登记制度；对于影响大、公益性强的组织，则实行公益认定制度。改善和加强以“年检”为核心的信息报告制度。采取公益举报制度，广泛动员公众参与环保公益组织的监管，并建立全国联网的公益举报受理机制。实行公益托管制度，加强政府在监管上的权威和刚性约束。二是要探索建立以政府采购为基础的环保公共服务新体制。政府采购公共服务是建设公共服务型政府的基本方向之一。在环境保护领域，可参照扶贫、养老、社区综合服务等其他领域的成功经验，由环保部委托成立专项基金，通过公开招投标的方式对环境保护领域的一系列公共服务进行基于市场机制的政府采购。也可参照英国的模式，选择比较成熟的环保公益组织进行政府委托。

2. 制定有利于环境保护的社会政策措施

加强环境保护，除了要制定相应的经济政策，如财政、税收、金融、产业等方面的政策之外，还需要制定相应的社会政策。社会政策更加广泛，涉及社会生活的各个方面。比如，家庭垃圾的分类回收管理，建筑垃圾的处理，家用轿车的使用和城市汽车交通管理，宾馆的空调使用管理，洗衣机的节水管理等，在各个方面都需要制定详细的包括引导、鼓励、奖励和处罚等手段的政策措施。甚至小到随手对垃圾的处理，也需要作出规定。日本在2011年地震海啸和核事故之后，出现电力普遍紧张，为鼓励全社会节电，就具体地提出号召每一个政府机关、每一间办公室、每一户居民家庭都开一半灯，虽然并没有强制规定，但人们都非常自觉地普遍实行。新加坡在早期治理随地吐痰和乱扔垃圾时，是处以重罚，以此纠正人们的不良行为，达到自觉遵守形成习惯。我国在夏季高温时间，“地球村”等环保组织积极倡导“26度空调节能行动”，受到政府的采纳，通过宣传变成人们在办公室里的自觉行动，收到良好的效果。因此，要有针对性地制定社会各方面的节能减排和环境保护的政策措施，以引导和奖励保护环境的行为，限制和惩罚破坏环境的行为，达到全社会人人参与环境保护的目的。

3. 培育和形成有利于环境保护的社会行为规则

环境保护要从改变人们的社会行为习惯做起，从一点一滴的小事做起，逐渐形成社会公众行为。比如，形成节约的意识和习惯。中国人口众多，能源资源有限，环境承载力弱，应该成为一个十分注重节约而不再是一个浪费的国家，使节约成为深入人心的观念和国民的自觉行动。尤其是中国人的大吃大喝的浪费，讲摆场比阔气的浪费，贪大求洋不计成本的浪费，都需要痛下决心加以治理。现在一些社会志愿者开展的“光盘行动”，提倡到饭店吃饭点菜够吃就好，吃完盘子里的饭菜，不要造成浪费。这一行动引起公众的响应，取得了较好的效果。可见，一些不良的消费习惯是可以改变的。要在社会公众中广泛宣传和倡导节约资源和保护环境的行为规则，以引导社会大众的行为方式和生活习惯，变成全社会的自觉行动。这方面具有“滴水穿石”的力量，只要持之以恒坚持下去，就可汇聚成全社会环境保护的“汪洋大海”，真正建成资源节约型和环境友好型社会。

4. 妥善处理公共环境突发事件

随着公众环境意识的提高，对环境问题越来越重视。一些公共环境事件往往成为群体性行为，并引起社会的广泛关注。这就需要做好事发前的沟通、事发中的管控、事发后的处理，可以利用环保组织和社区做好中间的协调工作，化解公众与企业、与政府之间的矛盾，形成最大化的利益汇合点。比如 2013 年以来北京等北方大部分地区发生的严重雾霾问题，引起了公众的高度关注，人们希望了解事情的真相，知晓严重雾霾对人体健康的影响，以及基本的防护措施，更进一步知道严重雾霾的成因，以及治理污染的具体措施。还有 2013 年 3 月上海黄浦江发生的上万头死猪漂流事件，也引起媒体和公众的很大关注。有关方面及时调查事情真相并向社会公布，并及时采取措施解决这一问题。在处理突发公共环境事件中，一方面尊重公众的知情权和监督权，及时告知公众并作出解释，采取坚决果断的应对措施；另一方面，可以发挥环保组织作为第三方中立地位的作用，包括参与调查、告知公众、履行监督、作出评价，提高处理突发公共环境事件的公信力和权威性，以赢得社会公众的理解和信任。

5. 建立环境保护的社会评价标准和监督机制

加强环境保护，需要建立起社会评价标准。一方面，是社会公众对环境的评价，通俗地说，就是社会公众是如何看待环境状况的，有什么样的标准。比如，城市的空气质量、水环境有没有改善，改善的程度如何，一些专业性的标准如 $PM_{2.5}$、化学需氧量、二氧化硫、二氧化碳含量等，如何变成公众的感受，如何与公众的感受相一致，等等。另一方面，是对社会每一个单位、企业、家庭、个人的评价，是否符合保护环境的标准，做得好坏，就像建立个人生活的“低碳标准”一样，有一个衡量的尺度。有了标准，才好监督制约。在建立相应的环境保护评价标准的基础上，进一步建立起监督制约机制。比如，一个人上班一个人开一辆车是不环保的，而不开车乘坐公共汽车或骑自行车是环保的，人们会逐步倾向于更环保的生活。现在，在一些发达国家，不开汽车而骑自行车的人越来越多，既环保又锻炼身体，受到人们的欢迎。人们的评价和监督，可以发挥重要的导向作用。同时，也要充分发挥新闻媒体和社会舆论在加强环境保护中的重要监督作用。所谓“众口烁金”，众志成城，依靠千百万群众的力量，就能够使环境保护成为全民参与的伟大事业，共同建设全人类的美好家园。

四、在全社会建立节约环保的现代生活方式

在经济发展和物质极大丰富的现代社会中，将可持续发展的理念融入人们的生活观念之中，再造科学的、节约环保的、可持续的现代生活方式，是推动环境保护与社会发展协调进步的重要研究内容。

（一）人类生活方式的由来及其与自然环境的关系

关于人类生活方式研究，已经有了相当长的历史，各国学者从不同的角度对人类的生活方式进行了阐释和解读，中国学术界对生活方式的研究主要从20世纪80年代开始，相对西方起步较晚，而且研究的数量也较少。在本报告中，我们从人类生产方式、生活方式和不同文明与文化三者之间的发展与互动关系，以及与自然环境的相互影响入手，研究如何建立能够与自然环境和谐的、与经济社会发展和文明进步相适应的现代生活方式。

1. 关于人类生活方式的定义

关于人类生活方式的提出，最早是马克思在其著作《德意志意识形态》中明确使用了“生活方式”。在马克思的理论体系中，“处于具体社会历史境遇中的‘现实的个人’”是其研究的对象和出发点。马克思在《德意志意识形态》中指出，人是一定社会生产力的使用者，处于一定社会关系之中，是一定社会意识形态的生产者和改造者；同时，马克思还强调人是其“个人能力、个人需要及个人意识和个人价值观念的有机统一体”。由此，在马克思看来，所谓生活方式，是指在一定的生产方式的基础上产生，在诸多主客观条件下形成和发展的人们生活活动的典型形式和总体特征。生活方式，在微观上是指个人能力的发挥方式、

个人需要的满足方式和个人思想道德价值观的思维方式。从社会宏观意义上看，则反映了社会生产力的发展水平、社会分工的发展程度、社会资源的配置状况、社会消费资料的分配状况、社会的文化观念等[①]。

从社会心理学的角度，Feldman 和 Thielbar（1971）概括了生活方式的四个特点[②]：（1）生活方式是一种群体现象。一个人的生活方式受到他所在的社会群体以及跟其他人之间的关系的影响。（2）生活方式覆盖了生活的各个方面。一个人的生活方式使他在行为上表现出连贯性。当我们知道一个人在生活的一个方面的行为方式，就可以推断他在其他方面的行为方式。（3）生活方式反映了一个人的核心生活利益。许多核心利益塑造了一个人的生活方式，比如家庭、工作、休闲和宗教等。（4）生活方式在不同人口统计变量上表现出差异。包括年龄、性别、民族、社会阶层、宗教和其他决定因素。另外，社会变迁也会导致生活方式的改变。

综合国内外学者的研究，我们认为生活方式的定义可以分为广义的生活方式和狭义的生活方式[③]。广义生活方式包括人们在劳动、物质消费、政治、精神文化、家庭及日常生活等一切社会生活领域中的活动方式。狭义的生活方式只包括人们在物质消费、精神文化、家庭及日常生活领域中的活动方式。在本报告中，我们主要研究狭义的生活方式，针对人们的衣食住行、劳动消费、娱乐社交等具体行动，开展具体研究。

我们认为，人类在社会上的存在主要包括三种形式，一是生产方式，二是生活方式，三是思维方式。这三者之间是相互作用和相互影响的。我们要研究某个国家某个阶段和某些特定人群的生活方式，必须要综合考虑到与之相对应的生产方式和思维方式，不能孤立地仅就生活方式而研究生活方式，否则一定是没有根基的，也是不具备指导性的。

非常有意思的是，在我们总课题的研究中始终贯彻一个思路，就是在现代社会提倡可持续发展，必须同等重视经济、社会、环境这三个重要因素。这与上述人类的生产方式、思维方式和生活方式形成一种奇妙的对应关系，即生产方式是经济的发展的内核之一，思维方式是社会文明进步的缩影，而环境变化则是生活

---

① 王云霞．环境问题的社会批判研究．中国社会科学出版社，2012.

② Aaron Ahuvia，阳翼．“生活方式”研究综述：一个消费者行为学的视角．商贸经济，2005，(11).

③ 国家统计局．浅谈社会经济发展状况对城市居民生活方式的基本影响．2008.

方式影响的结果。当然，这种对应并不是十分严格和准确的，比如生活方式同时也受到经济和社会发展程度的约束。我们想说明的是，经济、社会、环境这三个可持续发展的支柱是相互依存、相互制约的，而人们的生产方式、思维方式和生活方式也是相互依存、相互制约的。我们现在研究生活方式，必须同等地研究生产方式和思维方式对它的影响，最终形成生产方式、思维方式和生活方式的“三赢”，而对我们总课题的研究而言，最终也是要达到经济、社会、环境的“三赢”。

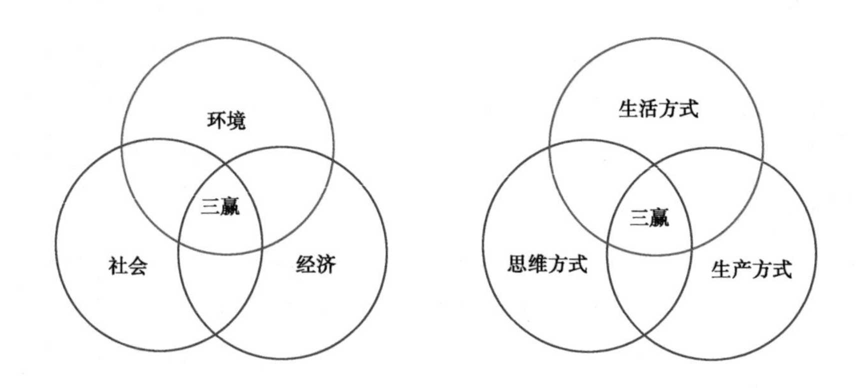

事实上，特定居民的生活方式与其所处国家的经济发展阶段是密切相关的，也是与其所属阶层的道德观、价值观、文化传统以及生活习俗的综合体现。从这个角度看，人类已经经历了原始社会、农业社会、工业社会和现代信息社会几个主要阶段，与此同时，人类也在不同的大陆上产生了多种多样的文明。这些因素交织起来，就诞生了太多种形形色色的生活理念和生活方式。因此，我们研究生活方式必须要承认这种差异性和多样性的存在，不能简单地判定高低优劣。以下，我们就从人类社会经济和生产发展的进化历程和各种文明文化发展的多样性来分析生活方式。

2. 人类生产方式的发展变迁与生活方式的关系及其影响

纵观人类发展的历史，人类生产和生活方式发生了几次重大的变迁。在这些变迁的背后，是人类与自然环境互动关系也随之发生了深刻的变化。这是因为，人类生产生活与自然环境所形成的互动关系，是人类生存所面临的基本关系，是人类文明的基本内容之一①。从人类文明演进的视角看，人类的生存和发展离不

---

① 李索清．对人类文明兴衰与生态环境关系的反思．生态环境与保护，2005，(2)．

开自然环境，一方面，人类的生存和发展必须以自然环境为依托，利用和改造自然，同时，自然环境也会因人的能动作用而不断发生变化，反过来从不同层次影响和制约人类的生存。从总体上看，伴随着人类科技文明的发展，人类认识自然、改造自然能力不断增强，人类也由对自然的依赖或以自然为主导逐渐过渡到人类对自然的控制和主导，人类的生产生活方式随之发生了巨大的改变。我们认为，随着人类社会生产力发展水平的不断提高和人类对客观自然规律认识的不断深化，人类社会的生产和生活方式有以下几个阶段：

（1）原始社会的生产和生活方式

在人类诞生以后的漫长岁月中，主要以狩猎和采集天然食物为生。在居住方面，在原始社会初期人类还不会建造房屋，以自然洞穴为栖身之所。之后，原始人走出洞穴，在地面上建造栖身之所，主要是利用树木和杂草搭在树冠上形成巢居。穴居与巢居同时并存了相当长的时期，直到今天，还有一些原始部落的人类居住在树上。在饮食方面，原始人类从茹毛饮血，到慢慢掌握了摩擦取火的方法，懂得保存火种，并用火做熟食。同时，一些地方的人类，已学会利用骨针来缝制苇、皮衣服，并逐步学会用葛、麻、棉花和羊毛纺织衣物。总体上讲，在这个时期，人类对自然的依赖性强，主要体现为依赖和适应，人类生产和生活受自然环境和自然资源的制约明显，人和自然之间保持了一种原始的和谐关系。

《孟子》："下者为巢，上者为营窟"，是说在地势低洼的地段作巢居，地势高亢的地段作穴居。北京人遗址发现大量的用火遗迹。其洞穴遗址从上至下有十三层文化堆积，其中第四层有较厚的灰烬堆积，最上面是鸽子堂，西侧第三层下面有一石灰岩巨石，其上有两大堆灰烬。第四层的灰烬厚达六米，灰烬中还发现了用火烧过的石头，动物骨骼及树籽，表明北京人已懂得用火，学会了保存火种，并已知道熟食。大约公元前5000年，世界各文明发祥地区都已就地取材开始了纺织生产。如北非尼罗河流域居民利用亚麻纺织；中国黄河、长江流域居民利用葛、麻纺织；南亚印度河流域居民和南美印加帝国人民均已利用棉花纺织；小亚细亚地区已有羊毛纺织。

（2）农业和畜牧业社会时期的生产和生活方式

随着农业、畜牧业的出现，人类的生产生活方式发生了改变。这个时期的主要特点是，农业和畜牧业是以家庭为基本生产单位、以手工为主要生产方式，生产的目的主要是为满足家庭生活需要而不是交换。同时，在这个时期社会的流动性较弱，人类的生活范围也不大。因此，这一时期人类在衣食住行方面的生活方式，基本上属于“自给自足”、“日出而作，日落而息。”由于农业和畜牧业生产规模小、强度低，因此农业社会中人们拥有一个优越的生态环境或生存空间，人类与自然保持一种顺应的、融合的非对立关系。当然，随着人口增长、生产力发展和人类活动范围的扩大，人类改造自然的能力也越来越强，对环境的影响也日益显著，也产生了破坏环境的副作用。如开荒、砍伐森林等引起的森林破坏、水土流失、沙漠蔓延、生态平衡失调、自然资源减少等，造成了局部性的环境破坏。还有，在一些地区为了争夺土地资源而频繁发动战争。总体看，农业和畜牧业时期，人类的生产和生活方式，使得人与自然关系表现出整体相对和谐，同时存在阶段性或区域性的不和谐的关系。

黄河流域曾是中国古代文明的发源地，据考证，商代时黄河流域的森林覆盖率曾达到50%以上。西汉末年和东汉时期进行了大规模的开垦，促进了当地农业生产的发展，但由于滥伐森林，水源得不到涵养，水土流失严重，结果造成水旱灾害频发，土地贫瘠；再如4000多年前，南亚的印度河流域曾经气候宜人、农业发达、物产丰富，但由于人们无休止的过度开发，使肥沃的土地变成了不毛之地，也形成了今天的塔尔沙漠；同样，发祥于幼发拉底河和底格里斯河的古巴比伦王国在创造灿烂文明的同时，也因为忽视对生态环境的保护，结果导致巴比伦王国的消失。如今伊拉克境内的古巴比伦遗址，已是满目荒凉，只有沙漠和盐渍化的土地。

（3）工业社会时期（18世纪60年代至20世纪50年代）的生产和生活方式

人类进入工业社会的标志，是两次科技革命。从18世纪60年代至19世纪中叶，从英国发明了以蒸汽机作为动力机并被广泛使用为标志的第一次科技革命。它开创了以机器代替手工工具，以工厂制代替了手工工场的“蒸汽时代”。这不仅是一次技术改革，更是一场深刻的社会变革，社会财富随着生产工具的机器化暴增，社会发展进入大工业经济时代。第二次科技革命起于19世纪70年代，主要表现在四个方面，即电力的广泛应用、内燃机和新交通工具的创制、新通讯手段的发明和化学工业的建立。在这一时期里，一些发达资本主义国家的工业总产值超过了农业总产值；工业重心由轻纺工业转为重工业，出现了电气、化学、石油等新兴工业部门。世界由“蒸汽时代”进入“电气时代”。

1866年德国人西门子（Siemens）制成发电机，1870年比利时人格拉姆（Gelam）发明电动机，电力开始用于带动机器，成为补充和取代蒸汽动力的新能源。电力工业和电器制造业迅速发展起来。人类跨入了电气时代。

内燃机的诞生和使用。19世纪七八十年代，以煤气和汽油为燃料的内燃机相继诞生，90年代柴油机创制成功。内燃机的发明解决了交通工具的发动机问题。1885年，德国人卡尔.本茨成功地制造了第一辆由内燃机驱动的汽车。1902年12月，以内燃机为动力的飞机飞上蓝天，实现了人类翱翔天空的梦想。内燃机车、远洋轮船、飞机等也得到迅速发展。

内燃机的发明，石油的开采量和提炼技术也大大提高，还推动了石油开采业的发展和石油化工工业的产生。1870年，全世界只生产了大约80万吨石油，到1900年已猛增到2000万吨。人们开始从煤炭中提炼氨、苯、人造燃料等化学产品，塑料、绝缘物质、人造纤维、无烟火药也相继发明并投入了生产和使用。

电讯事业的发展尤为迅速。继有线电报出现之后，电话、无线电报相继问世，为快速地传递信息提供了方便。从此，世界各地的经济、政治和文化联系进一步加强。

在工业社会时期，以农业与乡村为主体的经济体制变成了以工业以城市为主体的经济体制，大规模地改变了人类生活的方式。其中最突出的是人口由农村大量流向城市，城镇化的生活方式走上了人类发展的舞台，并成为主角，主要表现在：电灯的应用，使人类进入光明时代；汽车及其飞机的发明，大大提高了人们出行方式，使人类的活动范围更加宽广；有线电话和有线电报开始使用，人与人之间的交流更加便捷。塑料、人造纤维的生产和使用，使人们脱离了对传统纺织品的依赖。

在工业社会时期，人类的劳动生产率大幅度提高，增强了人类利用和改造自然环境的能力。但同时，人类的工业活动也大规模地改变了环境，导致人类所依赖的生态环境开始恶化。人们在尽情享受工业生产带来的物质文明大发展的成果时，环境污染问题也从天而降，给人们的日常生活甚至生存带来了巨大的威胁。

19世纪80年代开始，英国的伦敦就多次发生可怕的有毒烟雾事件。19世纪后期，日本的足尾铜矿区排出的废水污染了大片农田。1930年，比利时的马斯河谷工业区发生了严重的大气污染事件。1943年，在美国洛杉矶发生了多起可怕的光化学烟雾事件。1948年10月在美国还发生了多诺拉烟雾事件等。从20世纪50年代至80年代，环境问题更加突出，集中爆发，出现了一批震惊世界的公害事件。先后发生了1952年的伦敦烟雾事件、1953—1968 年日本熊本县的水俣病事件、1961 年日本四日市的哮喘病事件、1955—1972 年日本富山县的骨痛病事件以及 1968 年在日本九州爱知县等 23 个县府发生的米糠油事件。这些事件连同比利时的马斯河谷烟雾事件、美国洛杉矶的光化学烟雾事件和美国多诺拉烟雾事件一起构成了历史上震惊世界的"八大公害"事件，形成了人类历史上第一次环境问题的高潮。

（4）全球化时期的生产和生活方式

20 世纪 50 年代以后，原子能、电子计算机、宇航工程、生物工程等领域出现了重大发明和突破，随之产生了第三次科技革命。20 世纪后期，计算机网络技术、生物信息技术、基因工程技术、微电子集成技术等学科的高度融合并产业化。从此，蒸汽机、电能基础上的工业社会让步于计算机、电信基础上的知识经济社会，人类进入了全球化时代（Globalization）。在这个时期，科学技术的突破使得劳动生产率明显提高，产业结构明显变化使得城市人口猛增，社会居民结构和生活状况发生剧烈变化。科技进步和生产力显著提高，人类活动范围已扩张到全球的各个角落，并且不再局限于地球表层，已拓展到地球深部及外层空间，这使人类的生存领域和生活方式再次发生了根本性的变化。

在全球化时期，人们之间的联系不断增强，人类生活范围扩大到整个世界，全球意识逐步崛起。国与国之间在政治、经济贸易上互相依存。全球化亦可以解释为世界的压缩和视全球为一个整体，出现了"地球村"的概念。同时，具有共性的文化思想和道德观逐渐在全球通行。一个人可以住在美国的乡下，吃着法式面包，穿着英国产鞋子，耳朵里听着日本产的随身听，口袋里放着诺基亚手机，乘坐德国产的奔驰汽车，通过中国制造的掌上电脑用数种语言和任何一个国家的人联系工作。

1969 年，美国社会学家 M. 麦克吕亨在其畅销书《地球村里的战争与和平》中惊呼：美国的许多家庭通过电视直播，越南战争"亲临其境"，电视通讯卫星使地球变小了，地球犹如一个村庄。到了 80 年代，《哈佛商务评论》主编 X. 勒维特撰写的有关"商业全球化"的文章，则进一步完备了全球化的概念。联合国前秘书长加利在 1992 年联合国日致辞时说："一个真正的全球性的时代已经到来了。"

在全球化时期，世界已成为一个不可分割的整体，但全球人口的急剧膨胀、自然资源的日益短缺、生态环境的不断恶化，使人与自然的关系变得越来越不和谐，对人类生存安全构成了极其严峻的挑战。一方面，环境问题日益变成了全球性的问题。同时，地球上任何人任何地方人类的活动，对环境造成的危害也都会受到全球的瞩目，引起全球人的高度关注和警惕。

全球化时期，环境问题分为三类：一是全球性的大气污染如温室效应、臭氧层破坏和酸雨；二是大面积的生态破坏，如大面积森林被毁、草场退化、土壤侵蚀和荒漠化；三是突发性的严重污染事件，如1984年发生在印度博帕尔的农药泄漏事件，1986年发生在苏联的切尔诺贝利核电站泄漏事故等。

大范围乃至全球性的环境污染和大面积生态破坏，并威胁到人类在地球上的可持续生存和发展。以温室效应为例，众所周知，温室效应的一个直接后果就是全球气候变暖。全球气候变暖则会导致很多难以预料的不良后果。据科学家预测，全球变暖将会导致上亿人口面临粮食短缺，11亿—32亿人缺水，高山的积雪将加速融化。科学家还预计，全球变暖将导致世界上四分之一的陆地动植物在未来50年内灭绝。全球变暖还会导致极端天气频发、空气和水源污浊，引起更多同洪水相关的意外事故，威胁人类的食物供应，造成数以百万计的环境难民，给具有净化空气和水作用的许多生态系统带来压力，甚至导致生态系统的崩溃。

事实上，人与自然共处在地球生物圈之中，人类的繁衍与社会的发展离不开大自然。一方面，人类从自然界索取资源与空间，享受生态系统提供的服务功能，向环境排放废弃物；另一方面，自然也对人类产生影响与反作用，包括资源环境对人类生存发展的制约，自然灾害、环境污染与生态退化对人类的负面影响。当前，全球性的温室效应、臭氧层破坏、酸雨，以及全球性的物种灭绝、植被破坏、森林消失、草场退化、耕地减少、水土流失等使人们认识到必须变革现有的生活方式，保护环境，善待自然。

（5）低碳环保等可持续的现代生产和生活方式

20世纪60年代以来，可持续发展理念的广泛兴起与蓬勃发展，可以说是人类发展观的一次质的飞跃，它既是划时代的发展观，又是崭新的世界观、文明观和自然观，它深刻地揭示了经济社会繁荣背后的人与自然冲突，对传统的“征服自然”等不可持续发展观提出了挑战。1972年在斯德哥尔摩召开的联合国人类环境会议以来，可持续发展观逐渐被世界各国人民所接受。特别是1992年在巴西里约热内卢召开联合国环境与发展大会，183个国家和地区的代表、102位国家首脑出席了这次“地球高峰会议”，会议通过了《里约热内卢宣言》和《21世纪议程》两个纲领性文件，标志着可持续发展观被全球持不同发展理念的各类规矩所普遍认同。走可持续发展之路，实现人与自然和谐发展成为全世界的共识，促进人与自然和谐发展成为人类的共同使命。从此，以可持续发展为核心价值的理念正全面改变着人类的生产和生活方式，农业的有机化、工业的生态化、废弃物的减量化、产品和营销的绿化，已经成为生产发展的潮流；穿天然再生的生态服装，吃健康有益的绿色食品，住环保家居，出行使用公共交通等环保第一的理念全面融入衣、食、住、行各个方面，形成有利于环境的生产和生活方式已经成为全社会的共识和追求。

3. 人类不同文明、文化与生活方式的关系及其影响

文明是反映人类社会发展程度的标识，反映了一个国家或民族的经济、社会和文化的发展水平与整体面貌，文明也根深蒂固地影响着不同人群的生活方式。从历史上看，人类文明的发展大致经历了原始文明、农业文明和工业文明三个阶段。当前，很多学者又提出了生态文明的理念。我们这里从西方工业文明、东方文明和现代生态文明的角度，讨论其对人类生活方式的影响①。

（1）西方工业文明下的生活理念和生活方式

西方工业文明是以无限获取利润为动力、以现代资本主义体系为制度、以建立在工商业和金融业上的城市为载体的文明，它为人类社会带来了巨大财富。从文化上看，西方工业文明的基础就是人类中心主义，即将人视为自然万物的主宰和中心，一切以人为中心、一切以人为尺度、一切从人的利益出发，人的物质需求也是无止境的。同时，自然财富是无限的，自然就是不断满足人类无限欲望的对象。因此，人类只要不断地征服自然，就能促进经济发展，满足人们不断增长的物质需要。西方的工业文明从其产生的时候起就因其诸多弊端的存在而成为许多思想家反思和批评的对象。卢梭曾对使工业文明过分膨胀的工具理性侵蚀人的道德理性、破坏人与自然和谐的可能性和危险性发出警告。马克思、恩格斯更是对资本主义工业文明所导致的人与人、人与自然的异化进行过深刻的反思。事实上，人的欲望可以无限，但地球的资源是有限的。西方工业文明在 300 年间创造了空前的物质文明和社会财富，几乎等于传统社会的总和，但也消耗了亿万年的自然储备，带来了严重的经济危机和全球生态危机。

面对西方工业文明所暴露出来的内在困境，一些政治精英与学者试图从中寻找摆脱困境的路径，他们发现在世界各大文明和古老宗教中也蕴藏着生态智慧：

从基督教文明来看，人类对生物群的尊重和保护也应成为自然法的内容，包括生物自然多样化存在的权利、健康生活的权利、自由进化的权利、不受人类侵害的权利、共享于地球的权利等。为这些权利而奋斗，已成为二十年来西方环境

① 潘岳．在改革开放 30 周年环保高峰论坛上的演讲．2009.

运动的伦理基础。

从伊斯兰文明来看，伊斯兰教的基本教义倡导人与自然的和谐一体。《古兰经》提出，世间万物皆由真主所创，人是自然的一部分，代真主管理大地，人应该合理适度地利用自然，反对穷奢极欲和浪费，最具代表性的话是，“你们应当吃，你们应当喝，但不要浪费，真主的确不喜欢浪费者”。

从印度文明来看，印度教徒蔑视外在和物质的东西，重视内在精神。与其他民族崇拜精英不同，印度教社会的精英都崇拜那些隐居于森林之中的圣徒，认为其体现出的精神力量是永恒的。因此现代印度则有抱树运动，人们抱着树，防止林业工人砍伐森木，这被全世界看做应用印度教思想的一次环境起义。

（2）东方文明下的生活理念和生活方式

五千年中国传统文化的主流，是儒释道三家。在它们的共同作用下，中华民族形成了自己独特的文化体系，那就是“中”、“和”、“容”，即中庸之中、和谐之和、包容之容，讲究有序、平衡、包容、协调，倡导“道法自然”、“天人合一”的理念，追求“自谦”、“不争”、“适度”、“宽容”的生活态度，强调无论做什么事情都不要过度，要适可而止，最终达到返朴归真、恬淡自然的境界。这些朴素的思想指导中国人要修身养性，并爱万物、爱自然。中华传统文化是农业文明的产物，但其价值观仍然适用于今天工业化社会。

儒释道三家都在追求人与自然的和谐统一。儒家讲求“仁民爱物”，即人与人、人与物之间，犹如同胞手足，朋友兄弟，万物一体而相互仁爱。佛教虽为外来文化，但很好地实现了与中国本土文化的融合。佛教提出“佛性”为万物本原，“山川草木，悉皆成佛”，万物之差别仅是佛性的不同表现，本质乃是佛性的统一。众生平等，是佛家最根本的原则之一。道家崇尚“自然”，希望通过“道法自然”实现人道契合、人道为一。老子认为，万物与人既是平等又是相互联系的，主张顺道而为，复归于朴。庄子提出了“天地与我并生，而万物与我为一”的生命境界。

东方文明的价值观在现实制度和生活中具体落实为一个“度”字。“度”就是分寸，就是节制，就是礼数，就是平衡，就是和谐。概言之，“度”不仅是中国政治智慧，也是中国人的生活智慧，更是中国生态智慧的凝练表达。《朱子治家格言》从清代一直到民国都是五六岁儿童开蒙的课本。里面说到“一粥一饭，当思来之不易；半丝半缕，恒念物力维艰”“宜未雨而绸缪，毋临渴而掘井”，就是教育人们形成节制、有度、从容的生活态度和生活方式。

因此，中国的精神和道德层面里，有非常多的和可持续发展的理念可以相互契合的地方。现在，在很多人眼里中国人有着大手大脚、铺张浪费的现象，这在某种程度上是由于中国在过去30多年发展太快，刚刚富裕起来之后的一种“心态”的迷失，绝对不代表中国人的主流价值观念。相反，大力提倡节约环保的现代生活方式，引导人们在服饰、饮食、日用品、建筑、交通、行为等方面抵制旧有生活陋习和奢侈消费，从全新角度把人们对提高生活质量的需求引向正确的方向，是现在中国社会的主流心态和共识。

（3）现代生态文明下的生活理念和生活方式

20世纪60年代以来，随着全球环境污染的进一步恶化，人们开始了有意识地寻求新的发展模式，核心是处理好人和自然之间关系，以解决当前的经济危机与生态危机。在这个时期，东西方文明令人惊讶的殊途同归，共同提出了生态文明的理念。生态文明不追求物质享受的最大化，而是强调建立健康文明、从容有度的生活方式，通过节制人的无限欲望，追求充实饱满的精神追求，最终形成人与自然和谐相处的状态。

生态文明大大拓展了人类文明的含义和内容。在生态文明的理念中，人与自然的关系是天成的。人不能选择脱离自然的道路，只能选择某种有利于自身发展的与自然的关系。在人的能力空前提高的今天，人与自然的关系在很大程度上还要依赖人的价值观、生活方式、社会关系等诸多因素的协调和谐。具体来说，生态文明的基本内容，主要包括以下几方面：

第一，人类要尊重自身首先要尊重自然。人类与自然是一个相互依存的整体。以损害自然界的生物种群来满足人类无节制的需求，只能导致整个生态环境资源的破坏和枯竭，最终危害人类自身。人类必须重新认识自身与自然的关系。从自然的角度说，人与自然是平等关系，而不是主从关系，更不是征服与被征服的关系。

第二，价值观的革命。人类并不是自然的主宰，而是自然的一部分，人类的价值观并不能仅仅以人本身为最终目标，人类的功利和幸福不能逾越自然所允许的范围。人类只有在与自然协调和谐相处的前提下，才能获得真正持续、健康的功利与幸福。但是，功利与幸福及其程度的界定又是由人的价值观所决定的。生态文明是人类价值观必然的选择。

第三，保护生态环境是伦理道德的首要准则。生态文明的伦理道理是以维护地球生态环境系统正常运转，保护自然生态的良好状态为伦理道德的首要准则，人类其他的一切行为，首先必须以服从这一道德准则为前提。

第四，生态文明是社会结构的重要组成部分。生态文明发展结构理论把生态环境资源作为社会结构理论的重要组成部分，在经济、政治、文化“三领域”框架中加上“生态环境”，建立起“四领域”的总体框架，因为优美的地球生态环境是人类文明繁荣发展的基础和前提，人类文明必须把保持自然生态环境系统的正常运转作为其重要标志之一。

与此同时，很多学者提出了“生态人”的理念①，用来解决当前的生态危机、建立人与自然的和谐关系。在国内，徐嵩龄先生最早提出了“理性生态人”，随后还有学者使用“社会生态人”、“德性生态人”和“美学生态人”等概念。“生态人”的提出者和拥护者普遍把当代的生态环境问题归咎于“经济人”假设，坚信解决生态问题的出路就在于用“生态人”代替“经济人”假设，塑造新型人格模型。在他们看来，“生态人”是“当今理想的人格模式”，生态文明是代替工业文明的新型文明，而“生态人”是“生态文明的主体承负者”。由此，由“生态人”的假设模型出发，提出人的生活方式的“生态化”，形成促进人与自然和谐相处。

从以上人类生产方式和不同文明与文化的发展、演进以及变革，及其与人类的生态方式的关系研究中，我们看到，人类社会发展到今天，人与自然的关系已经到了一个一触即发的危急阶段。建立节约环保的现代生活方式是实现人与自然和谐相处、推动可持续发展的必然要求。与传统的农耕文明不同，在工业化生产大发展的今天，需要有节制的、环保的生活方式，以此来维系人居环境的可持续性。在经济发展和物质极大丰富的现代社会中，将可持续发展理念融入人们的生活观念之中，再造科学的、现代的生活方式，是推动环境保护与社会发展协调进步的基础工作。过去的 30 多年中国经济实现了高速增长，人民的物质和文化生活水平得到了较大的提升，已经进入了中等收入国家的行列。在这个时候，中国政府清醒地看到，中国经济的增长的另一面是对能源和资源的巨大需求，因此从国家层面看，已经提出了发展循环经济，建设资源节约型社会和环境友好型社会等决策。可以说，刚刚富裕和繁荣起来的中国，随即就面临着既要实现经济的稳定增长，又要保持良好的生态环境，这个非常现实的“两难”问题。因此，中国提出建设生态文明、实现可持续发展，绝对是对世界各国和全球人民负责任的态度和行动。这也要求在中国人民中倡导和普及低碳环保的现代生活方式。实事求是地讲，从目前中国国情和人民生产方式和生活方式来讲，实现低碳环保的现代生活还不是一蹴而就的事情。

---

① 袁云．从“生态人”假设到人的“生态化”生活方式．天津行政学院学报，2012，(3)．

（二）当代人类生活方式的研究

生活方式犹如一面镜子，不仅可以反映出各时期经济发展的程度和生活水平，而且可以真实地反映人类对环境的态度。当代人类在衣、食、住、行、用等的生活方式和生活态度上，仍然存在诸多不节约、不环保、不可持续的现象。我们必须带上“环境责任”的眼镜去重新梳理和审视这些问题。

1. 当代人类生活方式对环境的主要危害

（1）居住方式

当前，中国正在经历一场全球最大规模的城镇化过程，人口的大规模转移也推动了建筑产业的发展。中国已经成为世界上每年新建建筑量最大的国家，每年新增20亿平方米建筑面积，相当于消耗了全世界40%的水泥和钢材，仅上海、北京每年的建筑量差不多是整个欧洲每年的建筑量。建筑业占用土地面积巨大，消耗资源化能源的数量巨大。人类从自然界获取资源的50%用于建筑物，产生的固体废弃物的50%也来自建筑物，这对自然环境带来了巨大的威胁。

——建筑寿命过短。许多住宅远远未达到设计使用年限就被拆除，平均使用寿命不足30年，目前中国每年拆除的老旧建筑占新建建筑面积的40%。按照我国的强制标准，普通建筑规定的合理使用年限是50年。但统计结果显示中国建筑平均寿命不到30年；而欧洲建筑的平均生命周期则超过80年。英国建筑的平均寿命132年，美国的建筑寿命74年。

——户型过大。我国城镇人均住房面积2004年已达25平方米，2011年增加到30平方米，这么快的发展速度在全世界都是绝无仅有的。根据我国新的政策界定，消费者享受普通住宅政策的标准是120平方米以内，同时允许各地上浮20%，即144平方米以内。在我国当前的发展阶段，144平方米的住宅标准远远超过了日本、瑞典、德国等发达国家。

从1990年到2002年，日本新建住宅的平均户型基本在80至100平方米之间浮动。在发达国家，经过长期居住实践之后，已经确定了一个符合适度、合理、可持续发展的户型区间：单套住宅建筑面积在85平方米至100平方米之间。在上个世纪70年代到90年代，这些发达国家的新建住宅户型也曾越做越大，现在户型又重新回归到一个合理的区间。2002年，日本、瑞典和德国这3个发达国家新建住宅的平均建筑面积分别是：85平方米、90平方米和99平方米。

——建筑节能问题突出。我国建筑直接和间接消耗的能源已经占到全社会总能耗的46.7%，现有建筑95%达不到节能标准，新增建筑中节能不达标的超过8成；单位建筑面积能耗是发达国家的2至3倍。其中，我国住宅钢材的消耗高10%至25%，卫生洁具的耗水高30%，而污水回用率仅为发达国家的25%。

（2）汽车化生活方式

根据发达国家的经验，汽车时代的迅速发展，提高了人们的生活水平和质量，带动消费水平升级，促进了国民经济发展，但同时也消耗了大量自然资源、污染了空气，加剧了交通压力，给能源供给、道路交通、环境质量带来了极大的压力。

——巨大的能耗需求。美国是“汽车上的王国”，能源消耗的一半以上都用在了汽车上。这种生活方式让美国人在世界饱受诟病。2010年，我国汽车保有量已经超过了7000万辆，每年的销量接近2000万辆，到2020年将达到2.5亿辆。在一些大城市，汽车保有量更是爆发式的增长。以北京市为例，2012年北京常住人口超过2000万人，机动车保有量已达520万辆。按照目前我国汽车市场的增长速度，每年新增汽车消耗的成品油相当于新建一个2000万吨级炼油厂。

——带来空气污染，在大中城市，工业污染正让位于汽车带来的烟雾污染。上海城市规划人员估计，该城市90%的空气污染来自于机动车。覆盖在中国城市上空的烟雾，导致了庞大的医疗成本，中国肺病发病率在过去30年内翻了一番。据世界银行估计，因空气污染导致的医疗成本增加，以及人生病丧失生产力，使得中国GDP被抵消掉5%。以发生在2013年1月中国的雾霾天气为例，机动车、燃煤、工业污染和扬尘是形成雾霾和空气重污染的主要来源。其中，汽车尾气的占比最高达到了34%。

案例1：洛杉矶光化学烟雾事件。洛杉矶在20世纪40年代初就有汽车250万辆，每天消耗汽油1600万升。到70年代，汽车增加到400多万辆。市内高速公路纵横交错，占全市面积的30%，每条公路通行的汽车每天达16.8万次。由于汽车漏油、汽油挥发、不完全燃烧和汽车排气，每天向城市上空排放大量石油烃废气、一氧化碳、氧化氮和铅烟。这些排放物，在阳光的作用下，发生光化学反应，生成淡蓝色光化学烟雾。这种烟雾中含有臭氧、氧化氮、乙醛和其他氧化剂，滞留市区久久不散，造成许多人眼睛痛、头痛、呼吸困难。光化学烟雾还降低大气能见度，影响汽车、飞机安全运行，造成车祸、飞机坠落事件增多。

案例2：2013年1月，中国大部分地区出现了雾霾天气，特别是中东部地区还随之发生了空气重度污染。这次雾霾天气的特点：一是持续时间较长，是1961年以来全国平均雾霾天数最多的，其中，北京雾霾日数为25天，合肥雾霾日数为30天、南京和杭州29天。二是这次雾霾天气的分布范围也很广。根据国家气象卫星遥感监测显示，全国有30个省（区、市）先后出现雾霾天气，雾霾覆盖范围平均有71.6万平方公里，其中1月22日雾霾范围达222万平方公里。三是受大范围雾霾天气影响，空气中污染物不易扩散，全国17个省（区、市）的城市空气质量下降，部分城市空气质量达到严重污染程度，部分地区能见度不足200米，全国受影响人口约6亿人。

（3）饮食方式

目前，世界欠发达地区面临粮荒，全球还有几亿人在挨饿。但食品的浪费现象也很惊人。

——不良购物习惯。人们习惯一次性在超市购买大量食品，很多食品还没来得及食用就过期了。根据《纽约时报》报道，美国人每年购买的食品超过四分之一都被扔掉。调查发现，美国市场上供出售的食品每年有大约1.6亿吨，其中有4370万吨最后被扔掉，相当于市场上所有食品的27%。平均到人头，相当于每名美国人每天浪费大约400克食品。而英国政府最近发布的一份统计报告显示，英国平均每年在“从农场到冰箱”的过程当中要扔掉总价值约200亿英镑的食品。据统计，英国人每天扔掉440万个苹果、160万根香蕉、130万罐酸奶、66万个鸡蛋、55万只鸡、30万包土豆片和44万份熟食。浪费的食品价值相当于一个普通家庭一年多支出420英镑。

——追求包装良好的食品。在英国，由于包装的损坏，超市和食品店会将所有包装损坏的食品扔掉，餐厅会扔掉厨房里的边角余料，而普通人也会扔掉甚至只是外皮变成褐色的香蕉以及存放时间稍微长了一点的外卖食品。有时因为抽检的水果表皮有几个瑕疵，相关的整批水果就要遭受“连坐之苦”，白白烂掉。英国每年由于这个原因扔掉的食品有300万吨之多。

——餐桌上的浪费。据统计，中国人在餐桌上浪费的粮食一年高达2000亿元，被倒掉的食物相当于2亿多人一年的口粮。上海餐饮业每天产生的餐厨垃圾超过2000吨。重庆餐饮界近日提供了一个“比较保守”的估计，餐桌上的10%都被消费者浪费掉了。

——大量使用一次性筷子。一次性筷子是社会快节奏，节约资源的生活产物，也是导致森林资源急剧下降的产物。根据统计，中国市场各类木制筷子消耗量十分巨大，其中每年消耗一次性木筷子450亿双（约消耗木材166万立方米）。每加工5000双木制一次性筷子要消耗一棵生长30年杨树，全国每天生产一次性木制筷子要消耗森林100多亩，一年下来总计3.6万亩。

（4）过度包装

目前，对商品进行过度包装的现象日趋严重，不少包装已经背离了其应有的功能。一些专家认为，包装物的价值超过被包装产品价值的1至2倍，就称为过度包装。

——过度包装的主要方式：一是结构过度，有的商品故意增加包装层数，在内包装和外包装间增加中包装，外观漂亮，名不副实；有的商品包装体积过大，实际产品很小，喧宾夺主；还有的商品采用过厚的衬垫材料，保护功能过剩。二是材料过度，在一些食物的包装中，很多采用实木、金属制品，大大增加了包装成本。三是装潢过度，盲目采用上好的包装原材料，增加包装成本，有的甚至还在商品中附加几倍甚至几十倍于商品价值的礼品，提升商品价格。

——过度包装的危害表现在以下几方面。一是浪费大量资源。包装用的纸张、橡胶、玻璃、钢铁、塑料等，大都来源于木材、石油、钢铁等，造成了很大的浪费。二是污染环境。消费者抛弃大量包装废弃物，加重对环境的污染。包装废弃物的年排放量在重量上已占城市固体废弃物的1/3，而在体积上更达到1/2之多，且排放量以每年10%的惊人速度递增。过度包装产生的成本相当可观，而这些耗费大量资源的过度包装物，到了消费者手中全部变成了生活垃圾。

（5）电子废弃物

电子废弃物俗称“电子垃圾”（E-waste or Waste Electronic Equipment），是指被废弃不再使用的电气或电子设备，主要包括电冰箱、空调、洗衣机、电视机等家用电器和计算机等通信电子产品等的淘汰品。电子垃圾需要谨慎处理。在中国等一些发展中国家，电子垃圾的现象十分严重，造成的环境污染威胁着当地居民的身体健康。

电子废弃物的成分复杂，不少家电含有有毒化学物质，其中半数以上的材料对人体有害，有一些甚至是剧毒的。比如，一台电脑有700多个元件，其中有一半元件含有汞、砷、铬等各种有毒化学物质；电视机、电冰箱、手机等电子产品也都含有铅、铬、汞等重金属；激光打印机和复印机中含有碳粉等。电子废弃物被填埋或者焚烧时，其中的重金属渗人土壤，进入河流和地下水，将会造成当地土壤和地下水的污染，直接或间接地对当地的居民及其他的生物造成损伤；有机物经过焚烧，释放出大量的有害气体，如剧毒的二恶英、呋喃、多氯联苯类等致癌物质，对自然环境和人体造成危害。

据2010联合国环境规划署发布的报告，我国已成为世界第二大电子垃圾生产国，每年生产超过230万吨电子垃圾，仅次于美国的300万吨；到2020年，我国的废旧电脑将比2007年翻一番到两番，废弃手机将增长7倍。近年来，我国步入家电更新换代高峰期，也是电子垃圾高速增长期。粗略估计，我国平均每年需报废的电视机在500万台以上，洗衣机约600万台，电冰箱约400万台，每年淘汰1500多万台废旧家电，这还不包括保有量迅猛增长并迅速更新的电子及通讯器材，包括手机、DVD等。与此同时，随着全球电子垃圾产品以每年18%的速度迅猛增长，废旧电脑等电子垃圾产品大量进入中国。据统计，美国每年以废旧电脑为主的电子垃圾有80%被偷运进入了中国、印度和巴基斯坦。

（6）农村生活方式

对中国来讲，现代化的过程中，不能忽视9亿农民生活方式的变化。在我国改革开放以来，随着农村经济的发展，村民生活水平也不断提高。农村居民家庭平均每百户拥有的洗衣机、电冰箱、空调机、摩托车等耐用消费品不断增加。但这也使得农村居民生活能源的使用发生了巨大的变化，农村居民生产、生活废物、废气的排放也相应地增加。农村所受的污染日益严重。加之不少农村居民的生活方式却还相对落后，环保意识较差，废品随意丢弃，废气随意排放，生活污水乱流，禽畜粪便乱拉，导致农村环境污染严重。

——农业生活污染。自古以来，农村的能源消耗主要用于烧饭、取暖，用的主要是柴草、桔杆。尽管这也排放废气，但排放的规模、浓度，对大自然的影响不十分严重。20世纪下半叶以来，这种情况已发生了变化，许多地方农村居民的烧饭、取暖全部用上了方便、省力的煤，或比煤更好用的液化气、天然气。过去农村的炊烟，如今已成了“工业废气”。农村所用电力，许多也是火电，用电增加，也增加了矿物燃料的耗费。

——农村生活垃圾处理不当。农村生活垃圾来源，一是包括农村居民在日常生产生活中产生的煤渣、厨卫垃圾、人畜粪便、塑料废纸、废旧电器、织物橡胶等废弃物。二是相当多的是从城市转移来的生活垃圾，据相关数据显示，我国每年有90%以上的城市垃圾是在郊外填埋或堆放。随着经济的发展，农村人口居住由分散趋向集中，生活垃圾对环境造成的影响也逐渐凸显起来，而且在农村地区没有统一的规划和垃圾处理中心，村民随意丢弃生活垃圾。不少村民的生活污水都是直接排放到河流，将垃圾直接堆放在路边或者河沿，造成农村地区生活垃圾污染日益严重。由于资金、技术有限以及其他原因，村镇生活废弃物处理厂的建设及容量都不能满足实际需要。一份有关中国农村污染状况的调查表明，从农村污染的主要污染源来看，生活垃圾排在第一位，工矿业污染排在第二，化肥、农药等农业污染排在第三。

2. 变革现有的生活方式是中国乃至全球的共同话题

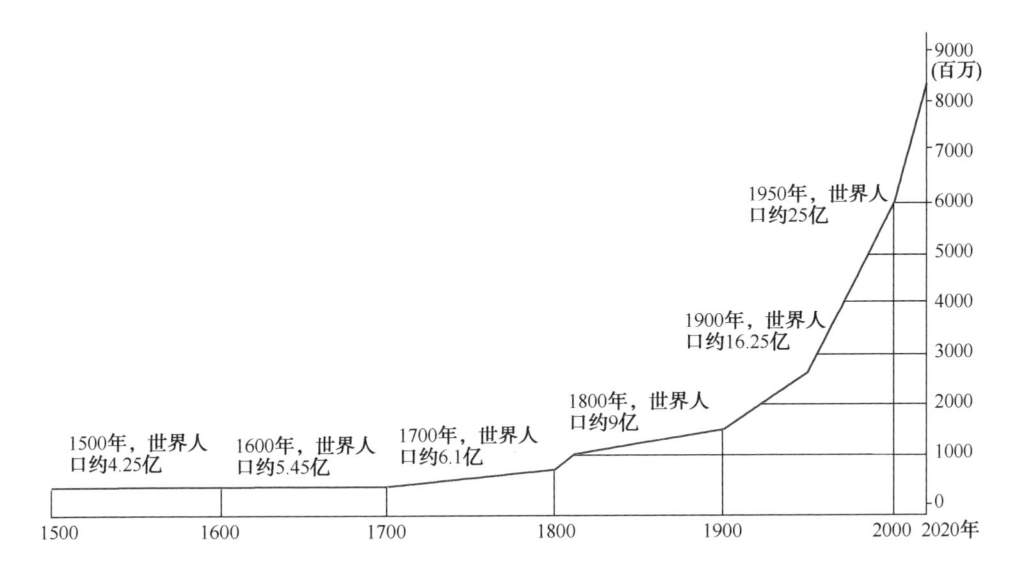

（1）地球到底能够养活多少人

根据2011年《纽约时报》的一篇文章①，全球人口将在很短的时间内突破70亿，人类发展迎来了一个新的里程碑。根据文章的分析，目前地球上人口已经有60多亿，自人类诞生以来经过几百万年，到1804年世界人口才有10亿，经过123年后，1927年才达到20亿，33年以后即1960年达到30亿，14年后即1974年达到40亿，13年后1987年达到50亿，又经过12年即1999年超过60亿，2007年2月25日已经达到65亿。事实上，2011年10月31日凌晨前2分钟，作为全球第70亿名人口的象征性成员丹妮卡·卡马乔已经在菲律宾降生了。由此看来，地球人口增长是随时间呈指数式增加，按目前增长速度，每年世界要增加7800万人口，真是“人口爆炸”。预测到2070年地球人口将近120亿。

然而，目前地球70多亿人口中，大多数人口处于比较贫困的状态，人均国民总产值在1万美元以上，高收入国家人口只占世界总人口14%，人均国民生产

① 《Can the earth support an ever-expanding rise in population?》，Newyork Times，2011.

总值在800美元以下的低收入国家，占世界总人口59%，人均产值在400美元以下的贫困国家，占世界总人口13%，许多国家人民在饥饿线上挣扎，没有解决温饱问题。地球人口爆炸性增长，而人类又要贪得无厌，大量消耗地球自然资源，破坏生态环境，现在70多亿人口就产生以下全球性难以解决的问题。

以今天的美国人的生活消耗为例，美国人每年消耗的电能是1950年的12.5倍，人均年耗能量是中国的20倍，如果13亿中国人都要达到美国的富裕程度和人均能耗水平，则全世界现在所生产的能源，包括煤炭、石油、电力，都供应我们中国还不够。有人测算过，至少还需要3个地球。这些情况说明，今天全球的发展模式必须作出重大的变革。

（2）从人均生态足迹看中国人生活方式的发展困境

2010年，受中国环境与发展国际合作委员会和WWF（中国）的联合委托，由全球生态足迹网络和中国科学院地理科学与资源研究所合作完成的《中国生态足迹报告》分析指出，作为一个国家，中国消耗了全球生物承载力的15%；尽管生物承载力不断增加，中国的需求已经超过其自身生态系统可持续供应能力的两倍多。报告指出，中国的人均生态足迹是1.6全球公顷，也就是说，平均每人需要1.6公顷具有生态生产力的土地，来满足其生活方式的需要。中国的人均生态足迹在147个国家中列第69位，这个数字低于2.2全球公顷的全球平均生态足迹，但仍然反映出中国面临的重要挑战。在Andrea Westall提供的《英国和欧洲的可持续消费政策》[①] 中，人类的生态足迹已经超过地球承载力的50%，中国超过了20%，而北京则超过了120%。同时，生态足迹还在不断增长。（Humanity’s ecological footprint（EF）exceeds Earth’s biocapacity by more than 50%（1.5）; for China the figure is 20%（1.2）and for Beijing（2.2））。事实上，根据一些专家的分析研究，中国是一个人均资源相对贫乏的国家，人均占有煤炭、石油、天然气、土地资源、淡水资源和森林资源，分别仅为世界人均占有量的1/2、1/9、1/23、1/3、1/4和1/6。

放眼今天的世界，在人类欲望、生产规模和资源消耗的竞争中，人类所面临的已经不仅仅是资源的“相对稀缺”，而是“绝对稀缺”问题。人类整体生存处于环境恶化的威胁之下，人类的健康与福利面临全面的危险，任何人都需要面对和关注生态环境问题。改善和挽救生态环境，需要人类的智慧和行动。为此，每个国家需要转型，整个人类社会需要转型。作为当代世界的每一个人，迫切需要

---

① 《UK & European Policy for Sustainable Consumption》，Andrea Westall.

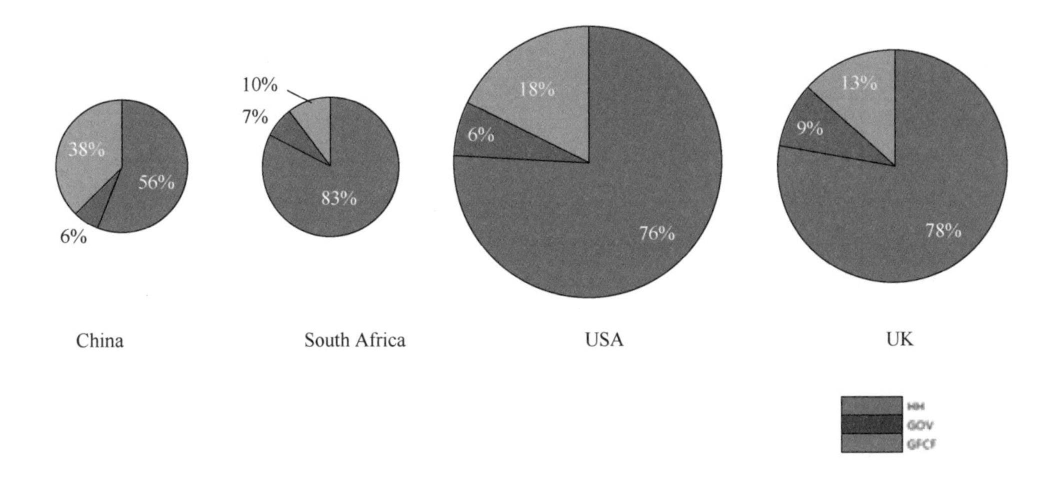

在经济行为、生活方式和价值体系上作出根本性的改变和调整。

（三）大力提倡节约环保的现代生活方式

我们推动可持续发展，建立完全新型的节约环保的现代生活方式，引导人们在服饰、饮食、日用品、建筑、交通、行为等方面抵制旧有生活陋习和奢侈消费，从全新角度把人们对提高生活质量的需求引向正确的方向。

1. 当代中国人理想的生活方式

改革开放以后，特别是进入 21 世纪，居民生活方式呈现跃进式发展。主要表现在：

（1）家庭规模缩小，居住迈向小康。根据国家统计局发布的 2010 年第六次全国人口普查主要数据，全国 31 个省、自治区、直辖市共有家庭户 40152 万户，家庭户人口 124461 万人，平均每个家庭户的人口为 3. 10 人，比 2000 年人口普查的 3. 44 人减少 0. 34 人。与此同时，全国家庭的平均住房面积为 116. 4 平方米，人均住房面积为 36. 0 平方米。

（2）消费水平快速提高。2012 年，恩格尔系数为 35. 8%。家庭饮食社会化趋势日趋明显。人们的生活消费已呈现多元化、社会化的基本特征。家庭设备用品日益表现出多功能、高科技、新样式的特点。技术水平先进、经久耐用的日用消费品不断进入普通百姓家庭，使得居民家庭生活中人们的出行变得更加方便、舒适、快捷。家务劳动也不再显得繁重，日常生活增添了更多的安乐感、享受感、富足感，居民生活方式现代化特征进一步显现。

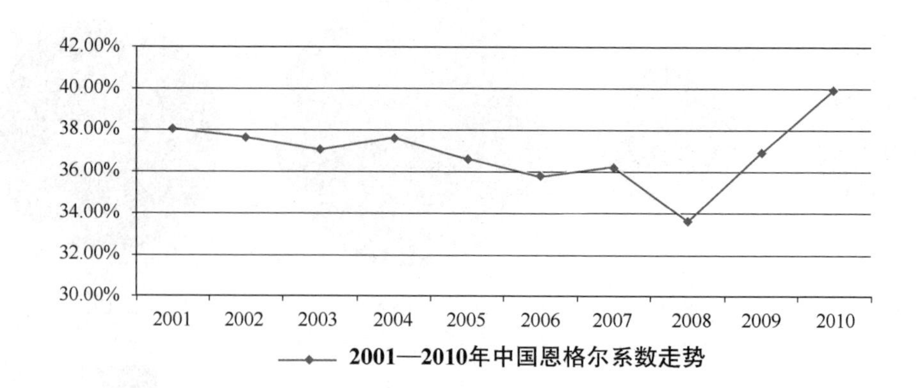

（3）城市居民业余生活丰富多彩、充满生气。20 世纪 90 年代，国家将每周职工的工作日由 6 天改为 5 天，并且国家法定节假日还增加到 11 天，实行了职工带薪年休假制度，居民闲暇时间明显增多。人均文娱费用和旅游花费大幅增长。

人们的生活方式是与所处的社会经济发展状况相对应的。有人曾把中产阶级美国梦用 6 个基本标准进行量化，包括有房、有一部好车、接受过大学教育、有退休保障、有医疗保险以及有休假时间。在中国的现实阶段，提出了到 2020 年全面建成小康社会的发展目标。事实上，这个目标体系由经济发展、社会和谐、生活质量、民主法制、文化教育、资源环境等 6 个方面（或子目标）构成，每个都与当代中国人的生活息息相关，要求我们逐步实现更为理想、健康、和谐的生活方式。

全面建成小康社会统计监测指标体系及综合评价方法

监 测 指 标	单位	权重（%）	标准值（2020 年）
一、经济发展		29	
1. 人均 GDP	元	12	≥31400
2. R&D 经费支出占 GDP 比重	%	4	≥2.5
3. 第三产业增加值占 GDP 比重	%	4	≥50
4. 城镇人口比重	%	5	≥60
5. 失业率（城镇）	%	4	≤6
二、社会和谐		15	

续表

监 测 指 标	单位	权重（%）	标准值（2020 年）
6. 基尼系数	-	2	≤0.4
7. 城乡居民收入比	以农为 1	2	≤2.80
8. 地区经济发展差异系数	%	2	≤60
9. 基本社会保险覆盖率	%	6	≥90
10. 高中阶段毕业生性别差异系数	%	3	=100
三、生活质量		19	
11. 居民人均可支配收入	元	6	≥15000
12. 恩格尔系数	%	3	≤40
13. 人均住房使用面积	平方米	5	≥27
14. 5 岁以下儿童死亡率	‰	2	≤12
15. 平均预期寿命	岁	3	≥75
四、民主法制		11	
16. 公民自身民主权利满意度	%	5	≥90
17. 社会安全指数	%	6	≥100
五、文化教育		14	
18. 文化产业增加值占 GDP 比重	%	6	≥5
19. 居民文教娱乐服务支出占家庭消费支出比重	%	2	≥16
20. 平均受教育年限	年	6	≥10.5
六、资源环境		12	
21. 单位 GDP 能耗	吨标准煤/万元	4	≤0.84
22. 耕地面积指数	%	2	≥94
23. 环境质量指数	%	6	=100

2. 在中国社会建立节约环保的生活方式的基本思路

20 世纪 90 年代以来，全球性的资源耗竭和环境污染迫使人类开始对自身生活方式进行反思。与此同时，可持续发展理论逐渐完善，将环境影响纳入经济社会发展的范围。按照可持续发展的要求，不仅要对生产模式进行变革，而且要对生活方式进行变革。因此，我们应当建立起中国特有的、符合中国经济社会发展实际的生态文明理念和核心价值体系，按照尊重传统、适应时代，保护个性、提倡多元，提升公平、健康有序的要求推行和建立一种现代社会生活方式。我们认为，遵循可持续发展的理念，建立现代生活方式应当遵循和体现以下三方面原则：

一是适度原则。即从人与自然的角度看，要以人的需要作为出发点，以人的健康生存作为目标，逐步减少无意义的浪费和对人类健康无益甚至有害的生活方式；承认地球资源的有限性，要把各种需求的水平控制在地球承载能力范围之内，保持资源和环境的“可持续性”。适度原则既反对过度浪费，也反对过分节约。过度浪费超出了人自身的正当需要，消耗了更多的资源、加剧了环境的破坏，本质上只是满足了一些不合理的社会与心理需求。过分节约虽然减少了消费中的物质消耗，但降低了生活质量，抑制了生活情趣，也不利于身心与个性的健康发展。

二是公平原则。生活方式所带来的环境问题的实质是人与人之间的环境利益冲突。由于资源的有限性，当一部分人过度使用环境资源的时候，其他人使用同类环境资源的利益就会受到挤压或损害。因此，从人与人之间的关系看，提倡面向全体公民的公平原则，谁消耗资源越多，谁应负的代价越大，体现出公平与公正性。公平的原则认为，追求生活质量的权利，对于当代的每一个人、对于后代的每一个人应该同等地享有；任何人都不应由于自身的生活方式而危及他人的生存（代内公平），当代人不应该由于本代人的生存方式而危及后代人的生存（代际公平）。

三是以人为本的原则。即在生活方式上要形成合理的结构，以实现人的本质需要以及人的全面发展为目标，提倡环境友好的生活方式，注重生活质量的提高而不仅仅是资源数量的消耗。为此，必须倡导在生活方式中，增大精神文化的比重。在物质生活中，增大绿色产品的比例，把对环境有害的各种生活方式控制在最低限度。

综合起来，我们认为可持续的、科学的、现代的生活方式是指，形成符合人的身心健康和全面发展要求、促进经济社会发展、追求人与自然之间、人与人之间和谐互利的生活观念、生活方式、生活行为，最终实现人类生活“发展性”与“持续性”的双赢。

3. 主要行动

（1）建立低碳环保的现代消费模式

联合国开发计划署早在1998年提出的《人类发展报告》里就总结了20世纪

人类消费的得失并提出警告：20 世纪人类消费急剧增长，但存在一些误区，出现了炫耀型、竞争型、摆阔型的消费。目前，我国政府提出“要在全社会形成崇尚节俭、合理消费、适度消费的理念”。改变长期形成的“面子消费”、“奢侈消费”、“便利消费”等落后消费方式。

——小型公务聚餐采用“份儿饭”。在澳大利亚，10 个人以下的小型宴请实行分餐制，由宾客根据菜单选择自己喜欢吃的食物。如果是 10 个人以上，那么在活动当中，参加者是没有座位的，需要站立着和别人进行沟通，而在饮食环节，往往是会采取自助式。

——针对食物浪费设处罚规定。德国被认为是处罚餐厅浪费最严的国家。在德国，无论自助餐还是点餐，都不能浪费，一旦发现有人浪费，任何见证人都可向相关机构举报，工作人员会立即赶到，按规定罚款。

——宣传“够吃就好，杜绝浪费”的消费理念。在北京，在微博和论坛上“晒光盘”也成为了节日期间新的时尚，不少网民晒出菜肴被吃得干干净净的餐盘照片，身体力行向中国式“剩”宴说“不”。在腾讯微博上，“中国光盘节宣言”阅读量超过 400 万，转发数千条，新浪微博上参与“光盘行动”微话题讨论的微博也超过了 1530 万条。

——少买不必要的衣服并且采取节能方式洗衣。人们每年购买的衣服数量繁多，款式多样，衣服洗涤也经常使用洗衣机。同时，服装在生产、加工和运输过程中，要消耗大量的能源，产生大量污染物。

（2）选择低能耗的绿色出行方式

一些发达国家，城市客流量的 50%—60% 由公共交通承担，东京达 90%，而中国城市公共交通承载的客流量还不足 10%，即使国内最为发达的北京也不足 1/3。

——居民在出行时，应充分利用方便快捷的公共交通系统，发挥水运、铁路、城市公共交通等比较优势，考虑多种交通运输方式的衔接和协调，选择方便、快捷、经济有效的出行路线。出行时注意控制好私家车的使用，尽量选择乘坐公共交通工具或者步行、骑自行车，环保驾车、文明驾车，用自己的实际行动来缓解道路交通压力，节约能源，减少环境污染，形成自行车、机动车和行人和谐发展。

——鼓励拼车上下班。在一些美国城市，政府要求在上下班高峰时期，进出城的主干道内侧设立一条专门快速路，只有车内至少有两人或三人的汽车方能驶入。

——提升油品质量，减少汽车尾气。一项权威的项目环境影响评价结果显示，由于油品质量升级，北京市汽车尾气中排放的二氧化硫每年将减少4000吨以上，二氧化碳每年也将减少4000吨以上。

——发展新能源汽车。新能源汽车是指除汽油、柴油发动机之外所有其他能源汽车，包括燃料电池汽车、混合动力汽车、氢能源动力汽车和太阳能汽车等。新能源汽车废气排放量比较低，已在世界各地推广。中国在新能源的研究上与发达国家差距不大，发展新能源汽车既能拉动消费，还能减少污染，是一举两得的好事情。

(3) 减少化石能源和薪柴消费

推广沼气使用、节柴改灶、太阳能热水器、太阳能光伏利用、秸秆优质化能源利用；尽量选择混合燃料、电力动机车及低排量、低耗量的机动车，减轻交通、出行对环境的污染；减少对塑料制品、一次性消费品、纺织品、皮革的需求量，不乱扔电池等电子废弃物。

(4) 建立现代化居住方式

持续大规模的住宅建设与人多地少、资源短缺的矛盾，是当前中国城镇化建设必须解决好的问题。

——合理选择户型。在不降低舒适度、不降低生活质量的前提下，把户型面积做小一点，既能减轻沉重的购房压力，又能减少长期住房消费中可观的费用支出。

——合理改造老旧建筑。通过维修更新，改善老旧建筑的居住环境，延长其使用年限。因建筑本身以外的其他原因需要拆除的，也应由相应部门和专家论证，重要地段的建筑甚至应进入听证程序。

——大力提倡“节能省地型”建筑，内涵是节能、节地、节水、节材。争取到2020年，我国大部分既有建筑完成节能改造，新建建筑实现节能65%的目标，实现建筑建造和使用过程中节水率在现有基础上提高30%以上，新建建筑对不可再生资源的总消耗比现在下降30%以上。

(5) 倡导并扶持农村低碳环保生活

——发展农村新能源。更新改造传统的省柴节煤炉灶和节能低碳炕，加快省柴煤灶炕的升级换代。有条件的地方，利用生物技术，将果皮菜叶、秸秆、人畜粪便、灰土、养殖场污水放入封闭的沼气池，进行厌氧发酵。沼气为居民提供生活燃料；沼液可以喂养蚯蚓等有益昆虫，改良土壤结构；沼渣入田增加土壤肥力，另外也可以利用沼渣种植蘑菇、金针菇等菌类。

——推广应用保温、省地、隔热新型建筑材料，发展节能低碳型住房，在北方地区引导农民建造太阳房和使用太阳能热水器。

——综合利用多种方式处理生活垃圾，如征用农民废弃宅基地修建中小型的垃圾回收站，分类集中收集居民的生活垃圾，将可燃类垃圾送往发电站，可回收利用的送往有能力生产的企业实现资源再生产。

——面向农户开展环保宣传工作，提高农民的环保意识，宣传工作中，可以发放图文结合的海报、书籍等宣传材料，指导大家形成低碳环保的生活方式。政府也可以适当给予农户奖励。

(6) 在公共生活中建立起推动全社会树立崇尚自然、崇尚环保、崇尚节约的社会风尚和文化氛围

——加强生态文明观念的教育。转变环境教育观念，从人与自然和谐统一的高度树立正确的自然观、环境价值观。利用各种媒体和舆论工具，大力宣传环境保护知识和环境法规，提高公民环保意识，大幅度提高社会公众参与可持续发展的力度。同时环境教育要求人们依据环境法则，逐步树立起“一个地球”的意识，树立起人与自然平等、国际间和代际间公平的思想。

——加强生态道德教育。增强人们对于生态环境的道德意识。生态道德是人类道德的重要方面，保护自然环境、维护生态平衡是人类为了自身的生存所应履行的道德义务与责任。生态道德既包括人对自然的道德，也包括人对人的道德。从“人是自然”的观念出发，人对人的道德亦是人对自然的道德的表现。

——加强生态法制教育。保护自然环境，建设生态文明，不仅需要人类的道德自觉，同时更需要社会法制的保障。

第五章　城镇化进程中改善人居环境和减少贫困

环境问题实际是人与环境的关系问题，而城市是人类生产生活的高度集中之地。中国在过去30年及未来相当一个时期内，正处于城镇化加速推进的阶段，如何解决城镇化进程中凸显的环境问题已经成为经济社会发展面临的重要课题。本章着重分析中国城镇化进程中带来的主要环境问题，结合中国的国情，提出解决城市环境问题的有关建议。

一、中国城镇化快速推进

中国是一个世界上人口最多的发展中国家，2012年底总人口达到13.5亿，其中城镇常住人口7.1亿人，城镇化率为52.57%。中国的城镇化在结构和分布上有以下几个特点，这些特点又是造成各种环境问题的客观因素。

（一）在短时期内城镇化加速推进

1979—2012年，中国城镇化率年均提高1个百分点，每年从农村转移到城市的人口超过1500万人，相当于一年创造一个特大城市，这样的城镇化规模世界上任何国家都是绝无仅有的（图5-1）。但目前中国的城镇化率不仅明显低于发达国家近80%的平均水平，也低于发展中国家60%左右的平均水平（图5-2）。如印尼的城镇化率为54%，墨西哥的城镇化率78%，巴西是86%。按照城镇化的一般规律，城镇化率在30%—70%之间时，城镇化处于加速发展的阶段。预计未来20年，中国的城镇化仍将保持年均提高1个百分点左右的推进速度。

（二）城镇化以小城镇为主体

尽管中国有北京、上海、天津等一些人口超过千万的特大城市。但总体上无论从城市布局还是人口分布，都以星罗棋布的中小城市为主。截至2010年底，全国地级及地级以上城市287个，其中400万人口以上城市14个，200万—400

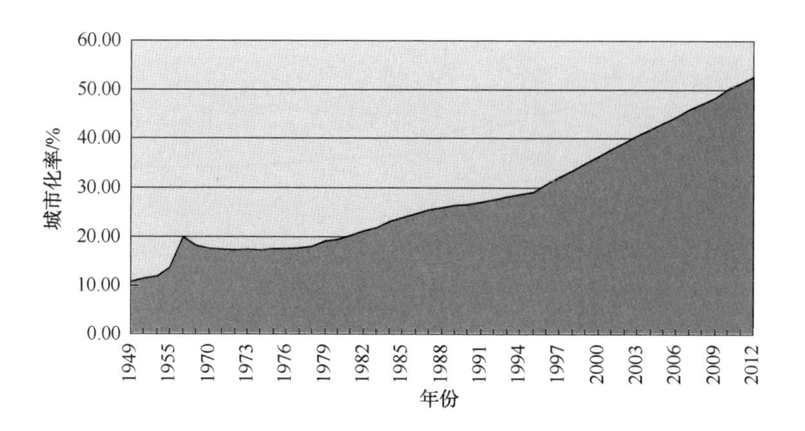

图 5-1　1949—2012 年中国城镇常住人口占总人口比重（%）

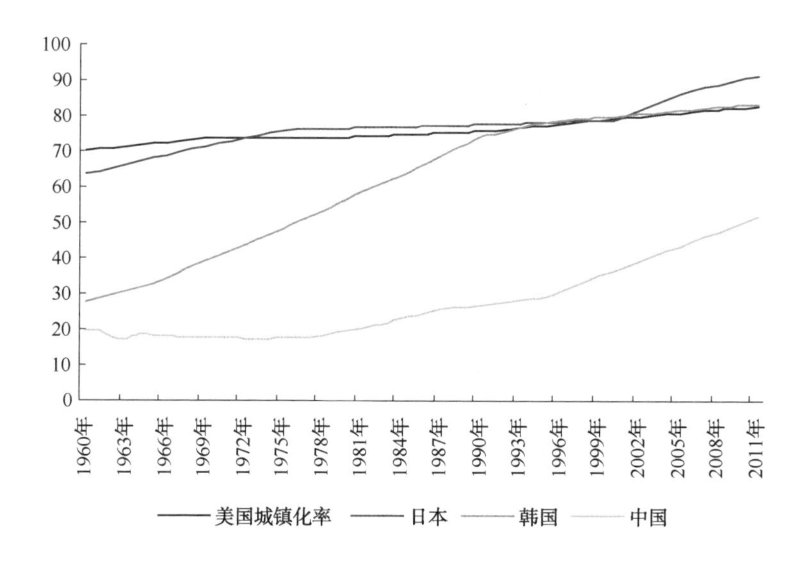

图 5-2　中国与美日韩城镇化率对比（%）

（资料来源：国家统计局）

万人口城市 30 个，100 万—200 万人口城市 81 个，50 万—100 万人口城市 109 个，20 万—50 万人口城市 49 个，此外还有近 2 万个建制镇（图 5-3）。其中，50 万人以下城镇人口占全部城市人口的 54%。

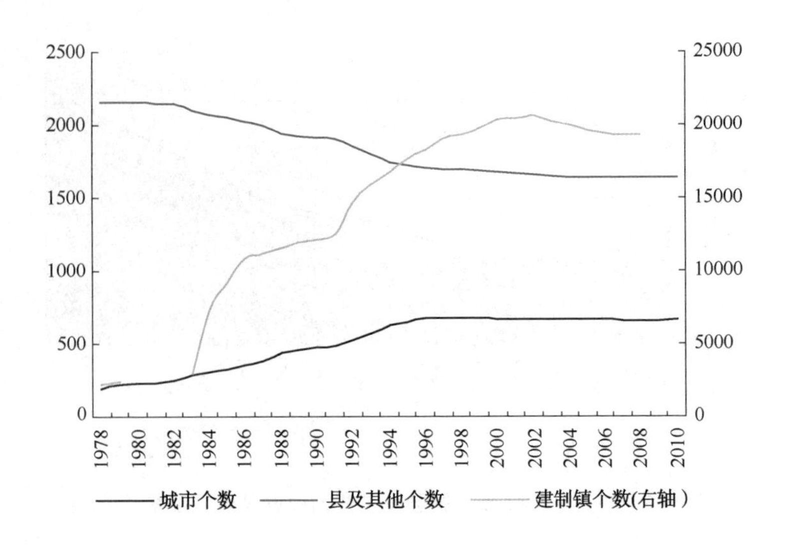

图 5-3　中国不同类型城市数量结构

（资源来源：中国城乡建设统计年鉴）

（三）人口转移更多依靠农民工进城

2012 年，中国的城镇化率如果按照户籍人口计算，仅有 35.29%。在城市和城镇中生活半年以上的农村人口达到 2.34 亿人，占整个城市人口的 33%（表 5-1），这部分人口主要是从农村到城市打工的群体（表 5-2）。

表 5-1　城市常住人口与户籍人口之差

单位：亿人、%		1978 年	1990 年	2000 年	2002 年	2007 年	2010 年	2011 年	2012 年
常住人口	城镇	1.72	3.02	4.59	5.02	6.06	6.70	6.91	7.12
	农村	7.90	8.41	8.08	7.82	7.15	7.71	6.57	6.42
	城镇化率	17.92	26.41	36.22	39.09	45.89	49.95	51.27	52.57
户籍人口	城镇	1.52	2.41	3.31	3.58	4.35	4.58	4.66	4.78
	农村	8.12	9.02	9.37	9.26	8.86	8.83	8.82	8.76
	城镇化率	15.80	21.10	26.08	27.89	32.93	34.17	34.55	35.29
城镇中无城镇户籍人口		0.20	0.61	1.28	1.44	1.71	2.12	2.25	2.34
占城镇常住人口比例		11.63	20.20	27.89	28.69	28.22	31.64	32.56	32.87

（数据来源：中国统计年鉴及公安部数据）

表 5-2　农民工规模

	2008 年	2009 年	2010 年	2011 年	2012 年
农民工总量（万人）	22542	22978	24223	25278	26261
1. 外出农民工	14041	14533	15335	15863	16336
（1）住户中外出农民工	11182	11567	12264	12584	
（2）举家外出农民工	2859	2966	3071	3279	
2. 本地农民工	8501	8445	8888	9415	9925

（资料来源：国家统计局农民工调查监测报告）

（四）城镇化水平在区域上存在很大差异

中国东部地区相对于广大的西部地区更适宜人居，也更符合以港口和交通枢纽为主的现代经济发展。西部地区 12 省区市国土面积占全国的 70%以上，而人口仅占 1/4。20 世纪 30 年代，中国著名地理学家胡焕庸画过一条线，从黑龙江的黑河到云南的腾冲，这条斜线的东南方向大约占国土的 1/3，聚集了 93%以上的人口。东西地区不仅人口密度有很大差距，由于经济发展水平不同，其内部城镇化水平也相差很大，东部地区城镇化率明显高于中西部地区（表 5-3）。

表 5-3　分区域城镇化率（单位：%）

地　区	2000 年	2007 年	2012 年
东部	45. 32	55. 20	61. 86
中部	29. 56	39. 41	47. 19
西部	28. 67	37. 01	44. 75
东北	52. 13	55. 81	59. 60

（资料来源：国家统计局）

（五）城市内部存在明显的“二元结构”

中国最大的差距在城乡之间，目前城乡之间收入分配差距达到 3. 23∶1。但目前城市间、城市内部的差距表现得更为突出，一面是高楼林立的新城区，一面是棚户区连片、困难群众聚居、基础设施陈旧的老城区，两者形成鲜明对比。中国一些偏远的城市特别是中小城市，大多是在推进工业化初期，依托资源富集地区，以资源开采加工为主导产业，“因矿设市”、“因厂设区”，迅速集中了一大批产业工人，形成了一批城市与独立工矿区。这些城市基础设施简单，当资源枯竭后，城市发展存在严重问题。2012 年底，全国仍有 5000 平方米以上集中连片棚户区 1200 万户，涉及 4000 多万人。再加上其他城市中没有管道自来水的住户，不同时拥有厨房和厕所的住户。据估算，涉及 1 亿以上的人口。住在棚户区的居民，大多为低收入或中等偏下收入群众，包括老企业职工及其家属、城市低保户、农民工及其家属，这是城市内部贫困问题的主要体现。

二、城镇化带来巨大环境压力

城镇化加速给中国经济持续快速发展带来了不竭的动力。1979—2012 年，中国年均经济增长速度达到 9.9%，其中近 20 年达到 10.5%，城镇化率提高发挥了重要的拉动作用。从未来发展来看，城镇化仍将是中国国内需求持续增长的最大潜力。但在城市快速发展的同时，也带来了前所未有环境压力，城市人居环境在一些地区明显恶化。

（一）城市建设运行耗费了大量资源

中国城镇化总体上是建立在高消耗、高排放、高扩张、低效率的基础之上，给资源环境带来巨大压力。土地资源占用过快。按建成区面积计算的人口密度，由 1981 年的每平方公里 2.71 万人，减少到 2010 年的每平方公里 1.67 万人。近 30 年时间，城市人均占用土地增加了 62%（表 5-4）。

表 5-4　城市建设用地面积

项　　目	1981 年	1990 年	1995 年	2000 年	2009 年	2010 年
城镇人口　（万人）	20171	30195	35174	45906	64512	66978
城区面积　（平方公里）	206684	1165970	1171698	878015	175464	178692
建成区面积　（平方公里）	7438	12856	19264	22439	38107	40058
城市建设用地面积　（平方公里）	6720	11608	22064	22114	38727	39758
城区人口密度　（万人/平方公里）	2.71	2.35	1.83	2.05	1.69	1.67

（数据来源：《2011 年中国统计年鉴》、《2010 年中国城乡建设统计年鉴》）

城市能源消耗急剧上升。在能源消费中，2010 年中国工交行业和城镇生活能源消费占 84.6%，交通能耗主要集中在城市。城镇人均生活能耗是农村人均水平的 1.54 倍。城镇每年单位建筑面积耗能约 13.2 千克标煤/平方米，是农村地区的 4.52 倍。城市和工业发展方式粗放，导致单位 GDP 能源资源消耗水平居高不下（图 5-4）。

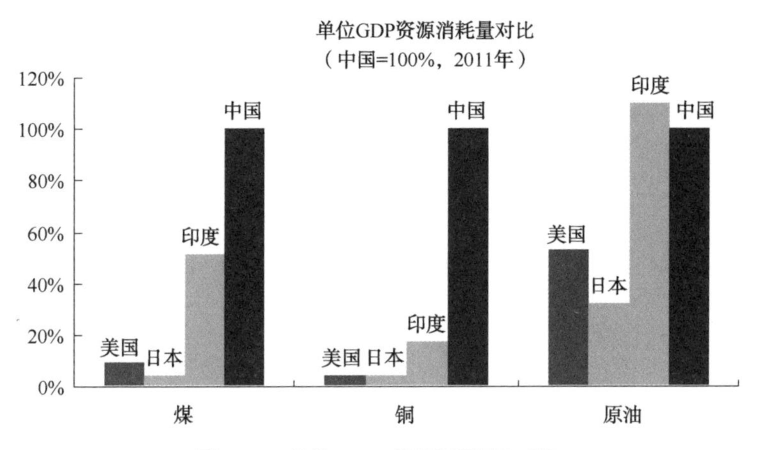

图 5-4　单位 GDP 资源消费量对比

城市水资源供需矛盾突出，越来越多的城市处于严重缺水状态。1979 年中国有 154 座城市缺水，到 2010 年 420 多座城市供水不足，其中 110 座严重缺水，缺水总量达 105 亿立方米。华北、东北、西北以及东部沿海地区的城市，缺水尤为严重。2010 年，北京、天津、上海、深圳、郑州、太原、石家庄等城市人均水资源不足 200 立方米，沈阳、唐山、青岛、烟台、济南、连云港、西安等城市人均水资源不足 500 立方米，处于极度缺水状态。随着城镇化的快速推进，今后城市水资源供需矛盾还将变得更加尖锐。预计人口城镇化率每提高 1 个百分点，全国生活用水净增加 5.62 亿立方米。

（二）城市空气污染日益严重

目前中国约有 1/5 的城市空气污染严重。颗粒物污染是影响城市空气质量的首要污染物。根据国家环境监测总站公布的《2013 年 1 月 74 个城市空气质量状况月报》，2013 年 1 月份，全国监测的 74 个城市空气质量总体的达标天数比例为 31.6%，超标天数比例占 68.4%，其中轻度污染占 24.7%，中度污染占 13.5%，重度污染占 20.2%，严重污染占 10.0%。其中 $PM_{2.5}$ 平均超标率为

68.9%，PM_{10}平均超标率46.9%，NO_2平均超标率为23.5%，SO_2平均超标率为14.9%。城市酸雨污染仍较重，分布区域主要集中在长江沿线及以南、青藏高原以东地区。2011年，在全国监测的468市（县）中，有227个市县出现过酸雨，占48.5%；140个市县酸雨频率在25%以上，占29.9%；有44个市县酸雨频率在75%以上，占9.4%。降水PH年均值低于5.6（酸雨）、低于5.0（较重酸雨）和低于4.5（重酸雨）的市（县）分别占31.8%、19.2%和6.4%。

（三）城市水污染状况得不到改善

2012年，全国七大水系的571个水质监测断面中，Ⅰ—Ⅲ类水质断面比例占63.9%，劣Ⅴ类水质占12.4%，七大水系水质总体为轻度污染。流经城市的河段普遍受到污染，一些地区已经出现了“有河皆干、有水皆污”的现象，近岸海域赤潮和三峡库区支流“水华”现象接连发生。2011年，监测的5个城市内湖中，东湖（武汉）、玄武湖（南京）和昆明湖（北京）为Ⅳ类水质，西湖（杭州）和大明湖（济南）为Ⅲ类水质。玄武湖、东湖和西湖为轻度富营养状态，大明湖和昆明湖为中营养状态。水污染已威胁城市饮用水安全。全国大中城市浅层地下水不同程度地遭受污染，约一半的城市市区地下水污染较为严重。在全国200个城市4727个地下水水质监测点中，水质较差级和极差的监测点分别占40.3%和14.7%，合计55.0%。大城市的中心地带、城镇周围区，以及排污河道两侧、引污灌溉区污染尤为严重。重要城市往往靠近大江大河，城市产生的大量污染物源源不断地排入其周围地表水体，甚至威胁城市饮用水源地。在全国113个环保重点城市监测的389个集中式饮用水源地中，不达标水量达21.3亿吨，占监测点取水总量的9.4%。

（四）城市固体垃圾处理跟不上

据统计，目前全国城市生活垃圾累计堆放存量达70多亿吨，占地80多万亩，并且以年平均4.8%的速度持续增长。全国2/3的大中城市陷入垃圾包围之中，1/4的城市已没有合适场所堆放垃圾。由于环保意识差和缺乏分类管理，已经形成“垃圾围城”之势，并将愈演愈烈。如，2012年中秋国庆“黄金周”，有着“东方夏威夷”之称的海南三亚，仅在中秋夜后，大东海景区3公里海滩共清理出50吨游人留下的生活垃圾，严重污染了海岸环境。就在“十一”当天，天安门广场共清理出生活垃圾7900多公斤，比往年同期大幅增加。受工业污染、固体废弃物、过度和不合理使用化肥等影响，土壤环境也呈现总体质量下降。近年来，不少地方出现重金属污染事件，严重影响人民群众尤其是儿童的身体健康。

（五）城市交通拥堵成为难题

交通拥堵已成为我国城市特别是大城市的常态。一些大城市即便采取了区域限行、尾号限行、购车限制等措施，也难以从根本上解决问题。目前中国约有2/3的城市在高峰时段出现拥堵。特别是在一些特大城市，交通拥堵现象十分严重，有的城市中心地区高峰时段几乎接近瘫痪状态。交通拥堵导致城市居民上下班通勤时间大幅增加。根据智联招聘和北京大学社会调查研究中心（2012）的联合调查，北京、上海、天津、沈阳、西安、成都等城市上下班平均通勤（往返）时间已经超过1小时，广州、青岛、武汉、重庆、郑州、南京、长春、深圳、杭州等城市也在0.85小时以上，其中北京高达1.32小时，上海达1.17小时，天津达1.15小时，分别居前三位。英国Regus（2011）的调查显示，中国的上班族平均每天在上班路上花费的时间领先全球，仅次于印度尼西亚，比世界平均水平高出31.7%，比加拿大和美国平均水平高出近乎1倍。

如此大规模长时间的通勤，既耗费了居民大量的时间和精力，也造成了无谓的经济损失、能源消耗以及环境污染。交通排放中的一氧化碳、碳氢化合物、氮氧化物（NO_x）和挥发性有机化合物（VOC_S）是城市大气的主要污染源，而城市空气中的浮尘（包括总悬浮物颗粒物TSP、PM_{10}、$PM_{2.5}$等）约有50%来源于汽车尾气排放及车辆行驶过程中带起的扬尘。城市交通能耗已经成为城市可持续发展面临的重要资源问题。国际能源机构（IEA）的研究报告显示，2000年全球约50%的石油消耗在运输部门，到2020年全球运输用油将占石油消耗的60%以上。中国每年因汽车增长而增加石油消费达3000万吨，过去五年国内新增炼油能力全部被新增汽车消耗掉。

（六）城市特色文化逐步消失

城市面貌是历史的积淀和文化的凝结，也是城市人居环境的重要体现。一座城市的文化发育越成熟，历史积淀越深厚，城市的个性就越强，品位就越高，特色就越鲜明。但是，一些城市在建设和发展中，城市面貌正在急速地走向趋同。由于抄袭、模仿、复制现象十分普遍，毫无特色的城市街区占据着越来越显著的位置，导致“南方北方一个样，大城小城一个样，城里城外一个样”的文化危机，“千城一面”的现象日趋严重。

三、在改善人居环境中推进新型城镇化

城镇化的核心是人的城镇化，而人的生存与发展必须建立在适宜的生态环境基础之上。快速发展的中国城镇化，需要改变原有的粗放扩张模式，在有限的资源环境支撑条件下，走出一条绿色、低碳、可持续的新型城镇化道路。

（一）集约利用土地

土地利用模式是影响城市可持续发展和绿色发展的首要因素。城市的空间结构、用地构成和区位关系，决定了居住、工作、购物、制造和消费等活动的空间分布、出行需要和能耗水平。中国是人均土地和有效耕地不足的国家，城镇化只能走集约利用土地之路。

一是形成紧凑的城市空间结构。城市空间结构是城市要素的空间分布和相互作用的内在机制，其不仅决定了整体城市形态，决定了城市的交通出行结构，对城市的资源能源消耗也起到了决定性的影响。紧凑的城市空间结构有利于步行、自行车、公共交通等低碳交通方式为主的交通出行方式，从而减少了由于城市空间过于分散而导致的大量小汽车出行带来的高能耗与大量的碳排放。所以，确定紧凑的城市空间结构应成为绿色城市发展空间系统规划的核心内容之一。

二是保持合理的土地开发强度。土地资源的有限性及社会经济发展对土地持续的需求，迫使人类社会必须节约用地，从而减少对土地资源的消耗，保证一定面积的自然生态空间。在城市土地利用中，合理确定土地的开发强度能够提高土地的使用效率与经济效益，延缓城市外延扩张的速度，有利于形成紧凑的城市空间结构，更好地保护农业生产及生态环境，促进生产、生活与环境的协调发展。

三是规划完善的城市绿地布局。城镇化快速发展占用了大量的农田、生态绿地，这些用地变成城市建设用地后，使自然生态系统的循环受到阻碍，进一步加剧了城市对生态环境的影响。因此，必须从生态学原理入手，合理安排城市的土地利用。其中最主要的任务之一就是完善城市绿地布局。通过城市绿地布局，将城市内部的生态源、生态斑块与城市外围的生态源联系起来，形成自然生态网络。完善的城市绿地布局能够减少城市建设用地对自然环境及自然生态循环的阻隔，通过城市各类绿地的布局，提高城市用地的生态化水平。尤其是在提高城市地下蓄水能力、促进城市水体的自然循环以及吸收城市中的碳排放方面，完善的绿地布局将发挥更大的作用，从而实现城市的低碳生态化。

（二）建设便捷交通

交通是城市运转与发展的基本保障。交通系统的布局和模式直接影响土地利用效能。传统城市交通系统由于以便利汽车通行为发展导向，对交通与人、环境、资源、城市社会的关系重视不够，常常盲目加建、拓宽道路，缺乏系统规划，带来交通拥堵、交通污染、交通能耗以及交通安全等一系列社会问题。因此，应当以建设人性化、多元化、生态化、现代化的高效可持续化发展的绿色交通系统为目标，从区域交通、市区交通、社区交通等多个层面出发，建设绿化、便捷、高效城市交通系统。

应实现交通模式低碳化。交通模式是城市交通系统中不同交通方式所承担的交通量的比例关系。从发达国家实践看，公交、慢行交通和环保型交通工具是绿色交通体系建设的重点。优先发展公共交通是提高交通资源利用效率、节能环保、缓解交通拥堵的重要手段，世界上许多大城市都努力打造"公交都市"，东京、纽约、伦敦、新加坡、香港等较为典型，轨道交通和快速公交是发展的重点。因此，绿色交通体系必须以公共交通为主导，通过发展轨道交通，形成公共交通主导型交通模式。同时，在公共交通发展中，还要不断提高公共交通与私人交通相比较的竞争优势，以及实现轨道交通网络与公共汽电车交通网络的有效衔接，让居民自愿选择公共交通出行方式，从而实现交通模式低碳化。

应实现道路网系统低碳化。形成具有低碳生态特性的交通系统，需要建设符合低碳生态特性的道路网系统。首先规划建设适宜出行的步行道路系统、非机动道路系统、公共交通道路系统，再规划建设私人小汽车、货运汽车的道路系统。通过合理规划，将多种功能在小范围聚集，使居民通过较短的交通出行就能够满足工作、购物、休闲等需要，在一定范围内倡导以步行或非机动车为主，满足日常生活出行需求，减少交通出行的碳排放。在城区之间主要通过公共交通满足居民日常跨区域出行。如果全面形成各种出行方式有效结合的道路网络系统，方便居民的步行、非机动车、公共交通等方式的出行，更多的居民会主动地选择低碳的交通出行模式，从而促进形成具有低碳生态特性的交通系统。

实现交通基础设施生态化。主要是指在交通基础设施规划建设过程中以生态学为基础，以人与自然的和谐为目标，以现代技术和生态技术为手段，最高效、最少量地使用资源和能源，最大可能地减少对环境的冲击，以营造和谐、健康、舒适的人居环境状态。

（三）推广清洁能源

实施清洁生产。不仅指生产过程要节约原材料、能源并减少排放物，同时也要求最大限度地减少整个生产周期对人的健康和自然生态的损害。传统生产是一种只强调物质生产而忽视生态环境保护的生产方式。改变这种生产方式，需要不断提高社会清洁生产意识，引导人们转变传统生产观念，让清洁生产的要求和方式深入人心，使采用清洁能源资源、预防和减少污染成为政府、企业和社会的自觉行为。

推广清洁能源。在确保充足和稳定的能源供应基础上，优化调整能源结构，落实节能技术措施，建立社会经济效益和生态环境效益双赢的能源供需系统。实现能源结构多元化，改变以煤炭为主的单一能源来源结构，加快发展可再生能源，减少煤炭等化石能源在能源消耗中的比重。完善能源供应体系，通过加大区域经济合作保障能源供应安全，加大能源设施的建设力度，保障能源特别是清洁能源的供应通道安全畅通。在城市发展中的各个领域贯彻节能原则。在产业发展上应发展节能型产业，通过产业和产品结构调整以及节能技术措施的推广减少能源消耗。

发展循环经济。传统经济是一种单向流动的线性经济，即“资源—产品—污染排放”，其特征是高开采、低利用、高排放。这种经济条件下，人们高强度地把地球上的物质和能源提取出来，然后又把污染和废物大量排放到水系、空气和土壤中，对资源的利用是粗放的和一次性的，通过把资源持续不断变成废物来实现经济的数量型增长。绿色城镇化要求物质流动要充分循环，按照“资源—产品—再生资源”的循环方式来组织经济活动，所有的物质和能量能在这个循环中得到最大限度的利用，以把经济活动对自然环境的影响降低到尽可能小的程度。西方发达国家发展循环经济一般侧重于废物再利用，而中国处于工业化高速发展阶段，能耗物耗过高，资源浪费严重，前端减量化的潜力很大。因此，应当改变过去以单一的经济增长为导向的发展方式，明确地区综合生态承载力，并以此作为重要依据，在生产、流通和消费的各个环节尤其是源头环节，注重资源能源利用的高效化、减量化和资源化，尽可能使得物质和能源在循环中得到合理和持久的利用。

（四）治理主要污染

推进减量化、资源化和无害化。加大政策支持力度，积极推进生活污水减量化和生活垃圾减量，推进固体废物的减量化、资源化和无害化，大力推广垃圾分类收集与清洁直运，减少污染物排放，破解“垃圾围城”现象。

解决空气污染严重问题。空气是流动的，要彻底解决空气污染问题，需要建立生态环境联合治理的合作机制，尤其是要建立跨区域的生态环境补偿机制。否则，单靠一个地方采取治理措施，收效受限。在监测方面，新修订的《环境空气质量标准》提出，$PM_{2.5}$监测分三步走：第一步，2012 年在京津冀、长三角、珠三角等重点区域及直辖市、省会城市开展监测并公布信息；第二步，2013 年在全国 113 个环境保护重点城市和国家环境保护模范城市开展监测并公布信息；第三步，2015 年在所有地级以上城市开展监测并公布信息。在治理方面，国家《重点区域大气污染防治“十二五”规划》，对全国 13 个重点区域的 $PM_{2.5}$治理提出了有针对性的措施，明确要求五年内京津冀、长三角、珠三角 $PM_{2.5}$浓度年均下降 6%，其他 10 个城市群年均下降 5%。2013 年 3 月，针对连续不断出现的雾霾天气，北京市政府办公厅印发了《关于分解实施北京市 2013 年清洁空气行动计划的通知》，计划从 8 方面实施 52 项大气污染治理措施，明确了主要污染物年均浓度平均下降 2%的年度目标。将在 2013 年严控增量，按照“以新代老、总量减少”原则审批建设项目，不再新建、扩建使用煤等高污染燃料的建设项目，制定发布更为严格的第二批不符合首都功能定位的高污染工业行业调整退出目录等。

加强城市防灾减灾。随着人口、产业、工程设施不断向城市集中，城市发展与防灾能力不足的矛盾日益突出，加强城镇防灾减灾能力迫在眉睫。应以城乡防灾规划制定和实施为先导，以房屋建筑和市政公用设施抗灾设防监管为主线，以应急基础设施建设为重点，以城乡建设防灾减灾法律法规、标准体系为依据，以应急管理队伍建设和防灾减灾技术进步为支撑，进一步完善城乡建设防灾减灾管理体系。全面提高城乡建设防灾减灾能力，最大限度地避免和减轻灾害中因房屋建筑、市政公用设施破坏造成的人员伤亡和经济损失。

加强城市环境基础设施建设。加快城市基础设施管理体制改革，打破多头管理的局面，实现城市基础设施管理体制的转型。完善规划管理体系和政策法律保障体系，加强城市污水处理设施的规划，加强城市垃圾处理设施建设的规划和管理。加强城市主要污染物的监测工作，提高监测标准，增加监测结果向社会公布的即时性与透明度，建立完善城市主要污染物总量减排预警制度和防控机制。

（五）恢复生态系统

维护和强化整体自然生态系统的联系性十分重要。破坏山水格局，就等于切断了城市与自然联系的通道，阻碍了包括风、水、物种、营养等自然要素的流动，必然会影响城市的整体生态环境，以致使城市生态系统“死亡”。因此，维护区域山水格局和大地机体的完整性，是维护城市生态安全的一大关键。

生态环境的循环和衍生都离不开自然水体，大量的改造自然水体及岸线是对自然生态系统的严重的破坏，将极大地影响自然生态环境的循环。城市建设应维护原有水体及岸线的自然形态，对于已经改造的水体及岸线，应尽可能地恢复其自然形态。湿地不仅是人类最重要的生存环境，也是众多野生动物、植物的重要生存环境之一，生物多样性极为丰富，被誉为“自然之肾”，对城市及居民具有多种生态服务功能和社会经济价值。湿地能够提供丰富多样的生物栖息地、调节局部小气候、减缓旱涝灾害、净化环境。在城镇化进程中，越来越多的湿地面积正逐渐变小，有些湿地已经消失。应保护、恢复城市湿地，积极发挥湿地的生态服务功能，改善城市环境质量，促进城市可持续发展。

防护林体系与城市绿地系统相结合。早在20世纪50年代，中国就开展了大规模的防护林实践，带状的农田防护林网成为一大特色。这些带状绿色林网与道路、水渠、河流相结合，具有很好的水土保持、防风固沙、调节气候等生态功能。但防护林建设工程目标比较单一，只关注于防护，缺少将防护功能与城市其他功能相结合。在城市建设中，一些沿河林带和沿路林带，往往在城市扩展中的河岸整治或道路拓宽过程中被伐去，降低了防护林体系的质量，使防护林体系的完整性遭到破坏。可以将防护林体系保留与城市绿地系统相结合，共同组成城市绿色基础设施。

生物多样性是体现生态环境良好的重要指标之一。由于城市建设的快速发展，越来越多的生物逐渐消失，城乡生态系统遭到严重破坏。因此，应在保证整体格局的前提下，尽可能地保护和建设城乡生态系统的多样性，从而促进城市与周边自然环境形成良性的生态循环。

建设社区非机动车绿色廊道。完整的绿色基础设施应当是一个完整的绿色网络，能够很好地促进城市内外生态系统的联系。在城市内部需要通过多类型的绿色通道联系绿色基础设施网络。因此，在社区内也需要规划绿色廊道，作为绿色基础设施的组成部分。社区内的绿色廊道的规划建设可以结合非机动车道，建立方便生活和工作及休闲的绿色步行道及非机动车道网络。绿色廊道与城市的绿色系统、学校、居住区及步行商业街相结合，不仅可为步行及非机动车使用者提供一个健康、安全、舒适的通道，也可大大改善城市内部自然生态环境，从而达到调节城市生态系统的作用。

将城市公园及农田作为城市的重要绿色元素。在现代城市中，公园应是居民日常生产与生活环境的有机组成部分，随着城市的更新改造和进一步向郊区化扩展，城市中建设了更多的公园或城市绿地。然而，传统的城市公园为了方便管理，通常都有围栏作为边界。这种做法正好是将公园这一绿地与城市周边自然生态环境分隔开。在城市绿色基础设施建设中，应当积极利用公园这一绿色基质，将其融入到城市内各种性质用地之间，并以简洁、生态化和开放的绿地形态，渗透到居住区、办公区、产业园区内，并与城郊自然景观相融合，从而将城市公园与绿色廊道、城乡自然生态绿地联系起来。保护农田是未来中国可持续发展的重大战略。随着城市的发展，在城乡空间格局中，农田渗透入市区，而城市机体延伸入农田之中，将农田作为绿色基础设施的有机组成部分，既保护了农田又构建了完整的绿色基础设施。

（六）传承历史文化

城市文化从城市诞生之日起，经过长期历史过程，在原有基础上不断积淀形成。一座城市能够延续和发展，越来越取决于城市文化的延续。城市的发展不能仅仅关注经济积累以及建设数量的增长，应从“功能城市”走向“文化城市”。重新认识人类社会复合系统中的现有文化资源，应成为新时期城市文化建设的重要任务。结合自然、区域、民族等特点，在规划布局、功能区划、建筑风格等方面，注重文化特色。通过不断丰富城市自身特有的文化内涵，找到属于城市各自的文化发展路径，形成各具特色、适宜人居的城市体系。

四、改善城市环境的相关机制

城镇化是中国发展的一个大战略，到 2020 年全面建成小康社会，到本世纪中叶基本实现现代化，关键看城市发展水平。而城镇化成败的关键，则要看城市环境质量和社会公平状况。在一个相当长的时期内，中国城镇化还将处于快速上升阶段，城市发展需要在解决历史遗留问题的同时，面向未来，强化规划设计、依靠机制引导、注重运行管理，实现城市与环境协调发展。

（一）协调推进城镇化和工业化

城市发展不能单方面突进，需要与工业化、信息化、农业现代化同步协调、同步推进。中国的城镇化率和城市发展水平整体上落后于工业化进程，这也是城乡之间、城市内部出现的严重失衡现象的主要原因。与世界发达国家城镇化发展历程不同的是，中国城镇化推进不仅可以与工业化进程同步，还可以与信息化同步。如果结合得好，这可以大幅度缩短城市现代化的时间，减少建设成本，直接进数字城市甚至智慧城市时代（图 5-5）。城镇化与农业现代化结合，是解决土地制约、确保粮食安全的有效之道。中国农村建设用地占城乡建设用地的一半，如果城市建设用地增加与农村建设用地减少持钩，仅此一项，就可以实现在城镇化进程中，不仅不增加建设用地，反而会节约土地、增加耕地面积。人口有序向城市转移，可以为农村土地流转创造条件，进而为推进农村现代化，奠定土地集中规模耕作的基础。

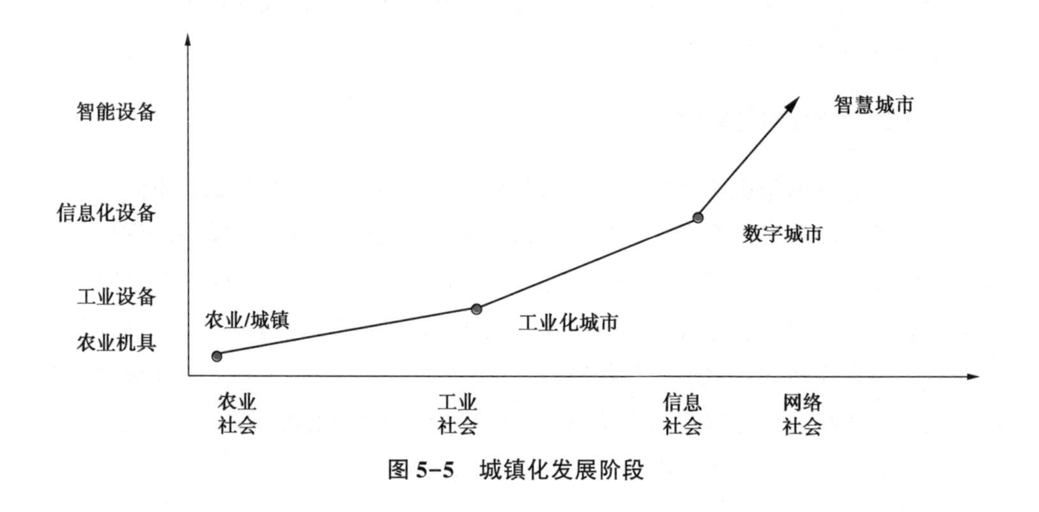

图 5-5 城镇化发展阶段

（二）依据资源环境资源承载力布局城市

中国总体上是人多、资源相对不足，各地自然条件差距也很大，东部宜居但缺地、西北地广但缺水、西南有水但多山。城市体系布局需要从国家层面进行总体规划，完全依靠市场力量，很难解决资源环境与人口的平衡问题。城市布局必须考虑一些硬性约束条件，如耕地保护、淡水资源、交通体系等。

中国的城市发展具有浓厚的政治色彩，政治、经济、金融、交通等资源高度向行政中心集中。这说明，政府在资源配置中发挥着重要的作用。然而，这种集中并不一定符合自然规律，它会导致城市扩张的盲目性，对局部的生态环境造成巨大压力，也会过早地出现“大城市病”。解决城市环境问题，需要弱化城市发展的行政色彩，引导行政中心与经济中心合理分离，形成功能相互协调、特色明显的城市体系。今后，应把城市发展的重点放在那些自然条件相对较好的中小城市，通过加强城市基础设施建设，完善居住功能，发展特色产业，解决中小城市发展明显滞后问题（图 5-6），疏缓特大城市的压力。

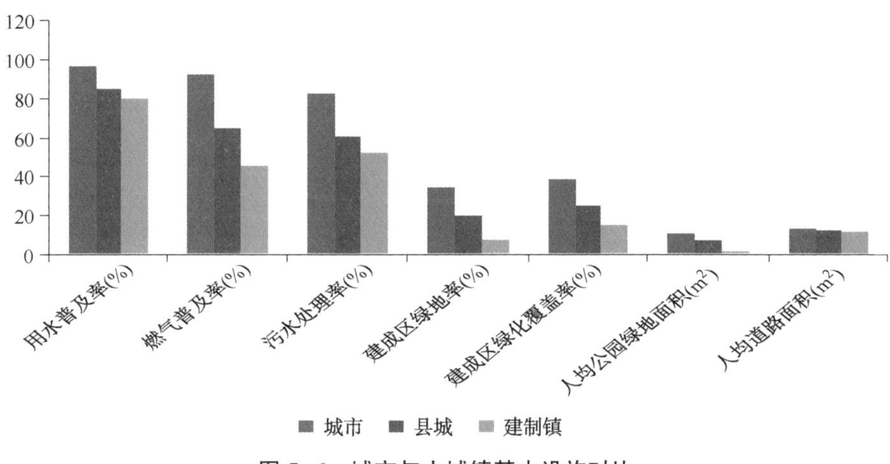

图 5-6　城市与小城镇基本设施对比

（三）提高城市自身规划建设水平

已有的特大城市需要调整功能区划，合理规划产业发展区、商业服务区与居住区，减少城市运行的交通压力。后发展的城市，需要提高规划的前瞻性，不能再犯同样的错误。任何城市建设，都应有基础工程先行、建筑标准先行，环保设施先行的理念。特别是支撑城市运行的“地下工程”，早规划、早建设，可以节约建设成本。城市建设应把节能环保作为重要的衡量标准，对原有建筑进行节能改造，对新建工程提高技术标准。加强垃圾归集管理，倡导绿色消费理念。城市绿化，应以生态功能为主，兼顾景观建设，多一些“森林城市”，少一些“花园城市”。

（四）建立城市发展新的融资机制

城市基础设施建设对城市经济生活和社会稳定的重要保障作用，受到了各地政府和社会各界越来越多的关注与重视，城市基础设施的投资逐年上升（表 5-5）。但城市建设投资来源单一，特别是过度依赖土地收入（图 5-7）。中国国际金融有限公司曾对未来十年中国城市基础设施建设投资规模作过预测，大概在 16 万—24 万亿元左右。依托土地的融资方式难以应对未来城市基础设施建设投资增长的需要，城市基础设施建设将存在巨大的资金缺口。

表 5-5　中国城市基础设施建设投资基本情况

年份	城市建设固定资产投资（亿元）	增速（%）	城镇固定资产投资（亿元）	比重（%）	国内生产总值（亿元）	比重（%）
1978	12.0	NA	NA	NA	3645.2	0.3
1981	19.5	35.4	711.1	2.7	4891.6	0.4
1991	170.9	41.0	4057.9	4.2	21781.5	0.8
2001	2351.9	24.4	30001.2	7.8	109655.2	2.1
2006	5765.1	2.9	93368.7	6.2	216314.4	2.7
2007	6418.9	11.3	117464.5	5.5	265810.3	2.4
2008	7368.2	14.8	148738.3	5.0	314045.4	2.3
2009	10641.5	44.4	193920.4	5.5	340902.8	3.1
2010	13363.9	25.6	243797.8	5.5	401512.8	3.3
2011	13934.2	4.3	302396.1	4.6	472881.6	2.9

（资料来源：中国城乡建设统计年鉴）

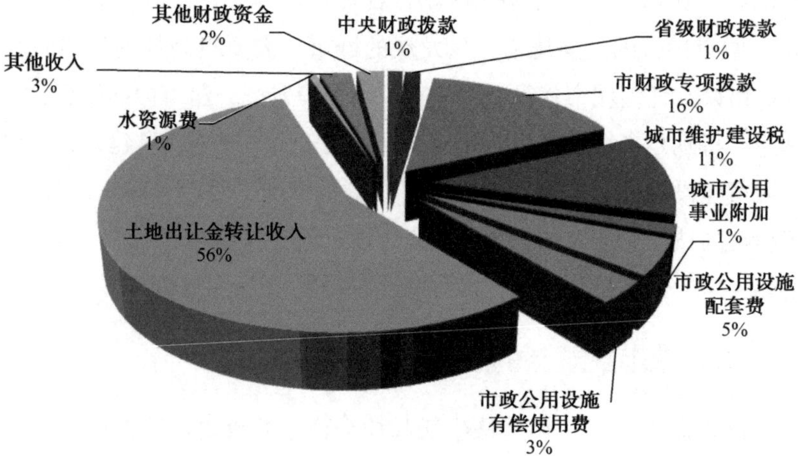

图 5-7　2011 年财政性城市维护建设资金构成比例

（资料来源：中国城乡建设统计年鉴）

需要建立政府与市场联动的城市基础设施建设投融资机制，根据城市基础设施具有的经营性、准经营性、非经营性的不同类型，建立不同性质的投融资体制，从而实现投融资主体多元化和投融资方式多样化。包括扩大地方政府发行市政建设债券的范围、改革与完善政策性金融、大力发展产业化基金等。还应完善以价格为核心的经济政策体系，按照行业平均成本+税费+合理利润的原则，制定和调整供水、供气、公交等城市基础设施行业的产品和服务价格，建立和健全适应价值规律和建设节约型社会要求、反映市场供求变化的经营性和准经营性基础设施价格运行体系和机制。

（五）消除城市贫困

城市贫困往往比农村贫困更可怕，也更难以解决。这要求政府提高基本公共服务均等化水平，在教育、医疗、养老以及就业等社会保障体系中，实现各地和城市内部的相对均衡发展。居住问题，是城市贫困的首要因素。中国的城市发展，不能走房地产行业完全市场化的路子。政府有责任建设保障性安居工程，解决那些低收入群体的居住困难。2012 年保障房的城市人口覆盖率已经达到 11%，规划到 2015 年提高到 20%。安居工程应融入城市整体规划，防止形成新的贫民区，还需要建立可持续的后续运行管理制度。

第六章　政策实践专题

一、社会组织在社会与环境事务中的作用研究

（一）概述

社会组织是政府和企业之外的一个重要的部门，在研究国家、市场和社会三者关系中，人们往往一定程度地强调各部门的作用，而忽略了三者之间互动、组合和相互补充、支持和互相监督功能。在研究社会组织的社会作用时，人们常常注重其社会影响力和经济地位，而忽略了社会组织在社会文化建设、环境保护、平衡社会各阶层之间矛盾、寻求实现个人价值的途径、关注弱势群体和建立和谐社会等方面的独特作用。本章将重点阐述社会组织在社会和环境事物中的地位和作用，描述环保组织参与国际环保活动的情况，设计社会发展与环境保护互动机制中社会组织参与的模型，并提出社会组织参与社会与环境事务的政策建议。

（二）研究背景

改革开放 30 多年来，中国经济发展取得长足进步。目前，中国经济总量已跃居世界第二，多项经济指标位居世界第一。然而，中国经济发展存在着过度消耗资源环境的问题，特别是随着工业化和城镇化进程不断加快，经济发展与资源环境的矛盾日益突出。一段时期以来，对于环境与经济关系的研究成为社会各界关注的重点，国家提出了转变经济发展方式、推进绿色发展和生态文明建设等理念，开始在实践中探索和深化经济的转型发展，强调在坚持以经济建设为中心的基础上，把节约资源、保护环境作为基本国策，大力实施可持续发展战略。可持续发展有三个重要因素：经济、社会、环境。这三个要素相互依存、相互制约。但是目前在中国，相比于环境与经济的关系所受到的重视程度，环境与社会的关系在整个中国可持续发展框架中，还没有受到应有的重视，人们对于可持续发展战略的认识也往往着重于环境和经济的层面，环境与社会的相互作用与影响等内在联系未得到很好的认识、关注以及深入研究。通常情况下，社会发展又分为三

个方面，一是教育、科技、文化、卫生等社会领域；二是社会组织行为；三是社会的行为方式。这三方面与环境都有直接或潜在的联系，需要人们去认真挖掘和研究。

另一方面，21 世纪以来，中国进入到环境事故与环境风险的一个集聚期和高发期，环境污染和生态退化对社会发展的影响日益凸显，已经成为引致当前中国一些社会问题的重要因素和导火索，一些因环境利益冲突引发的社会冲突、环境利益分配的不均衡和不公平、环境正义或者公平受到侵害、民众健康受到损害、环境引发的贫困问题、环境恶化带来的社会成本增加、贫困群体的基本公共服务水平不高乃至各类社会冲突加剧等问题成为社会各界热议的焦点。有专家认为，中国由环境问题引发的社会危机，有可能超越环境问题对经济的影响，成为影响中国和谐发展的重要因素。人们越来越深刻地认识到，经济的高速发展应当也必须有社会的进步作为保障，在资源环境的保护与经济社会的协调发展之间寻找出一条平衡、可持续的道路是关系到中国未来发展的重大战略问题。

从政府层面看，从最早提出循环经济开始，到建立节约型社会，再到环境友好型社会，再到低碳社会，已经体现出了对可持续发展三要素认识的进步，环境与社会的重要性越来越受到重视。中国政府在“十二五”规划纲要中明确提出要把转变经济发展方式贯穿于经济社会发展的全过程和各领域，建设和谐社会，共享发展成果，维护社会公平和正义。环境保护与社会发展问题正是保障和改善民生、建设和谐社会的具体体现，体现了国家改善环境与社会关系的意愿。另一方面，社会公众的环境意识也日益增长，参与环境保护与社会发展进程的热情不断高涨。例如，社会公众热议的 $PM_{2.5}$问题以及后续的相关政策和标准出台，反映出国家环境公共治理模式的转变，说明在中国环境保护建设与社会管理转型已经具备了现实的条件。事实上，根据我们的调研，包括经济发展的绿色转型等理念，实际上就是来源于或者借助于社会学领域，很多具体行动也已经不可避免地、越来越多地涉及社会发展的领域。因此，中国环境保护建设的未来发展方向，不仅限于经济领域，也将逐渐集中在社会领域，致力于推进社会进步和发展。

我们认为，环境保护与社会发展紧密联系，相互影响，互为因果。良好的环境是社会发展的重要条件，也是衡量社会文明程度的重要标准，同时也是社会民众的基本权利。保护环境，使社会全体成员共享环境保护的成果，才能增强整个社会的可持续发展能力。另一方面，社会发展是环境保护的重要基础和外部条件，一个法治、公平、正义、诚信、有序的社会，可以维护和促进环境的改善与可持续性。反之，环境污染或者生态退化将直接冲击长期经济增长潜力，对人群健康和生态系统健康带来长期影响，并冲击社会关系与结构，进而引发严重的社会问题；而紊乱或者扭曲的社会制度、结构和关系会引致和加剧环境的破坏，如果社会关系和结构得不到及时调整，那么就会形成环境与社会关系的恶性循环。

从我国环境保护的政策层面看，“自上而下”的行政手段以及基于市场的经济政策工具采用较多，“自下而上”的社会公众参与的治理手段相对较少，社会公众参与国家与社区环境治理的渠道和作用还没有充分发掘，成为国家环境治理的短板。中国政府在“十二五”规划纲要中明确提出要坚持多方参与、共同治理，统筹兼顾、动态协调的原则，完善社会管理格局，创新社会管理机制，形成社会管理和服务合力。这为未来中国如何处理环境与社会关系，如何运用社会公众手段参与环境治理指明了发展方向。解决中国的环境保护与社会发展问题，促进环境保护与社会发展的和谐，应当更加依靠社会力量和公众的参与，改进环境与社会管理的模式、方式与机制，推动环境与社会管理的转型。

当前，我国正处在发展转型的关键时期，改善民生、保障社会公众权益、加强社会管理和创新将是未来一段时期的国家战略重点。在新的历史条件和背景下，开展环境保护建设与社会转型发展研究，认真审视和认识当前及未来中国环境与社会关系的状况、存在问题、发展趋势和相关影响，进一步完善相关政策，实现环境保护建设与社会转型发展可谓正当其时，这不仅能促进环境的改善与可持续，而且是实现社会稳定和国家长治久安的重要基础，对国家未来长期平稳健康发展具有重大意义。

本章按照国家与社会关系理论、可持续发展理论、公民社会发展理论、全球治理理论、公共管理理论等进行比较研究，提出中国环保社会组织的发展现状，存在的问题和发展的基本规律。

（三）中国社会组织在社会和环境事务中的作用

1. 社会组织的发展概况

广义上讲，社会组织是指非官方的按照一定的宗旨和系统建立起来并为实现其目标而形成的集体，是社会的重要组成部分；就民法划分是社团法人，当然还有政府机关法人、企业法人和事业单位法人。而作为社会组织是影响社会事务的主体，具有民间性、组织性和自发性等特征。其形式包括法律范围内登记和未登记的各类社会组织。从中国目前政府文件认可的社会组织的范畴看，社会组织主要包括社会团体、基金会和民办非企业单位。据民政部统计，截止 2012 年底全国（不含港、澳、台）共在民政部登记注册的社会组织有49.2万个，其中社会团体26.8万个，民办非企业单位22.1万个，基金会2961个。在各级民政部门备案的城乡社区社会组织和农村专业经济协会有30万个。目前社会组织还未包括公益类事业单位、非营利性的公司以及未登记的非营利组织。

2. 社会组织在社会发展中的地位

（1）社会组织的经济地位

由于中国目前的社会组织种类繁杂，良莠不齐，到目前为止，还没有一个相对准确的统计。根据国家统计局公布的2003年数据，中国群众社团、社会团体和宗教组织职工人数为18.1万。从专家分析来看，这一数据显然低估了社团的就业规模与贡献。根据有关专家推测，2002年，中国社团的专职人员数为47.5万人；兼职人员为70.9万人，志愿者为140.3万人，志愿者每月奉献的时间为79万小时。根据中国民促会和民政部社会组织服务中心合作研究《社会团体专职人员职业化问题研究》课题组专家预测，2003年，中国社会组织专兼职人员就业规模为300万人。截止到2012年底，我国社会组织提供的就业岗位有1200多万个。从国际上看，根据约翰·霍普金斯大学非营利部门比较项目的调研结果，1995年美国、德国、英国、法国、日本、瑞典、意大利等22个国家，即使排除了宗教团体，22个国家的非营利部门是一个1.1万亿美元的产业。它雇用了近1900万个的全职工作人员。这些国家的非营利部门的支出因而平均达到国内生产总值的4.6%，非营利组织就业占所有非农就业的近5%，占所有服务行业就业的10%，占所有公共部门就业的27%。就业是衡量社会发展的重要参数指标，从以上中外就业情况比较，中国的社会组织在促进就业方面仍然具有相当大的发展空间。

（2）社会组织的治理地位

社会组织力量来源既不同于国家，也不同于商业部门。国家所依托的是法律、征税权、军队和警察等强制力量，商业部门所依靠的是自己的经济力量、创新机会和企业家的智慧；而社会组织所依靠的是由规范上、道义、知识、信仰、价值观和可靠的信息等而产生的意志推动力，是一种“软权力”。随着全球化浪潮的步步推动，从经济全球化、文化全球化逐步转向社团革命全球化。全球化治理过程中，除了国家作为行为主体之外，还出现了政府间国际组织、跨国公司、地方政府、非正式社团、公民运动以及由他们所组成的松散的联盟等多种行为体。随着全球性问题的增多，国家和政府组织在处理这些问题时所表现出来的能力上的缺陷，以及全球公民社会力量的壮大，从不同方面增加了社会组织作为一种非国家行为体，对世界政治的参与及影响力在逐渐增大，在全球治理中正在发挥着重要的、独特的其他行为主体不可替代的作用。

(3) 社会组织的国际地位

随着社会组织参与国际事务的能力逐步提高，社会组织已从传统的国际人道主义援助已涉及到国际事务的各个方面，尤其是在人权、环境、发展、和平倡导等诸多领域，许多社会组织已经显示出参与国际事务的协调和倡导的能力。因此，社会组织在国际事务的舞台上活动于国家、国际组织和跨国公司力所不及的领域，从全球社会经济发展过程中，敏捷地发现各类热点课题，利用信息技术分享和传播信息，积极倡导新的价值观和参与全球治理的规划，在各种层面上参与政策制定，参与政策和制度的实施并监督各利益相关者的实施情况。社会组织在国际事务中的作用已逐步显现，民间外交、公共外交在联合国的重要决策中都有社会组织的声音。

(4) 社会组织的社会地位

由于社会组织在其发展进程中与社会各阶层的联系极为密切，从公民的自我治理和自我价值追求，公民之间的社会互信、互助和关爱，公民社会之间人生理念和伦理价值的影响，到公民社会组织的规范形成、运作、参与社会事务和运筹社会资本能力都表现了较高的社会地位。众所周知，中国改革开放 30 多年的变化是举世瞩目的，经济高速增长，堪称经济学上的一大奇迹。但我们不能过于乐观，由于中国像大海航行中的一艘巨轮，在转弯和前进方向过程中很难保持相对稳定，这也是令世人不可想象的。经济发展背后的社会问题愈凸显，那么社会组织作为社会的有机主体，愈应在国家与公民之间建立起一条缓冲社会矛盾、推进社会改革、寻求社会公正的宽松地带。当然要建立这一缓冲宽松地带的基本条件，还面临着如何处理政府与企业、政府与社会组织、社会组织与企业互相补充、互相支持和互相制约的平衡关系。社会组织已经是构建和谐社会的重要力量。

(5) 社会组织的互补地位

在当今全球化的时代浪潮中，人们面临着生存、变革、创新和发展的新挑战。社会组织就其本身发展也面临着巨大的挑战。从世界范围内的社会组织由于法律地位、可控资源和政治倾向等原因，造成了社会组织在全球化治理中，必须与政府、企业、国际组织和利益集团形成互动。在国际关系中虽然处于边缘地位，从国际法的角度看，社会组织并不构成国际法的主体；从可控资源角度来讲，社会组织越来越多地依赖公共部门和私人部门的资助而积极寻求通过政府、国际组织、利益集团和企业来发挥作用；从政治倾向来看，社会组织在一定程度上并没有处于一个完全平等的地位上。从中国社会组织的情况来分析，社会组织是政府有力助手，政府主导、社会组织参与，真正做到政社分开，是社会管理的创新。

3. 中国环保 NGO 的发展

在 1994 年草根环保 NGO 诞生后的最初八九年中，他们的活动以开展环境教育为主，后来有人将这些活动戏谑地总结为“种树、观鸟、捡垃圾”老三样。不过这一时期，草根环保 NGO 也开展了不少重要的、以自然保育为特征的专项环保行动，如 1995—1999 年间保护滇金丝猴、“拯救藏羚羊”的行动，1998—1999 年间阻止四川射洪县原始森林滥伐、保护川西天然林的事件等。通过这一系列的行动，中国开始形成一批具有相同认知和使命感的环保群体。这一时期另一个重要的发展是，始于 1996 年的大学生“绿色营”催生了一大批大学生环保社团并构成一个松散的网络，并成为环保 NGO 重要的后备力量。总体来看，上述时期 NGO 的行动较为温和，以生态和环境保护为旗帜，以唤起民众的环境意识为目标。

进入 21 世纪以后，经过前一时期的行动积累和认识的变化，部分环保 NGO 开始转向权利话语和政策倡导。尤其是 2003 年以后，在一些重大公共事件中，中国的环保 NGO 以其鲜明的个性和高调姿态呈现在公众视野中。2003 年，与都江堰相邻的杨柳湖水库、贡嘎山下的木格错水坝以及怒江大坝，先后遭遇环保 NGO 的阻击；2004 年，环保 NGO 延续对怒江大坝的公开论争，并掀起北京动物园搬迁风波和对虎跳峡电站的质疑；2005 年，对圆明园防渗工程作出群体响应和听证参与。在这一时期，污染受害者的环境权、重大工程决策的公众知情权、参与权和监督权成为环保 NGO 的另一面大旗。

根据民政部统计，中国大陆的生态环境类社会组织从 2007 年的社会组织 5709 个，到 2012 年底增加到 7928 个，6 年内增长了 38. 9%，平均年增长 7. 5%。但从 2007—2008 年有一个较大的增长，达到 34%，而在 2010—2011 年出现了负增长，减少了 2. 1%，其他年度基本是微增长现象（表 6-1）。

与此同时，影响政府的环保政策，尤其是推动公众在环保政策制定过程中的参与机制，促进政府信息公开，成为环保 NGO 的重要目标。

4. 环保组织的角色和地位

环保 NGO 在环境保护领域具有十分独特的作用，能够弥补政府与市场的不足。不过，本次调查表明，仍然有很多受访者对环保 NGO 不了解，甚至 30%的受访对象认为环保 NGO“作用较小”或“几乎没有作用”。因此，让更多的人了解环保 NGO，认识到环保 NGO 的作用仍然任重道远。

表 6-1　民政部门登记环境生态类社会组织（2007—2012）

类型（年）	社会团体（个）	民办非企业（个）	基金会（个）	总数（个）	占总量比例（%）
2007	5530	345	34	5709	1.48
2008	6716	908	28	7652	1.85
2009	6702	1049	35	7806	1.81
2010	6961	1070	47	8078	1.81
2011	6999	846	64	7909	1.73
2012	6790	1078	60	7928	1.61

环保 NGO 的作用主要体现在四个方面。其一，提升公众的环保意识，改变人们传统的环保观念，例如改变人们重 GDP、轻环保的理念；其二，促进公众的环保行为，例如，组织公众参与种树的活动；其三，完善公众参与机制，例如，为公众参与环境保护搭建平台；其四，影响政策，例如，提高政策制定的科学性和实效性，通过多种渠道倡导减塑行动，最终推动政府出台“限塑”政策。这四个方面的作用是一个有机整体。其中，提升公众环保意识是基础，公众环保意识的提升又有助于公众环保行为的改善，公众环保行为的增强又会直接推动公众参与机制的完善，而公众参与机制的完善又会进一步推动环保政策的完善。意识指导行动，行动实践意识，没有行动的意识不产生社会效应，没有意识指导的行动将偏离环境保护的初衷。而没有形成机制的行动将不具有持续性。而政策的影响力是巨大的，很多问题的根本解决都取决于是否能够推动政策的调整与制度的变革。

（1）提升公众环保意识

环保需要每个人的参与，而环保行为的产生是以环保观念的建立和环保知识的普及为基础的。环保 NGO 通过出版书籍、印刷资料、举办讲座、组织培训，以及媒体的报道等方式，帮助公众了解到环境保护的重要性和迫切性，通过知识的普及，提升公众的环境保护意识。

从调查情况看，91.2%的环保 NGO 都开展过环保宣传教育活动。与政府部门环保宣传教育相比，环保 NGO 在提升公众环保意识方面最重要的特点是探索与创新，不断丰富环保宣传教育的内容，探索新的环保宣传教育方式，扩大环保宣传的范围。

（2）促进公众环保行为改善

环保宣传教育可以唤醒公众的环保意识，但是，环境的改善，污染的减少最终还是要落实到公众的行动上。然而，由于专业化分工，普通人往往没有太多的时间、能力来研究怎样环保，普通企业也苦于缺乏环保节能技术。但是，环保 NGO 通过宣传教育，特别是通过一些专业化的服务，帮助和促进了个人、企事业单位、社区环保行为的改善。

（3）完善公众参与的机制

我国《环境保护法》第 6 条规定：一切单位和个人都有保护环境的义务，并有权对污染和破坏环境的单位和个人进行检举和控告。《环境影响评价法》中对公众参与的程序、责任人员作了进一步明确规定：专项规划的编制机关对可能造成不良环境影响并直接涉及公众权益的规划，应当在该规划报送审批前，举行论证会、听证会，或者其他形式，征求有关单位、专家和公众对环境影响报告书草案的意见。除国家规定的保密的情形外，对环境可能造成重大影响，应当编制环境影响报告书的建设项目，建设单位应当在报批建设项目环境影响报告书前，举行论证会、听证会，或者采取其他形式，征求有关单位、专家和公众的意见。

然而，由于我国环保法律体系还不健全，法条停留在应然性规定阶段，缺少可以操作的具体规范。因此，在实践中，公众往往缺乏参与的渠道。由于个人处于分散的状态，不仅利益表达的力度有限，而且作为政府，收集分散化的个人信息，成本也很高，因此在这方面，环保 NGO 可以为公众参与提供组织化的途径。同时，通过环保 NGO 的努力，可以促进政府不断完善公众参与的机制。

（4）政策倡导

在环保公共政策的制定、执行和评估的各个环节，环保 NGO 都起着推动环境政策的出台、推动环境政策的有效执行、作为独立的第三方参与环境政策的咨询与评估等重要作用。

（四）社会组织参与社会与环境事务中存在的主要问题

1. 环保 NGO 发展的环境有待完善

环保 NGO 在促进人与自然和谐发展、加快我国政治文明建设、构建社会主义和谐社会方面都发挥着积极重要的作用，政府、社会等各方面应该大力促进环保 NGO 的发展。但是，迄今为止，我国环保 NGO 发展的外部环境并不尽如人意，主要表现在公众认知程度与支持程度不高；法制不健全，参与渠道不足；政府支持力度不够，管理方式不完善。

2. 环保 NGO 的能力亟待提升

环保 NGO 在反应民众环境诉求，维护社会稳定，促进公民有序政治参与方面有着重要意义。但是，目前我国环保 NGO 还存在能力弱、影响力小、草根环保 NGO 对外依存度较高等问题，环保 NGO 的能力亟待提升。

3. 环保 NGO 的发展难以满足社会的需求

环保 NGO 是致力于环境保护的重要民间力量，在提高公民环保意识，维护公众环境权益，参与公共决策方面发挥着积极作用。此外，实现科学发展，提高民主政治水平，构建和谐社会都亟待环保 NGO 发挥更大的作用。然而，当前中国环保 NGO 的发展还处于较低的水平，能力弱、影响小，还远远无法满足中国社会经济与政治发展的需求。

（五）在环境与社会互动发展中社会组织参与的模型分析

模型分析是对客观事物或现象的一种描述。模型是被研究对象的一种抽象。客观事物或现象，是一个多因素综合体。因素之间存在着相互依赖又相互制约的关系，通常是复杂的非线性关系。为了分析其相互作用机制，揭示内部发展与运作的基本规律，可根据理论推导，或对观测数据的分析，或依据实践经验，设计一种模型来代表所研究的对象。

模型反映了对象最本质的东西，略去了枝节，是被研究对象实质性的描述和某种程度的简化，目的在于便于分析研究。借助模型进行分析，是一种有效的科学方法。

本章拟采用德菲尔法，并辅以其他分析方法，分析社会组织参与社会发展与环境保护互动机制。

德菲尔法（Delphi Technique），即函询调查法，将提出的问题和必要的背景材料，用通信的方式向有经验的专家提出，然后把他们答复的意见进行综合，再反馈给他们，如此反复多次，直到认为合适的意见为止。

这是一种匿名的专家问卷调查法。选择数位的专家，设计出问卷，寄出问卷，回收整理相同意见的部分，不同意见的部分再次设计并寄出问卷，如此反复进行直到意见一致。

目前，课题已经进入到问卷设计的环节。而主体行为互动体系中，社会发展的主体是政府、企业和社会组织，那谁是主要影响因子，它们的结构和权重比例是如何体现的。而环境变迁的体系中的主体可能是企业、政府和社会组织，那谁又是主要影响因子，它们之间的关系又是如何的？这需要通过问卷设计和专家咨询、专业人士的讨论和利益相关方共同协商。

从民促会与欧盟合作研究社会组织与企业互动模型中我们得到启示，社会组织与企业的互动体系之中还处于一个劣势状态，但是社会组织对企业的影响因子在发生变化，尤其是信息技术与新媒体的影响度在明显上升。

（六）中国环保组织积极参与国际交流与合作

1. 涉外项目合作情况

从组织类型看，在参加国际交流与合作的环保民间组织中，50%的由政府发起成立的环保组织、46.7%的高校环保社团、80.8%的草根环保民间组织参与了国际合作项目。草根环保民间组织在参与国际合作项目上表现十分积极。

从资金来源看，大部分的国际合作项目资金都来源于国际基金会。其中，由政府发起的环保民间组织获得国际基金会的资助最多，其次是草根环保民间组织。

最近几年，国际合作项目数量不断增加，领域明显拓宽，涉及动物保护、湿地保护、气候变化、节能减排、水地图、绿色选择、江河保护等领域。从区域范围上看，国际合作项目多集中在北京、四川、重庆和云南等地。

在国际合作项目中，外方合作伙伴主要包括基金会、国际（境外）环保NGO、政府、企业、大学、驻华使馆等。主要表现以下三种合作方式：（1）国外合作伙伴仅提供资金支持，不参与项目合作。（2）国外合作伙伴提供资金支持和方案策划，不直接参与合作项目。（3）国外合作伙伴提供资金支持、方案策划、技术设备和人员，直接参与项目合作。

2. 参加国际会议情况

我国过半以上的环保民间组织曾参加过国际会议，所参加国际会议涉及的领域包括气候变化、能源、河流保护、NGO 经验交流、公众参与、生物多样性、节能减排、法律、动物保护、化学品安全等。

3. 出国访问情况

在开展国际交流与合作的环保民间组织中，50%的由政府发起成立的环保民间组织和57.7%的草根环保民间组织组团出访，说明这两种类型的环保民间组织在出访活动方面较为活跃。

4. 加入国际组织情况

目前，加入国际组织的中国环保民间组织数量较少。中国环保民间组织参加的国际环保组织主要有：白鳍豚组织、国际雪豹组织、世界自然保护联盟、国际红树林联盟、太平洋环境、新能源一代、东北亚青年环境网络、国际植物园保护联盟等。

从加入国际环保组织的成员看，绝大部分是由政府发起成立的环保民间组织和专业性较强的草根环保民间组织。

从加入的组织类型来看，目前中国环保民间组织参加的国际组织类型主要有人道主义救援和发展机构、私人基金会、宣传机构、专业协会、专家型非营利的咨询和项目执行机构。其中，以参加专业协会为最多，且由政府发起成立的环保民间组织和草根环保民间组织最为活跃。

从以上几项初略的统计可以看出，环境保护与促进社会发展是世界各国面临的共同课题，为实现可持续发展，走绿色低碳的发展道路是世界各国的共同选择，因此加强国际间的交流与合作不是一个国家发展的需要，而是世界各国的共同需求，不仅是政府的行为，也不仅是仅靠政府的力量而实现的，必须要社会组织的积极参与。社会组织参与的深度越深、广度越广，越有利于环保问题的解决。政府引导社会组织积极与政府形成互动，加强国际间的交流与合作，是促进绿色可持续发展的必由之路。目前，中国社会组织参与在环保与社会发展领域中态度是积极的，已取得可喜的成绩，已得到国际组织与各个国家社会组织的支持，获得了一定的经验，今后加强与深入继续做下去。最近十几年，尽管中国的经济取得了很大的发展，经济总量也大幅度的增加，但中国仍然是世界上最大的发展中国家，在环境保护与社会发展领域无论从资金和技术上，和发达国家仍然有很大差距，希望继续得到联合国和环保类国际组织的帮助和支持。

（七）社会组织参与社会与环境事务的策略研究

改革开放 30 多年来，社会组织虽然取得一定的发展，但是无论在参与社会事务，还是参与环境事务方面都还不够深入，因此有必要对社会组织参与社会与环境事务的策略进行研究。

1. 政府如何改善社会组织生存环境的策略

政府应提高社会组织生存的政策环境，落实中央提出的政社分开，推动环保类社会组织的独立性和自主性。政府应加大购买社会组织公共服务的力度，支持环保类社会组织的能力建设。政府需要对环保类社会组织登记、税收优惠和发展提供支持政策，促进公民参与环境事务政策体系。政府在国家重点和重大环境政策制定过程中，听取社会组织的建议与意见。政府授权部分环保类社会组织参与国家重点项目的立项、实施和监督。

2. 社会组织参与社会和环境事务的策略

提高自身的专业能力，吸引一批有专业能力的专家参与专业服务和咨询工作。完善内部治理结构，制定规范的管理制度，引导不同利益相关者参与社会组织的管理事务。扩大国际交流，学习外国先进知识和经验，尤其学习政府、企业和社会组织之间的互动模式和有益方法。积极、主动参与政府在环境领域的事务，提高对政府在制定社会政策与环境政策的影响度。推动与影响环境主体之间互动，加强社会组织与企业对抗、倡导与合作模式的研究。

（八）主要发现和结论

社会与环境事务的治理，是一项公共事务，需要动员全社会的各种资源和力量广泛参与，其中社会组织的力量应得到更好的使用。社会组织不仅自己要积极参与社会和环境事务的治理，还要协助政府做好相关工作。改革开放后，中国的社会组织取得了一定的发展，但是较短的发展历史决定了中国社会组织的理念文化和自身实力还远未成熟，甚至是比较脆弱的。在当前政策环境下，结合社会组织自身能力，通过课题组调研、分析，得出主要发现和结论如下：

1. 主要发现

（1）影响社会组织在社会与环境事务合作的外部挑战之一

- 中国 NGO 缺乏对环境管理事务的知情权益。
- 中国 NGO 缺乏对重大项目设计和可行性研究的参与权益，资格问题。
- 中国 NGO 缺乏对重大项目实施过程的监督权益。
- 中国 NGO 缺乏参与国际重大环境事务的能力。
- 中国 NGO 缺乏国家救济和支持政策。

（2）影响社会组织在社会与环境事务合作的外部挑战之二

- 环保类社会组织的社会认可度低，在机构的透明管理，问责体系和开放过程缺乏系统的组织和协同。
- 社会组织缺乏合作的资源和资源公平分享的平台，没有将社会资本扩大化，产生最大社会效益。
- 缺乏社会组织合作的支持实施体系，包括媒体、学界对社会组织的专业化运作的指导和影响。

- 企业对社会组织的影响度在发生变化，企业在关注社会组织的发展，尤其是非公募基金会的发展。

（3）影响社会组织在社会与环境事务合作的内部挑战

- 部分环保社会组织尚未具备合作的基本条件，从治理结构，到人力资源，从筹资体系到会员服务等。
- 社会组织之间沟通渠道不畅，例如社会团体、民办非企业单位和基金会之间的互动和相互支撑体系和文化还没有建立。
- 社会组织之间的合作存在资金配置不合理以及合作成本偏高，缺乏行业意识和行业规范。
- 环保类社会组织缺乏联盟意识，没有形成专业分工、行业互动和相互支持体系。

2. 结论

结合我国目前社会组织在社会和环境事务中的发展以及课题组的调研、分析，得出以下几点结论：

（1）完善社会组织登记、注册、实施和监督体系，推动环保类社会组织的健康发展。

（2）创新“政社关系”，为环保类社会组织参与社会与环境事务让渡空间。

（3）应建立环保类社会组织有效沟通机制及合作制度。

（4）用科学发展引领社会组织参与社会和环境事务方向。

（5）适时建立环保类社会组织联盟，以解决不同社会组织之间缺少互帮互助的关系。

（6）推动政府在环境管理事务的信息透明和公开。

（7）中国社会组织争取获得对参与国家重大项目设计和可行性研究的咨询地位。

（8）中国社会组织有对重大项目实施过程的监督权。

（9）加强国际合作，促进环境管理的经验交流和合作行动。

二、陕西西咸新区创新城镇化发展研究

党的十八大报告提出：“坚持走中国特色新型工业化、信息化、城镇化、农业现代化道路，推动工业化和城镇化良性互动、城镇化和农业现代化相互协调，促进四化同步发展。”西咸新区是推进西咸一体化、共建大西安的重要板块，作为陕西省城镇化的重要抓手和承接产业转移进而推动经济转型的重要平台，承担

着“建设大西安、带动大关中、引领大西北”的历史使命。在总结和反思传统城镇化利弊的基础上，西咸新区以建设现代田园城市为目标，创新城市发展方式，构建点状布局的市镇体系，发展新产业，探索新业态，以社会建设为先导，探索城乡一体、产城一体，走新型城镇化道路，期望为中国城镇化进程提供一种新的样本。

（一）我国传统城镇化模式及其利弊

我国城镇化伴随着工业化、现代化的发展。建国后经历了两个阶段，前30年城镇化率由大约7%提高到17%，平均每3年提高1个百分点；改革开放30年平均每年提高1个百分点，2011年城镇化率已达到51.3%。这两个阶段城镇化主要模式分别是企业带动模式和产业园区带动模式。

“一五”、“二五”时期，我国城镇建设普遍采用企业带动城市发展模式。先建生产线，围绕生产线建设生活区，进而配套建设子弟学校、职工医院、合作社等。先生产、后生活；先“治坡”、后“治窝”。比如，一汽之于长春市；油田之于大庆。西安的发展也与纺织城、军工城、电子城的带动分不开。我国用15年左右时间，搭建起一大批大中城市的骨架，形成新中国的工业基地和中心城市，为全国城镇化推进打下重要基础。

改革开放以后特别是90年代以来，城市建设更多采用园区模式，依托各类产业开发区发展城市。西安市通过建设各类开发区发展城市。长三角、珠三角城市群也都是依托大大小小的产业园区发展来的。

这种工业化带动城镇化的模式，好处是基础设施建设多快好省，城市建设迅速形成规模，生产力快速形成。但这种快速发展也带来一些问题，我们将其概括为“三个透支”：

1. 透支环境红利

这种模式“多快好省”主要省的是环境建设部分，基本建设投入不足，污水处理、垃圾处理因陋就简，往往污染了山川河流，积以时日，破坏了自然生态系统。

2. 透支土地红利

在城市建设方面，政府财政投入有限，主要依靠土地级差地租和溢价，沿城市边际扩张，投资小、见效快，以预期土地收益作为商业担保融资推动城市扩张，在商业模式上无疑是成功的。1999—2009年，全国土地出让金从595.85亿元增加到1.59万亿元，占地方财政收入比重由9.2%提高到48.8%。政府投资城市建设的能力极大提高，拉动了GDP增长，支撑了中国的快速城镇化。同时这种商业模式必然推动城市边界无限制扩张，“摊大饼”使城市越来越大，从而丧失效率，由此产生各种严重的“城市病”。

3. 透支人口红利

城镇化过程劳动力要素以吸收农村廉价劳动力为主，农民工成为“中国制造”最大的成本优势，支撑了中国的快速工业化和城镇化。大量农民工工作在城镇，而社会福利和养老保险还系于农村和农地，约2亿人工作生活在城镇但不享受市民待遇，成为城市的“他者”，造成城市内部的二元结构。形成第一代农民工留不住、第二代农民工不回去、第三代农民工回不去。城市快速发展成为吸附农村的土地、劳动力、资金等生产要素和资源的“黑洞”，城市兴起背后却是农村相对衰败，城乡二元结构进一步强化，城乡差距进一步拉大，农业生产水平停滞。“城市像欧洲，农村像非洲”，出现了“留守儿童”、“空巢老人”等农村社会问题。

这种透支环境红利、土地红利、人口红利的模式已经不可持续。转变城市发展方式，走新型城镇化道路，是新一轮城镇化的必然选择。

（二）新型城镇化要以社会建设为根本

1. 新视角：城镇化是伟大的社会变迁

过去，我们讲城镇化主要注重经济视角，从生产力发展的角度看城市怎么规划、怎么建。综合城市的市政、交通、人口等因素，看规划是否合理，生产力怎么组织、布局。新型城镇化首先应当是新的视角，从社会建设高度认识城镇化。从社会建设看，中国的城镇化是人类历史上的最大规模的人口迁徙，也是中国历史上最伟大的社会变革。2011年，我国城镇人口达到了6.9亿。按照国际经验，未来20年还将以每年1%的增速快速增长，也就是说，从改革开放开始算起，要用50年左右时间完成10亿人从农村到城市的迁徙。可以说，多年来我国政府最伟大的成就就是领导了人类历史上最大规模的人口迁徙，在稳定发展的环境中推动了巨大的社会变革。由此看来，推动城镇化不仅是经济发展的重中之重，也是社会发展的重中之重。创新城市发展方式，社会建设是最重要的环节。

2. 城市是社会有机体

从社会的角度认识城市，城镇化在人口聚集和面积扩张的同时，更重要的是实现产业结构、就业方式、人居环境、社会保障等一系列由“乡”到“城”的重要转变。城市不是物质的无机堆砌，而是社会有机体，人是城市的主体，城市的本质是市民社会。因此，我们不单要从社会生产力发展的角度，从基础设施建设、产业布局、经济发展等方面来研究城镇化，也要以人为本，从社会生产方式转变、社会发展的角度研究城镇化。赵乐际同志强调西咸新区要“三保三新”，即“保护历史文化、保护耕地、保护农民利益，发展新产业、形成新业态、建设新城市”。这不仅是对西咸新区建设原则的高度概括，也是在“四化并举”思路下处理城乡关系、产城关系的全新视角。“四化并举”中城镇化相对其他“三化”具有综合带动的特点，这就意味着新型城镇化要做到三个带动：带动新型工业化，促进产城一体；带动农业现代化，促进城乡融合、城乡一体；带动信息化，建设智慧城市。

3. 城乡一体是新型城镇化的核心问题

新型城镇化不是城市“吃掉”农村，而是城乡融合共生、实现城乡一体。而城乡一体的关键是“两转化”，即在城镇化过程中解决“农民转化为市民、农业转化为城市产业”的问题。

一是让农民带着劳动力和土地两个资本进城。一部分农民通过教育、培训转化成城市产业从业者，进城后仍然保留所承包农田，和拥有铺面一样，可以自耕，也可以出租入股经营。一部分农民通过农业专业合作组织成为职业农民，少量的专业户可转为农庄业主。让城镇化进程中的农民不再游离于城市产业之外，既能共享土地增值的红利，又能分享城市产业发展的红利。我们正在探索建立新型农村合作组织，创新土地流转方式，通过复合型田园农业建设实现村民到职业农民的转变。

二是将城市资本、消费引入农村，加快发展新业态。我们提出现代田园农业，既不是以户为经营单位的小农经济，也不是以规模经济为特征的大田农业，而是集一、二、三产特色于一体的复合农业。小集中、大分散；高就业、高附加值。法规明确城市边界，城镇周边就是农田，而且是法定不得侵占的农田，在这种预期下，鼓励城市资本、消费流向农村和农业，农民也会对农业进行长期投资；通过发展田园农业，创新农村的社会组织、经济组织，使得民俗文化、休闲旅游、绿色有机等要素与农业相结合，更多依托于服务城市功能，使农业成为复合型的高附加值产业，达到或接近城市产业平均收益水平。这种田园农业不仅是城市农产品的生产基地，也成为多种城市功能的载体。

4. 产城一体是新型城镇化的必然选择

城市建设要优先考虑就业导向，将城市生活半径与就业半径相结合。因地制宜，从资源环境、区位特点和解决就业出发，在城市规划中将产业与就业统筹考虑，合理布局和“集约、集群、集成”发展产业。同时，把信息化作为提升城市产业和优化城市功能的重要手段，实现数据的规模化集中吞吐、深层次整合分析、多领域社会应用、高效益持续增值，广泛应用物联网、云计算、下一代互联网等新技术，不断提高城市智能水平，建设“智慧城市”。

5. 新布局：建立城乡融合共生的市镇体系

新型城镇化，不是城市“吃掉”农村成为水泥森林，而是形成城乡兼容、互为依托的市镇体系。就空间布局而言，要以城镇群为主体形态，构建点状布局的市镇体系。由特大城市—中等组团城市—小城镇—村落（优美小镇）形成完整的市镇体系，以开阔田园和山川水系衬托其中。其中，特大城市如西安，作为区域中心，聚集数百万人口；若干中等组团城市形成30-50平方公里不等的城市板块，可容纳人口30万—50万人左右；围绕大中城市布局若干小城镇，人口3万-5万人左右；村落包括保留一部分自然村和兴建特色优美小镇，星罗棋布散落在城乡之间。按照这种空间格局，市镇各有明确的边界，在城市规划建设上停止“摊大饼”的老路，以克服传统城镇化的“城市病”，在形态布局上有效承载产城一体、城乡一体的功能，形成“开敞田园、紧凑城市”的“大开大合”空间布局。特别是农田、河流、山川、湿地，既能发展复合农业，同时又作为城市的生态功能区。赵正永同志将现代田园城市形态概括为“核心板块支撑、快捷交通连接、优美小镇点缀、都市农业衬托”，这是对现代田园城市市镇体系空间布局的高度概括。

（三）西咸新区新型城镇化道路的探索与实践

西咸新区位于西安、咸阳两市建成区之间，规划控制范围882平方公里，现有总人口89.3万人，城镇化水平仅有23%。作为陕西省城镇化建设的重要平台，西咸新区以现代田园城市为目标，按照新型工业化、信息化、城镇化和农业现代“四化并举”的思路，探索新型城镇化道路，创新城市发展方式，着力推进城乡一体、产城一体，同时实现提高人口承载力和集约用地的两个目标，成为陕西省加快城镇化和承接新一轮产业转移进而推动经济转型的龙头。

1. 构建现代田园城市市镇体系

西咸新区规划建设用地272平方公里，仅占规划范围的三分之一。按照“现代田园城市”大开大合的规划建设理念，构建新区“一河、两带、四轴、五组团”的城市发展格局，发展“核心城区板块、小城镇、村落小镇”的现代田园城市市镇体系。布局了空港新城、沣东新城、秦汉新城、沣西新城、泾河新城五大组团，每个新城开发建设面积从30—60平方公里不等，均有明确的产业布局

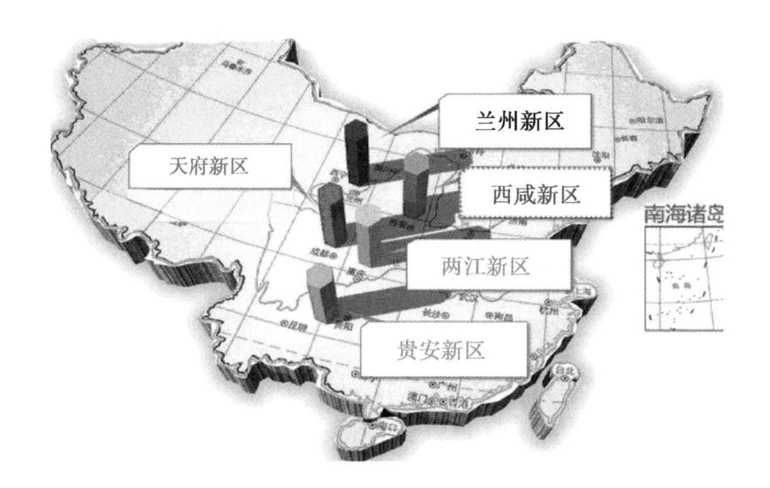

和功能定位。通过两条帝陵遗址带，渭河、泾河、沣河三条生态景观廊道，以及组团间楔形绿地为分隔，形成“廊道贯穿、组团布局”的田园城市总体空间形态。在城市组团间的开敞空间发展复合型田园农业，形成特色鲜明的优美小镇。

2. 以就业为核心布局产业，实现产城一体

西咸新区综合资源、环境条件和劳动力特点选择产业，更加注重以就业为核心来布局产业，以信息服务业、现代物流业、文化产业为重点，大力发展战略性新兴产业。规划建设沣东科技统筹示范园区、空港临空产业园区、沣西信息产业园区、秦汉文化旅游产业园区、泾河物流产业园区等十大产业园区，形成五个新城产业互补、错位布局、协同发展之势。利用西安中端人才聚集、地区区位居中的优势，抓住新一轮产业转移重大机遇，把信息服务业作为基础产业，做北京、上海等国内一线城市的后台服务中心。大力发展数据存储、呼叫中心、IDC 中心、灾备中心、数据交换共享平台等业态，积极创新产业模式。在沣西新城规划了 20 多平方公里的信息产业园，推动和国家部委、世界顶尖的高科技企业的战略合作。已有中国电信、中国移动和中国联通三大通讯运营商数据中心、全国人口数据中心、美特斯邦威西北电子商务中心等项目签约落户，国家林业灾备中心初步达成入驻意向。万通集团在西咸新区秦汉新城建设“立体城市”，探索产业、住宅、服务融合一体的城市发展新业态。现代服务业蓬勃发展，空港综合保税区起步区和华润万象城、西部飞机维修基地、秦龙现代农业园区等项目已开工建设。

3. 发展复合型田园农业

在城市的开敞田园空间规划了若干特色小镇，以田园为映衬布局休闲餐饮、艺术画廊、会议中心等现代服务设施，形成复合都市服务功能的现代农庄。西咸新区秦汉新城的张裕葡萄酒庄基本建成，不仅有葡萄种植，更有葡萄酒酿制、鉴赏，还有餐饮、休闲娱乐功能，使得田园农业成为都市产业。应用先进技术和现代经营组织方式，探索农业产业化、现代化、园区化的新路径，改造提升传统农业，推广设施农业示范工程，既要成为西安“菜篮子”，增强城市农副产品自给能力，又探索形成结合休闲养生、观光体验等功能的新兴田园农业，把农业提升为复合型的都市产业，加快泾河新城的秦龙现代生态智能创意农业园区、森禾现代花卉科技产业园建设。

4. 探索城乡一体化发展新机制

西咸新区正研究在农村集体建设用地有序流转制度和农村人口有序转移制度两个方面推进改革。探索农民带着劳动力和土地“两个资本”进城的模式。完善保障农民集体土地财产权和收益权的制度，试行土地权益股份化、土地承包经营权转包出租抵押等流转新模式。争取把新区列入全国征地制度改革试点单位，创建集体农用地流转补偿机制，盘活城乡存量建设用地，逐步建立“同地、同价、同权”的城乡统一建设用地市场。率先建立城乡统一的户籍管理、就业服务、社会保障以及教育、医疗等制度，实行居民居住证制度，使“农民”成为一种职业而非身份的称谓。

5. 重视社会建设，注重社会服务和配套

西咸新区引进了第四军医大学医教园，不仅会解决西咸新区内部的，而且会解决西安、咸阳两市综合、高端的医疗服务。此外，还规划建设集养老、康复于一体的西咸新区健康城，布局了很多社区医院。规划建设国际文化教育园区，已经和多家国际知名教育机构达成合作协议。通过加强社会配套，推动城市产业功能和社会服务功能的一体化。同时，大力发展社会组织，中介服务组织，促进社会发育，建设公民社会。借鉴农村村民自治制度，改良后引入城市社会管理，实现城市居民社区自治。

（四）陕西西咸新区开发建设现况

西咸新区自2011年6月成立以来，围绕“创新城市发展方式、建设现代田园城市”的目标，以项目为核心全力推进各项工作，“拉开骨架、对接主城、提升环境、产业起步、体现概念”，发展战略逐步清晰，各类生产要素加快聚集，一批体现现代田园城市核心理念的重点项目相继实施，产业园区建设全面启动，新区发展取得了显著成就。2011年至2013年4月底，共完成固定资产投资523.8亿元，招商引资到位资金240亿元。

1. 创新城市发展方式思路清晰

加快西咸新区发展战略研究，提出以建设现代田园城市为目标，创新城市发展方式，构建点状布局市镇体系，以社会建设为先导，以就业为核心布局产业，探索城乡一体、产城一体的新型城镇化道路。《关于将西咸新区设为国家创新城市发展方式实验区的建议》在全国政协6700多份提案中被列为5个重点提案之一，全国政协副主席陈宗兴带队的重点提案调研组专门赴西咸新区调研。温家宝、贾庆林、李克强等中央领导同志对西咸新区有关工作作出重要批示。西咸新区发展战略和实践探索得到了各方面的高度评价，认为陕西在新型城镇化建设上见事早、立意高、进展快，希望西咸新区能为全国创新城市发展方式提供样本和范式。

2. 基础建设快速推进

按照“快捷交通连接”的要求，编制完成了西咸新区轨道交通规划和各组团分区规划、道路专项规划、历史文化遗产保护规划，公共设施布局、市政设施等11个专项规划，计划10年投资1000亿元，构建“南岸成网、北岸连线”的轨道交通网络。目前，连通西安主城的渭河横桥和西宝高速西安至兴平段已竣工通车，已建成的兰池大道被誉为“大西安最美城市道路”，三桥新街、后围寨立交即将建成，西咸北环线和西安北客站至机场轨道交通正在加快推进，西安地铁一号线延伸线地勘等工作全面启动实施；沣泾大道、正阳大道、红光大道、红光大桥、富裕桥等“五路四桥”建设已完成投资58.4亿元，其中不含桥梁的道路项目在2014年底前基本竣工，跨河桥梁、跨高速和铁路桥涵项目在2015年底前基本竣工；总投资216亿元、总长度约220公里的空港南环路、秦皇路等56个组团路网项目也已经全面启动，交通条件大为改善，畅通西咸、快捷西咸的雏形已成。坚持把保障性住房作为民生工程、发展工程和改善投资条件的环境工程，两年来共启动42996套保障房的建设，累计完成投资44.8亿元，已有6797套基本建成。

3. 环境治理初见成效

秦汉新城段渭河综合整治18公里防洪工程、堤顶道路、景观工程全面完成，绿化面积达4000余亩，成为陕西省渭河治理的亮点工程；泾河治理帷幕全面拉开；沣河综合改造已完成投资约2亿元，面积1200亩的沣河湿地公园已建成开放；长陵绿化全面完成，栽植油松1万多棵；与陕西林业厅合作投资80亿的中国暖温带森林文化博览园建设项目已获国家批准，建成后将成为国内面积最大的森林文化博览展示园区。

4. 产业培育卓有成效

一是空港新城综合保税区和西部飞机维修基地完成投资5.58亿元，绿地新城项目完成投资13.44亿元，空港国际物联城项目正在加快施工，嘉民物流园项目正式落户。空港国际酒店项目主体顺利封顶。空港临空产业园完成投资1亿元，启动了“电子商务离境结算”试点工作。二是沣东新城三桥商业街区改造快速推进，启航佳苑安置小区和时代广场完成投资10.5亿元，华润万象城和大明宫沣东国际项目开工实施。阿房宫遗址公园项目规划获得国家文物局审批，一期项目建设启动。昆明池项目概念性规划完成，征地拆迁启动。统筹科技资源示范区建设启动，5000亩都市农业博览园项目计划5月开园。六村堡工业园、新加坡产业园加快建设。三是秦汉新城周陵产业园路网形成，新竹防灾救生设备等13个项目已开工，亚洲最大的张裕葡萄酒庄基本完工，秦汉清华中学主体工程竣工。陕煤建总部基地、秦汉大酒店和财富中心写字楼等项目正在加紧施工。成功引进投资37亿元的OCA创意产业园项目、投资38亿元的海荣集团新能源汽车电机装置生产项目和投资17亿元的交大二附院秦汉新城分院项目，特别是总投资50亿元的第四军医大学医教研综合园区落户秦汉新城，将建成西北地区规模最大的医药航母及科研中心，优化大西安医疗资源配置。四是沣西新城信息产业园成功引进了中国移动、联通、电信三大通讯运营商数据中心和全国人口信息处理与备份中心、美特斯邦威西北电子商务中心、省医疗器械检测中心、陕西广电网络中心、“西部云谷”等重大项目，在全国首家举起大数据产业旗帜，快速形成产业聚集。投资20亿元的中国联通西北数据基地项目已开工建设。五是泾河新城以华轩饰品城为示范的泾河湾田园城市示范区、以崇文重点镇为亮点的崇文文化旅游景区、以中国锂产业园为龙头的新能源新材料工业园区初具规模，秦龙现代生态智能创意农业园区、森禾现代花卉园区、温州电气生产基地等总投资1200亿元的45个项目开工建设。成功引进了乐华欢乐世界、西咸农庄等合同金额458亿元的项目，全国首个省级云计算示范平台落户泾河新城。

5. 现代田园城市核心概念得到体现

加快推进体现现代田园城市理念的核心概念项目，投资300亿元的万通立体城市项目落户秦汉新城，集约用地、产城一体的新型城镇化模式初步形成。昆明池生态文化景区项目启动实施，将恢复10.7平方公里水面，重现中国第一人工蓄水工程的胜景。集养老、康复于一体的西咸健康城开工建设，阿房宫遗址公园、西咸能源金融中心、国际文教园区等项目正在快速推进，西咸崇文庄园小镇已完成用地流转、项目部搭建和景观、市政道路设计等工作，即将全面开工建设。

6. 招商引资成效明显

一是狠抓招商引资。报请陕西省政府出台了《关于加快西咸新区发展的若干支持政策》，并据此制定了共三十三条的《西咸新区投资优惠政策》，积极参加西洽会、陕港澳经济合作活动周、央企进陕、民企进陕等大型招商推介活动，大力开展叩门招商、驻点招商、以商招商等招商工作，两年来共签约招商引资项目110多个，合同引资3750多亿元，目前已到位资金240亿元，香港利星行国际汽车城、普洛斯国际物流园区、宜家家居等一批投资大、前景广、带动作用强的项目落户西咸，西咸新区成为投资热土。二是搭建融资平台。在香港设立了西咸发展基金，首期募集金额20亿美元，基金总规模共800亿元，引起海内外投资商的高度关注。编制完成了《西咸新区系统性融资规划》，与国开行、农发行、长安银行、北京银行、交通银行等金融机构建立了战略合作关系，积极争取省金融办支持设立小额贷款公司1家、另有多家在筹建，搭建了更为广阔的融资平台。三是组建企业集团。成功组建了西咸文旅集团、西咸金融控股集团和西咸置业集团，特别是由西咸集团与陕旅集团、西影集团联合组建的西咸文旅集团，探索出了西咸新区与省属大型国企通过资本纽带合作开发的新机制，为西咸新区的长远发展赢得了更多的外部支持。

7. 体制机制进一步理顺

针对面临的耕地占补平衡这一迫切问题，先后与延安、榆林、铜川等市签署协议，建立了耕地占补平衡异地指标收购长期合作机制，在全国同类新区中率先完成了跨多个行政区的土地利用规划编制，西咸新区成为国家认可的全省第十二个建设用地报批单位，建设用地报批体制全面理顺，用地制约逐步解除。针对开发建设涉及两市七县，各种权利和义务有交叉现象等问题，建立了西咸新区“共建共享机制”，制定了西咸新区与西安、咸阳两市事权划分的基本原则，设立了空港、沣西、泾河和秦汉四个新城的财政金库，管理体制机制进一步理顺，发展活力逐步彰显。

8. 品牌影响持续提升

创办了西咸周刊，建设并开通了西咸新区门户网站和官方微博，首开陕西省政务“微访谈”先例，被新浪网评为“陕西政府机构微博影响力”前十名。开展了西咸新区名片有奖征集、有奖征文、知识竞赛等专题活动和首届现代田园城市高峰论坛、西咸新区发展战略研讨会、渭河时代研讨会、城市名片征集研讨会、创新城市发展方式国际研讨会等系列活动，在凤凰卫视、《经济日报》、《中国日报》、新华网等媒体发布了西咸新区城市名片宣传语和LOGO，西咸新区对外影响逐步扩大，现代田园城市联盟总部落户西咸。受到主流媒体持续关注，人民日报和中央电视台《新闻30分》、《经济半小时》等名牌栏目多次聚焦西咸新区开发建设，《陕西日报》、《华商报》等省级媒体均在头版头条刊发西咸新区的新闻报道，西咸新区的知名度和影响力不断提升。荣获“2011年陕西推动力区域”、“2012年陕西最具推动力区域”和“香港回归十五周年大型评选活动”内地投资热点奖，成为陕西对外开放和加快发展的新亮点。

三、亿利资源集团履行环境社会责任研究

（一）亿利资源企业概况

亿利资源企业主要发展沙漠绿色经济，包括生态修复、清洁能源（生物能源）、天然药业、现代农业、城镇化建设。企业的发展使命是“引领沙漠绿色经济、开拓人类生存空间”。企业的核心价值观是“厚道、共赢、领导力”。

1988年创业以来，亿利资源在发展企业的同时，致力于中国第七大沙漠库布其沙漠的治理，实施了5000多平方公里的沙漠生态绿化和修复工程，并大规模、高技术实施了沙漠产业，开创了“市场化、产业化、公益化”相结合的沙漠绿色经济发展机制，走出了一条抗争荒漠化、整体消除贫困、改善区域生态环境、整治沙漠土地的绿色发展之路。2012年6月，库布其沙漠生态文明被列为联合国“里约+20”峰会重要成果向世界推广，企业负责人王文彪获得了联合国颁发的“全球环境与发展”奖。2012年10月，库布其沙漠生态文明被中央确定为国家生态文明典型。

（二）亿利资源企业沙漠生态文明事业成果

1. 沙漠生态文明事业投入时间

25 年。即从 1988 年到 2013 年。

2. 沙漠生态绿化与修复公益总投入

23.34 亿元人民币。

3. 沙漠绿色经济投资

目标规划 1000 亿元，已完成投资 300 亿元。

4. 沙漠生态绿化与修复工程实施区域

库布其沙漠（占 90%的面积）、毛乌素沙漠（上海庙）、科尔沁沙地。

5. 沙漠生态修复总面积

5000 多平方公里，是全球荒漠化面积的 1/7000。

6. 控制荒漠化面积

1.1 万平方公里。

7. 创造沙漠生态系统生产总值（GEP）

经国际组织和北京大学科学评估，库布其沙漠 25 年间在不毛之地上创造出了 305.91 亿元的生态系统生产总值。GEP，生态系统生产总值，是一套与国内生产总值（GDP）相对应的、能够衡量生态系统生产和服务总值的量化考核机制，主要指标是生态供给价值、生态调节价值、生态文化价值和生态支持价值，如生态系统的土壤、肥力、生态环境、生态产品及服务的价值。GEP 和 GDP 相比较，GDP 注重经济方面的指标，而 GEP 更加注重整个生态系统的服务和产品的总值，着眼长效的、可持续发展。

8. 库布其沙漠沙尘暴明显减少

库布其每年的沙尘暴由过去的七八十次减少到了现在的三至五次，显著改善了沙漠的生态环境，而且有效遏制了沙尘暴对北京、天津、河北等地区的影响。

9. 库布其沙漠降雨量明显增加

降雨量从过去的 70 毫米增长到现在的 300 多毫米。

10. 库布其沙漠土壤结构

出现了大面积厘米级的土壤迹象，专家称沙漠中每厘米土壤的自然形成时间至少需要 1 万年。

11. 库布其沙漠生物多样性

生物多样性得到了明显恢复，出现了绝迹多年的狼、狐狸等野生动物。

12. 牧民收入大幅增加

沙漠农牧民或以土地入股、或以有偿租赁、或以劳务投入参与企业的生态建设和沙产业项目，1990 年沙区牧民的人均收入不足 2000 元，2012 年沙区牧民的人均收入增长到了 30000 多元。企业的生态建设和沙产业项目 20 多年来累计为当地以及四川、青海、宁夏等地的农民工提供了近 10 万个绿色就业岗位，目前每年创造的绿色就业岗位可达到 1 万人（次）。

2009 年，企业投资 1.2 亿元为沙区牧民子弟建设了一所国际一流学校，解决了沙区孩子的上学难题。同时建设了“库布其农牧民培训学校”，通过引进以色列等世界先进沙漠产业技术，开办农牧民免费培训学习班，帮助沙区牧民脱贫致富。

（三）库布其沙漠生态绿化与修复方式

1. 规模化治沙

按照“以路化区、分割治理”的治沙策略，一是在库布其沙漠修筑多条纵横交错的穿沙公路；二是建设了长 240 多公里的防沙护河锁边林；三是建设大漠腹地生态修复工程。

2. 科学化治沙

经过 20 多年的实践，企业成功研发出了“水气法种树新技术”。该项技术种植一棵树的时间只有短短十几秒钟，两人配合每天可种植 20 多亩树，较以前锹挖植树效率提高了 30 多倍，每亩成本降低了 2000 多元，成活率也提高到 90%以上。库布其沙漠 20 多年的成功实践证明，我国有 10 亿亩左右的沙漠是可以通过生态绿化来改造治理的。如将该技术推广应到这部分可治理的沙漠中，可节省投资近 20000 亿元人民币，而且可保证大面积种植成活。

3. 产业化治沙

亿利资源经过反复实践，探索出了防沙与经济效益并存的复合生态模式，在地下大规模种植甘草等中药材植物，在地上大规模种植有经济价值的灌木乔木，并采取“公司+农户，企业+基地”的联盟发展模式，大力发展了千亿规模的“有机肥料、有机材料、有机能源、有机食品”沙漠绿色经济。通过持续的沙漠生态建设，整理和改良了数百万亩的沙漠绿色良田，极大地提升了沙漠土地的价值空间。

（四）亿利资源沙漠生态文明事业五年规划

1. 沙漠生态绿化与修复工程总面积达到 10000 平方公里

2. 沙漠绿色经济总投资达到1000亿元

重点发展“有机肥料、有机材料、有机能源、有机食品”四个有机产业、并加大力度实施寒旱物种生态修复产业。

有机肥料：利用库布其沙漠草炭资源和有机废渣，建设500万吨/年碳基复混肥。

有机材料：充分利用沙漠资源、沙漠经济基地园区及周边工业废弃物，引进国际、国内先进石油压裂支撑剂、纳米铀材料新技术，生产沙质透水新材料、石油压裂剂、沙漠工艺品和保温、防火建筑环保材料。

有机能源：与以色列凯伊玛公司合作开发100万亩蓖麻、扁桃种植、30万吨蓖麻油加工项目。

有机食品：以市场为龙头，建设现代农业物流配送中心及冷链储运中心，带动沙漠大面积种植绿色、无公害、有机果蔬、花生等经济植物；同时积极与以色列凯伊玛公司合作，对沙漠苁蓉、甘草、大黄、马齿苋、黄芩等生物资源进行规范化、标准化种植和提取加工，开发治疗粉刺、异位性皮肤炎、脂溢性皮肤炎和干癣产品等。

实施寒旱植物生态修复产业。一是培育和发展寒旱林草种苗产业，建设10万亩苗圃基地；二是实施沙漠地区和城郊地区生态修复与绿化。

3. 办好两年一届的库布其国际沙漠论坛

4. 建立全球首个国际沙漠生态系统科学家联盟

2013年3月19日，由国际沙漠协会联合中国亿利公益基金以及来自全球各国的荒漠化防治科学家共同在北京发起成立全球首个沙漠生态系统科学家联盟。该联盟旨在汇聚全球国家最多、地域最广、学术最权威、影响力最大的100多个国家的300多位杰出科学家，共同致力于中国沙漠生态建设，促进沙漠绿色经济和生态文明发展，推动世界生态环境的改善。

（五）亿利资源企业治沙实践总结

亿利25年的沙漠事业发展实践总结为“一个模式、两层循环、三化机制、四种方式和五项成果”，解决了防沙治沙“钱从哪里来”、“利从哪里得”、“如何可持续”的问题。

一个模式：

提出了“沙漠也是资源”的新理念，在这一理念指导下创立了“可持续公益商业治沙模式”。通过技术创新，变沙害为沙利，变劣势为优势，修复沙漠生态、发展沙漠经济，实现了“由求生存到谋发展、由输血治沙到造血治沙”的两大跨越，走出了一条“治沙、生态、民生、经济”平衡驱动的绿色可持续发展之路。这是亿利资源治沙实践的焦点、重点和难点。

两层循环：

沙漠生态经济循环：防沙治沙—生态修复—产业开发—土壤改良—防沙治沙。

沙漠生态社会循环：防沙治沙—生态经济—民生改善—人与自然和谐—防沙治沙。

三化治沙机制：

市场化：通过市场行为，变国家投资治沙为企业产业与公益投资治沙，变输血治沙为造血治沙，科学合理利用沙漠资源，采取多元化方式，引导农牧民市场化行为承包种树、种草、种药材，吸引社会力量参与，共同推动沙漠事业。

产业化：沙漠生态培育沙漠经济，沙漠经济反哺沙漠治理。一是大面积种植耐旱型的沙柳、柠条等平茬刈割植物，既能绿化沙漠，又能搞饲料、肥料产业。二是大面积种植甘草等耐旱药材，发展天然健康产业。三是发展沙漠生态修复和园林绿化产业。

公益化：先公益、后生意，先生态、后经济。25 年前，从生产 1 吨盐提取 5 元治沙育林基金，到现在捐赠总资产 30%永续收益持续治沙，长效公益性投入保障了沙漠生态公益事业的可持续发展。

四种治沙方式：

一是生态工程治沙。

实施了大面积人工围栏封育和飞播工程，围封固沙。

实施了全长 200 多公里长的锁边林工程，锁住四周，渗透腹部。

建成了 500 多公里纵横交错穿沙公路，把库布其沙漠化整为零，并沿路通电、通水、扎网格，大规模种树、种草、种药材，以路划区，分割治理。

最终形成了“南围、北堵、中切”的“孙子兵法”式治沙格局。

二是生态移民治沙。

把分散在几千平方公里沙漠里的农牧民转移出去，实施了集“党建、教育、文化、产业”于一体的、宜居宜业的沙漠生态移民工程。帮助引导农牧民参与沙漠生态建设和沙产业，让农牧民多渠道就业创业，增收致富。让沙漠得以休养生息，实现了绿富同兴。

三是生态技术治沙。

20 多年来，企业自主研发了 100 多项沙漠生态技术，培育、改良了沙柳、甘草等 20 多种免耕无灌溉的耐寒、耐旱经济植物。发明了“水气法植树新技术”

种树，每亩成本降低了1800多元，成活率由20%提高到85%以上，而且能在高高的沙丘上种活树，这项技术如能应用到我国西部有条件的沙漠里，可以大大提高治沙绿化效率，大幅减少治沙投资。

四是生态产业治沙。

企业在种下第一棵树的时候就埋下了产业经济的种子。构建了一个“生态+公益”互动共赢的生态产业发展机制，形成了“防沙治沙—生态修复—产业开发—土壤改良”一体化生态循环产业链。防沙治沙和生态修复促进了土地整治和产业开发，土壤改良和生态产业开发带动了防沙治沙和生态修复。

五项治沙成果：

创新了一个理念。在20多年的实践中，我们首创了“沙漠生态经济”的新理念，把沙漠当成一种宝贵资源，科学巧妙进行改善和利用，实现了人与自然的和谐，实现了“治沙、生态、民生、经济”的平衡驱动发展。

绿化了一座沙漠。20多年的努力，让昔日的“死亡之海”变成了富裕文明的生态绿洲，沙尘暴天气由过去每年七八十次减少到现在的三五次，降雨量也由过去的70毫米增长到现在的300多毫米。沙漠生物多样性得到了明显恢复，绝迹多年的仙鹤、天鹅、野兔、沙冬青、胡杨等100多种野生动植物出现在沙漠。沙漠中改良出了大规模的厘米级厚的土壤，被专家们称为“沙漠奇迹”，这也为国家拓展可利用国土空间找到了重要途径。

培育了一批产业。25年来，企业变沙漠劣势为优势，利用沙漠土地、阳光、生物质等沙漠资源，培育和发展了沙漠生态修复产业、沙漠天然健康产业、沙漠材料开发应用和沙漠再生能源等极具发展潜力的沙漠生态经济产业。

改善了一方民生。沙漠农牧民成为治理沙漠最大的受益者。经过20多年对沙漠生态建设和沙漠生态经济的持续投入，十多万沙漠农牧民的生活发生了翻天覆地的变化，人均年收入由十几年前的几百元增长到了现在的30000多元。企业捐资为沙漠里的孩子建设了沙漠国际化学校，让沙漠里的孩子们在家门口接受良好的教育。

凝聚了一域民心。亿利资源的生态建设和沙漠产业主要依靠当地各族百姓。多年来，我们先后组建了220个“生态建设民工联队”，并把党组织建在民工联队，依托党的组织优势和凝聚力，把散居在沙漠中几万农牧民汇聚在有组织、有体系的沙漠事业中，实现了绿富同兴，密切了党群、干群、企群关系，特别是密切了民族关系。

亿利资源提出到2020年，使沙漠生态修复和绿化总面积达到15000平方公里，相当于全国 2020 年沙漠土地整治规划总面积的 1/13，把库布其沙漠建设成为“美丽中国”的窗口，为实现中国梦贡献力量。

库布其沙漠2000年卫星遥感图

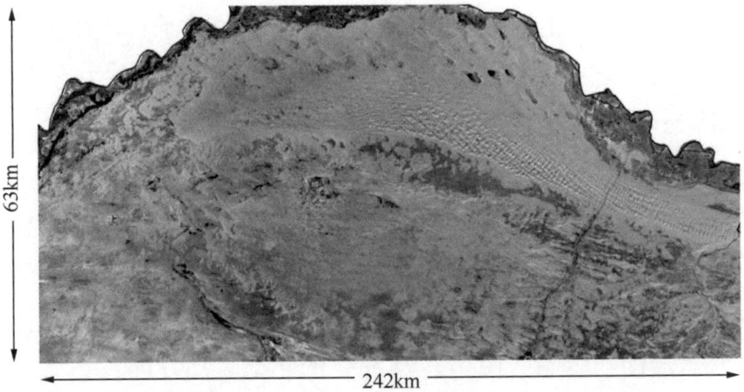

63km

242km

库布其沙漠是中国第七大沙漠，总面积1.86万平方公里，位于北京正西部，距北京直线距离800公里，是北京三大风沙源之一。

库布其沙漠2009年卫星遥感图

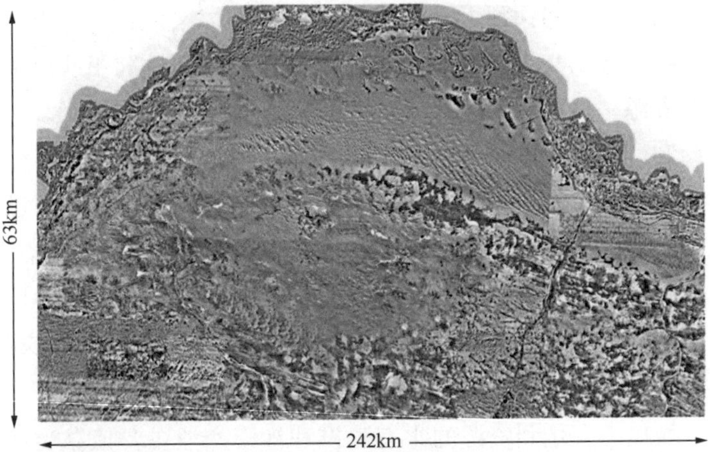

63km

242km

库布其沙漠是中国第七大沙漠，总面积1.86万平方公里，位于北京正西部，距北京直线距离800公里，是北京三大风沙源之一。

20世纪80年代的库布其沙漠

20世纪90年代的库布其沙漠

2010年的库布其沙漠

库布其沙变

党和政府改革开放爱民惠民

昔日黄沙滚滚

今日生机盎然

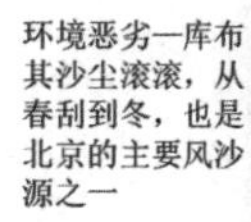
环境恶劣—库布其沙尘滚滚，从春刮到冬，也是北京的主要风沙源之一

沙区牧民变为生态产业股东、工人、新牧民。现在农牧民的人均年收入三万元左右。巴雅尔和他家的宝贝骆驼，在七星湖景区每天旅游收入千元

行路难——沙漠里没有路，到离家最近的医院要走几天

思维方式转变

沙漠变通途——纵横交错的穿沙公路让沙区的牧民走出了大漠、沙区农牧民的思维生活生产方式发生了改变

生产难——牧民以沙漠过渡游牧为主要的生产方式，草场退化，导致生态环境恶化

库布其沙漠变绿洲，成为中国北方的生态绿色屏障，沙区农牧民生存环境发生了巨变。对北京以及东北亚等周边地区的生态环境产生了积极影响

生活方式转变

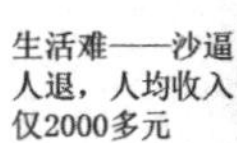
生活难——沙逼人退，人均收入仅2000多元

文化生活丰富——牧民们自发组织的互助式的文艺队，为游客、为社区演出，吹拉弹唱有滋有味

上学难——有些孩子到了十几岁才能上学

生产方式转变

沙漠里有了现代化学校——亿利东方学校的孩子们坐在明亮整洁的教室里学习

文化生活难——没有电、没有通讯、没有电视、没有文化生活

散落在沙漠里的土矮房变成了现代沙漠小镇—— 36岁的牧民斯仁巴布走出了低矮破旧的土房，住进现代化的新村，还搞起“牧家乐”，年收入二十多万元

参 考 文 献

[1] 工业革命以来西方主要国家环境污染与治理的历史考察. 世界历史，2000，(6).

[2] 工业革命以来西方国家城市化进程与规划启示，2012 中国城市规划年会论文集（城市化与区域规划研究）. 北京：北京大学出版社，2012.

[3] 工业革命：英国世界霸权形成的前提，大国崛起. 北京：人民出版社，2006.

[4] 产业革命以来西方发达国家经济增长方式的变革及启示. 科学网，2008-2-14.

[5] 净化空气：世界 5 大洲 14 个国家空气污染控制的法律与实践（Clean ALr Around the World，The Law and Practice of Air Pollution Control in 14 Countries in 5 Continents），国际空气污染防治协会联盟 1988，摘自新浪网.

[6] 特里弗·梅. 1760—1970 年英国经济和社会史（Trevor May，An E- conomic and Social History of Britain 1760-1970），1987，摘自新浪网.

[7] 曲格平. 第二座里程碑，迈向 21 世纪——联合国环境与发展大会文献汇编. 北京：中国环境科学出版社，1992.

[8] 宁大同，王华东. 关于西方国家环境污染阶段的划分，全球环境导论. 山东科学技术出版社，1996.

[9] 李京文，方汉中. 国际技术经济比较——大国的过去、现在和未来 .

[10] 杨朝飞. 环境保护与环境文化. 北京：中国政法大学出版社，1994.

[11] D. 斯特拉德林，P. 索赛姆. 大城市的烟雾：1860—1914 年英美控制空气污染的努力. 环境历史（D. Stradling & P. Thorsheim，“The Smoke of Great Cities，British and American Efforts to Control Air Pollution，1860—1914”，摘自新浪网；Environmental History），第 4 卷，1999，(1)，摘自新浪网.

[12] 拉蒙得·多米尼加. 资本主义、共产主义与环境保护：德国人的经验教训. 环境历史（Raymond Dominick，“Capitalism，Communism and En-vironmental Protection，Lessons from the German Experience”，Environmental History），第 3 卷，1998，(3)，摘自新浪网.

[13] 2003—2012 年政府工作报告.

[14] 中国科学院. 中国可持续发展战略报告. 2006.

[15] 环境保护部. 中国环境法规全书.

[16] 环境保护部. 中国环境政策全书.

[17] 环境保护部. 新时期环境保护重要文献选编.

[18] 中国环境与发展国际合作委员会年度政策报告. 2006—2012.

[19] 中国科学可持续发展战略研究组. 建设资源节约型和环境友好型社会.

[20] 中国共产党第十八次全国代表大会报告.

[21] 第十二届全国人民代表大会第一次会议. 国务院机构改革和职能转变方案.

[22] 2013 年政府工作报告.

[23] 国家环境保护部. 关于培育引导环保社会组织有序发展的指导意见.

[24] 中华环保联合会. 中国环保民间组织发展状况报告（2008）.

[25] 中华环保联会. 中国环保 NGO 蓝皮书.

[26] 黄浩明. 环保民间组织现况、挑战和对策.

[27] 刘应杰. 深刻认识中国与日本发展的显著差距——赴日考察报告.

[28] 刘应杰，邓文奎，龚维斌. 中国生态环境安全. 合肥：安徽人民出版社，2004.

[29] 刘应杰. 中国的发展战略和基本国策读本. 北京：中央党校出版社，2008.

[30] 宋言奇. 国外生态环境保护中社区“自组织”的发展态势. 国外社会科学，2009，(4).

[31] 孙志祥. 北京市民间环保组织个案研究.

[32] 王飞. 我国环保民间组织的运作与发展趋势.

[33] 秦洪良. 国外公众参与环境保护的主要做法.

[34] 卢莎，石昌智等. 环境保护公众参与调查与对策.

[35] 崔凤，唐国建. 环境社会学. 北京师范大学出版社，2010.

[36] 葛察忠，钟晓红，毕军. 建设环境友好型社会经济政策. 中国环境科学出版社，2007.

[37] 洪大用. 中国环境社会学. 社会科学文献出版社，2007.

[38] 吕胜利，吕晓英. 资源、环境与经济社会发展的模拟研究. 中国环境科学出版社，2008.

[39] 彭峰. 法国公众参与环境保护原则的实施及其对我国的借鉴.

[40] 王名，陶传进. 公共管理新课题：政府与非政府组织的关系.

[41] 黄炳元. 试论环保非政府组织（NGO）在我国的转型变化和未来作用.

[42] 王云霞. 环境问题的社会批判研究. 中国社会科学出版社，2012.

[43] [美] 德尼·古莱. 残酷的选择——发展理念与伦理价值. 社会科学文献出版社，2008.

[44] 袁云. “从生态人”假设到人的“生态化”生活方式. 天津行政学院学报，2012，(3).

[45] 徐嵩龄. 环境伦理学进展：评论与阐释. 北京：社会科学文献出版社，1999.

[46] 李承宗. 从价值论看“生态人”的合法性. 自然辩证法研究，2006.

[47] 顾智明. 论“生态人”之维——对人类新文明的一种解读. 社会科学，2004.

[48] 马克思恩格斯全集. 北京：人民出版社，1960.

[49] 李霞. 生活方式的变迁与选择. 人民出版社，2012.